Creo Parametric 5.0 for Novices

(Learn By Doing)

CADSoft Technologies

Creo Parametric 5.0 for Novices: Learn By Doing

CADSoft Technologies

CADSoft Technologies

ISBN 978-1-64057-025-2

NOTICE TO THE READER

Publisher does not warrant or guarantee any of the products described in the text or perform any independent analysis in connection with any of the product information contained in the text. Publisher does not assume, and expressly disclaims,any obligation to obtain and include information other than that provided to it by the manufacturer.

The reader is expressly warned to consider and adopt all safety precautions that might be indicated by the activities herein and to avoid all potential hazards. By following the instructions contained herein, the reader willingly assumes all risks in connection with such instructions.

The Publisher makes no representation or warranties of any kind, including but not limited to, the warranties of fitness for particular purpose or merchantability, nor are any such representations implied with respect to the material set forth herein, and the publisher takes no responsibility with respect to such material. The publisher shall not be liable for any special, consequential, or exemplary damages resulting, in whole or part, from the reader's use of, or reliance upon, this material.

DEDICATION

To students, who are dedicated to learning new technologies

THANKS

To the faculty and employees of
CADSoft Technologies for their cooperation

Training Programs offered by CADSoft Technologies

CADSoft Technologies provides quality training on various software packages including Computer Aided Design, Manufacturing and Engineering (CAD/CAM/CAE), Web and Programming Languages, animation, architecture, and GIS. The training is delivered at its training centers or online at any time, place, and pace to individuals, corporates as well as the students of colleges, universities, and CAD/CAM/CAE training centers. The main features of this program are:

Training in Classroom Settings and Online Training

A team of highly experienced instructors and qualified engineers at CADSoft Technologies conduct the classes in classroom environment at its centers. This team has authored several textbooks that are rated "one of the best" in their categories and are used in various colleges, universities, and training centers in North America, Europe, and in other parts of the world.

CADSoft Technologies with its cost effective and time saving initiative strives to deliver online training in the comfort of your home or work place, thereby relieving you from the hassles of traveling to training centers. CADSoft Technologies strives to be the best training institute in USA. CADSoft Technologies has developed career oriented programs by following state-of-the-art teaching and learning methodologies.

Courses Offered by CADSoft Technologies

CADSoft Technologies conducts classes on the following software packages:

CAD/CAM/CAE: *CATIA, Pro/ENGINEER Wildfire, Creo Parametric, Creo Direct, SOLIDWORKS, Autodesk Inventor, Solid Edge, NX, AutoCAD, AutoCAD LT, AutoCAD Plant 3D, Customizing AutoCAD, SolidCAM, NX CAM, NX Mold, Creo Mold, Alias Design, Alias Automotive, Ansys Fluent, and ANSYS*

Architecture and GIS: *Autodesk Revit Architecture, AutoCAD Civil 3D, Autodesk Revit Structure, AutoCAD Map 3D, Revit MEP, Navisworks, Primavera, and Bentley STAAD Pro, MS Project, MX Road, ArcGIS, and Raster Design*

Animation and VFX: *Autodesk 3ds Max, Chaosgroup VRay, Autodesk 3ds Max Design, Autodesk Maya, Blackmagic Design Fusion Studio, Adobe Premiere, Adobe Photoshop, Adobe Indesign, Adobe Illustrator, Corel Graphic Design CorelDraw, The Foundry NukeX, and MAXON CINEMA 4D*

Web and Programming: *C/C++, HTML5/CSS3, JavaScript, jQuery, Bootstrap, PHP, MySQL, Dreamweaver, VB.NET, Oracle, AJAX, and Java*

Personality Development: *Personality Development and Engineering Ethics/Soft Skills Course*

Table of Contents

Chapter 6: Options Aiding Construction of Parts-I

Chapter 7: Options Aiding Construction of Parts-II

Chapter 8: Options Aiding Construction of Parts-III

Chapter 9: Advanced Modeling Tools

Chapter 10: Assembly Modeling

Chapter 11: Generating, Editing, and Modifying the Drawing Views

Chapter 12: Dimensioning the Drawing Views

Chapter 13: Other Drawing Options

Chapter 14: Working with Sheetmetal Components

This page is intentionally left blank

Preface

Creo Parametric 5.0

Creo Parametric is developed by Parametric Technology Corporation. It provides a broad range of powerful and flexible CAD capabilities that can address even the most tedious design challenges. Being a parametric feature-based solid modeling tool, it not only integrates the 3D parametric features with 2D tools, but also assists in every design-through-manufacturing process. This software is remarkably user-friendly.

This solid modeling software allows you to easily import the standard format files with an amazing compatibility. The 2D drawing views of the components are automatically generated in the Drawing mode. Using this software, you can generate detailed, orthographic, isometric, auxiliary, and section views. Additionally, you can use any predefined drawing standard files for generating the drawing views. You can display the model dimensions in the drawing views or add reference dimensions whenever you want. The bidirectionally associative nature of this software ensures that any modification made in the model is automatically reflected in the drawing views. Similarly, any modification made in the dimensions of the drawing views is automatically updated in the model.

The ***Creo Parametric 5.0 for Novices: Learn By Doing*** textbook has been written to enable the readers to use the modeling power of Creo Parametric 5.0 effectively. The textbook also covers the Sheetmetal module with the help of relevant examples and illustrations. The mechanical engineering industry examples and tutorials are used in this textbook to ensure that the users can relate the knowledge of this book with the actual mechanical industry designs. The salient features of this textbook are as follows:

- **Tutorial Approach**
  The author has adopted the tutorial point-of-view and the learn-by-doing approach throughout the textbook. This approach guides the users through the process of creating the models in the tutorials.

- **Real-World Mechanical Engineering Projects as Tutorials**
  The author has used the real-world mechanical engineering projects as tutorials in this textbook so that the readers can correlate them with the real-time models in the mechanical engineering industry.

- **Tips and Notes**
  Additional information related to various topics is provided in the form of tips and notes.

- **Learning Objectives**
  The first page of every chapter summarizes the topics that will be covered in that chapter. This helps the users to easily refer to a topic.

- **Exercises**
  Exercises are given at the end of the chapters and they can be used by the instructor as test questions and exercises.

Symbols Used in the Textbook

Note
The author has provided additional information to the users about the topic being discussed in the form of notes.

Tip
Special information and techniques are provided in the form of tips that helps in increasing the efficiency of the users.

Formatting Conventions Used in the Textbook

Please refer to the following list for the formatting conventions used in this textbook.

- Names of tools, buttons, options, groups, tabs, slide-down panels, and Ribbon are written in boldface.

  Example: The **Extrude** tool, the **OK** button, the **Editing** group, the **Sketch** tab, and so on.

- Names of dialog boxes, drop-downs, drop-down lists, dashboards, areas, edit boxes, check boxes, and radio buttons are written in boldface.

  Example: The **Revolve** dashboard, the **Chamfer** drop-down of **Engineering** group in the **Model** tab, the **Thickness** drop-down of the **Shell** dashboard, the **Extended intersect surfaces** check box in the **Options** slide-down panel of the **Draft** dashboard, and so on.

- Values entered in edit boxes are written in boldface.

  Example: Enter **5** in the **Radius** edit box.

- Names and paths of the files are written in italics.

  Example: *C:\Creo-5.0\c03*, *c03tut01.prt*, and so on.

- Different options available for invoking a tool are given in a shaded command box.

  Ribbon: Model > Datum > Sketch

Naming Conventions Used in the Textbook

Tool

If you click on an item in a toolbar or a group of the **Ribbon** and a dashboard or dialog box is invoked to create/edit an object or perform some action, then that item is termed as **tool**.

For example:
Line tool, **Normal** tool, **Extrude** tool
Fillet tool, **Draft** tool, **Delete Segment** tool

If you click on an item in a toolbar or a group of the **Ribbon** and a dialog box is invoked wherein you can set the properties to create/edit an object, then that item is also termed as **tool**, refer to Figure 1.

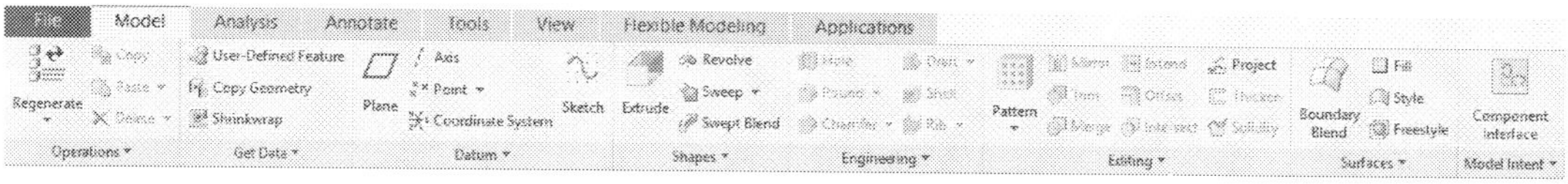

Figure 1 *Various tools in the* ***Ribbon***

Button

The item in a dialog box that has a 3D shape like a button is termed as **Button**. For example, **OK** button, **Cancel** button, **Apply** button, and so on.

Dialog Box

In this textbook, different terms are used for referring to the components of a dialog box. Refer to Figure 2 for the terminology used.

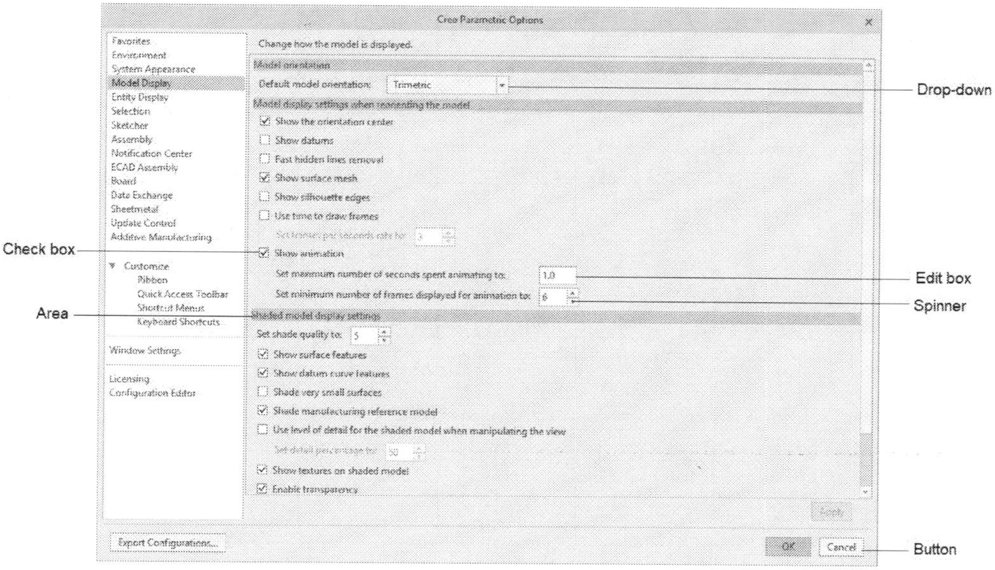

Figure 2 *The components of a dialog box*

This page is intentionally left blank

Chapter 1

Introduction to Creo Parametric 5.0

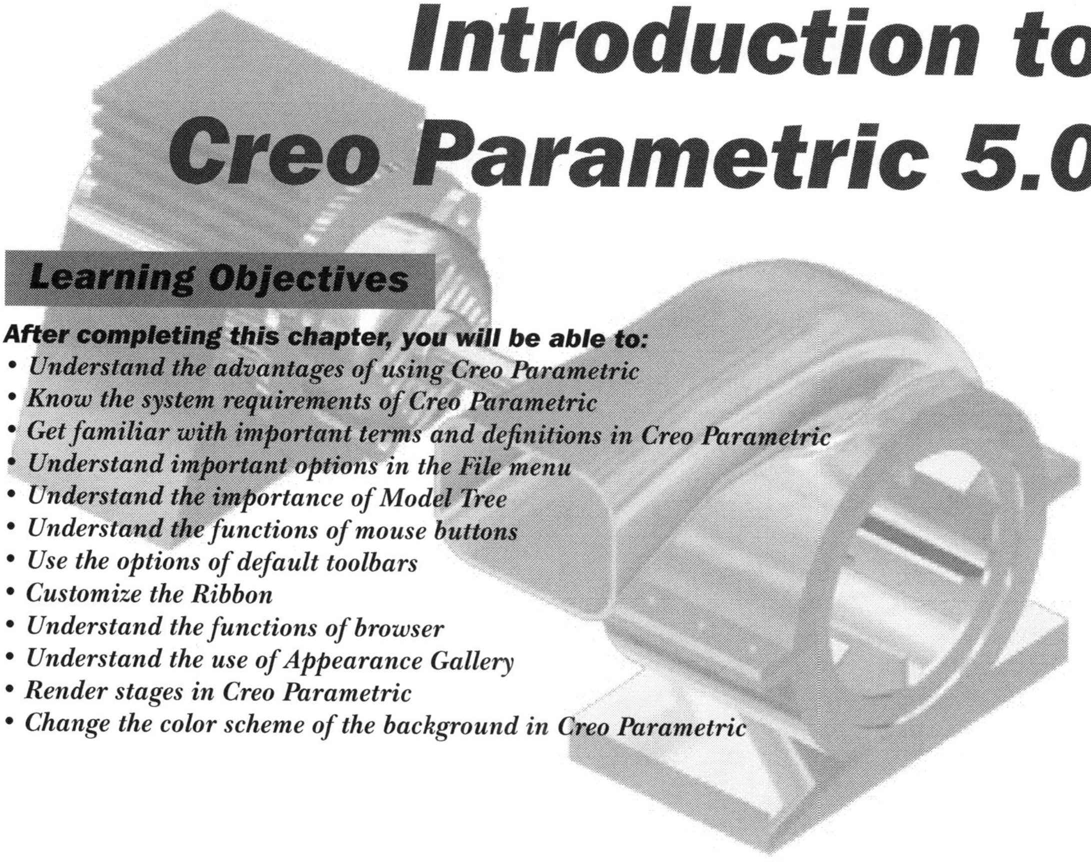

Learning Objectives

After completing this chapter, you will be able to:

- *Understand the advantages of using Creo Parametric*
- *Know the system requirements of Creo Parametric*
- *Get familiar with important terms and definitions in Creo Parametric*
- *Understand important options in the File menu*
- *Understand the importance of Model Tree*
- *Understand the functions of mouse buttons*
- *Use the options of default toolbars*
- *Customize the Ribbon*
- *Understand the functions of browser*
- *Understand the use of Appearance Gallery*
- *Render stages in Creo Parametric*
- *Change the color scheme of the background in Creo Parametric*

INTRODUCTION TO Creo Parametric 5.0

Creo Parametric is a powerful software used to create complex designs with great precision. The design intent of a three-dimensional (3D) model or an assembly is defined by its specification and its use. You can use the powerful tools of Creo Parametric to capture the design intent of a complex model by incorporating intelligence into the design. Once you understand the feature-based, associative, and parametric nature of Creo Parametric, you can appreciate its power as a solid modeling tool.

To make the designing process simple and quick, the designing process have been divided into different modules in this software package. This means each step of the designing is completed in a different module. For example, generally a design process consists of the following steps:

- Sketching using the basic sketch entities
- Converting the sketch into features and parts
- Assembling different parts and analyzing them
- Documenting parts and the assembly in terms of drawing views
- Manufacturing the final part and assembly

All these steps are divided into different modes of Creo Parametric namely, the **Sketch** mode, **Part** mode, **Assembly** mode, **Drawing** mode, and **Manufacturing** mode.

Despite making various modifications in a design, the parametric nature of this software helps preserve the design intent of a model with tremendous ease. Creo Parametric allows you to work in a 3D environment and calculates the mass properties directly from the created geometry. You can also switch to various display modes like wireframe, shaded, hidden, and no hidden at any time with ease as it does not affect the model but only changes its appearance.

FEATURES OF CREO PARAMETRIC

Different features of the software are discussed next.

Feature-Based Nature

Creo Parametric is a feature-based solid modeling tool. A feature is defined as the smallest building block and a solid model created in Creo Parametric is an integration of a number of these building blocks. Each feature can be edited individually to bring in the desired change in the solid model. The use of feature-based property provides greater flexibility to the parts created. For example, consider the part shown in Figure 1-1. It consists of one counterbore hole at the center and eight counterbore holes around the Bolt Circle Diameter (BCD).

Now, consider a case where you need to change all the outer counterbore holes to drill holes keeping the central counterbore hole and the BCD for the outer holes same. In a non feature-based software package, you will have to delete the entire part and then create a new part based on the new specifications. Whereas, Creo Parametric allows you to make this modification by just modifying some values in the same part, see Figure 1-2. This shows that the solid parts created in Creo Parametric are a combination of various features that can be modified individually at any time.

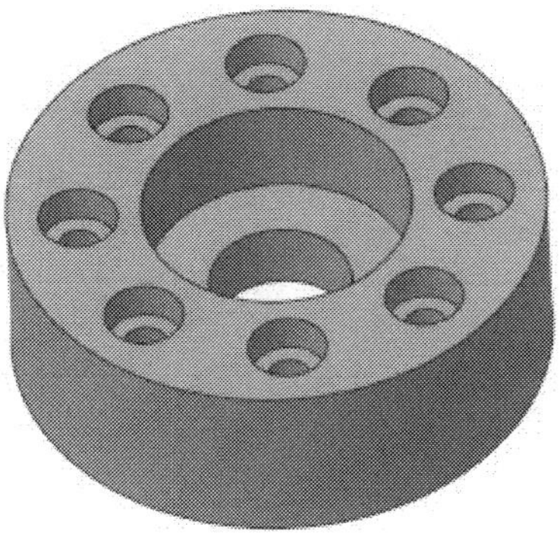

Figure 1-1 Model with the counterbore holes

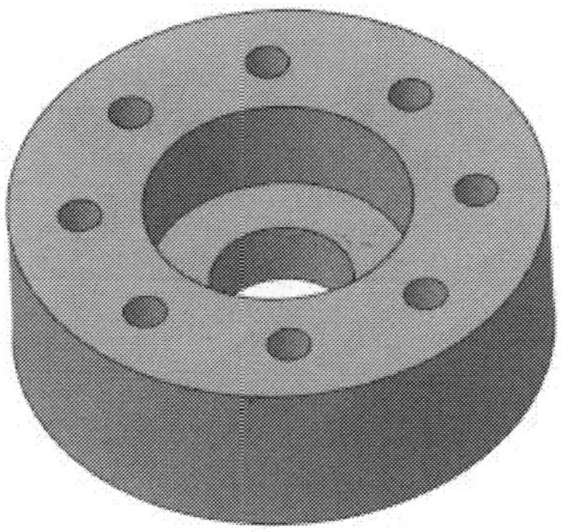

Figure 1-2 Model after making the modifications

Bidirectional Associative Property

There is a bidirectional associativity between all modes of Creo Parametric. The bidirectional associative nature of a software package is defined as its ability to ensure that if any modifications are made in a particular model in one mode, then those modifications are also reflected in the same model in other modes. For example, if you make any change in a model in the **Part** mode and regenerate it, the changes will also be highlighted in the **Assembly** mode. Similarly, if you make a change in a part in the **Assembly** mode, after regeneration, the change will also be highlighted in the **Part** mode. This bidirectional associativity also correlates the two-dimensional (2D) drawing views generated in the **Drawing** mode and the solid model created in the **Part** mode of Creo Parametric. This means that if you modify the dimensions of the 2D drawing views in the **Drawing** mode, the change will be automatically reflected in the solid model and also in the assembly after regeneration. Likewise, if you modify the solid model in the **Part** mode, the changes will also be seen in the 2D drawing views of that model in the **Drawing** mode. Thus, bidirectional associativity means that if a modification is made to one mode, it changes the output of all the other modes related to the model. This bidirectional associative nature relates various modes in Creo Parametric.

Figure 1-3 shows the drawing views of the part shown in Figure 1-1 generated in the **Drawing** mode. The views show that the part consists of a counterbore hole at the center and eight counterbore holes around it.

Now, when the part is modified in the **Part** mode, the modifications are automatically reflected in the **Drawing** mode, as shown in Figure 1-4. The views in this figure show that all outer counterbore holes are converted into drilled holes.

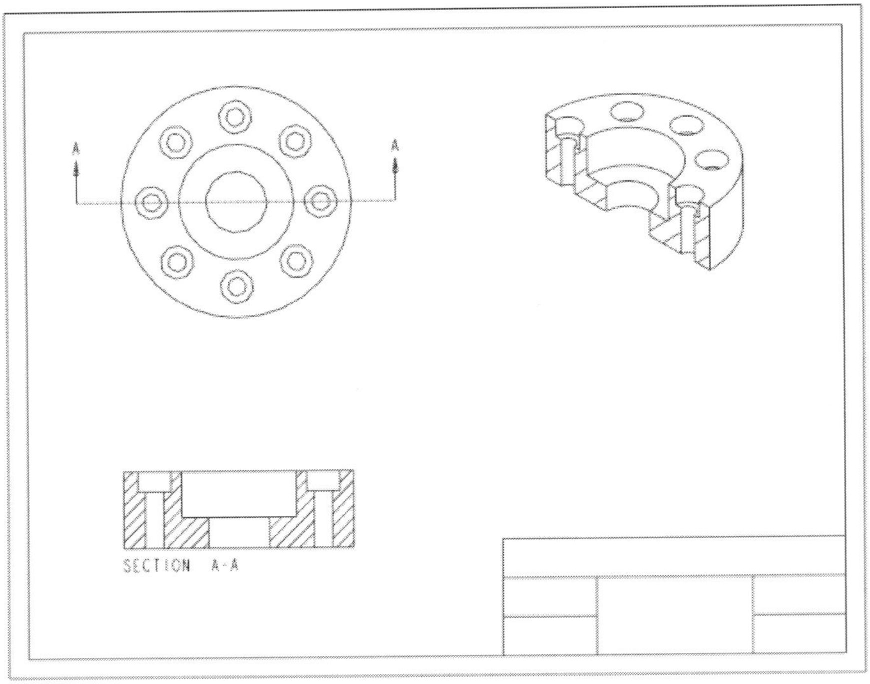

Figure 1-3 Drawing views of the model before modifications

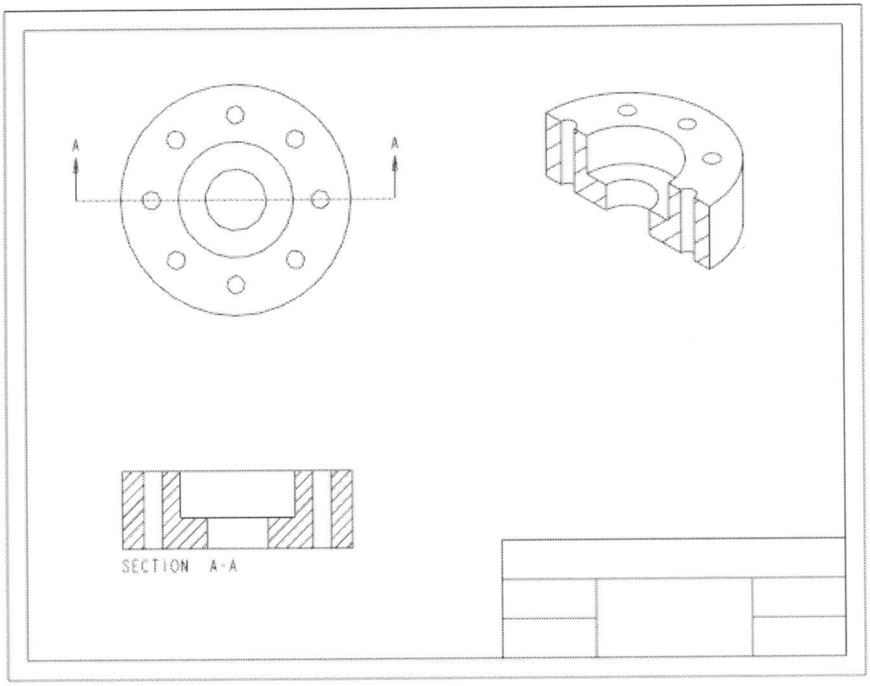

Figure 1-4 Drawing views of the model after modifications

Figure 1-5 shows the Crosshead assembly. It is clear from the assembly that the diameter of the hole is more than what is required (shown using filleted rectangles). In an ideal case, the diameter of the hole should be equal to the diameter of the bolt.

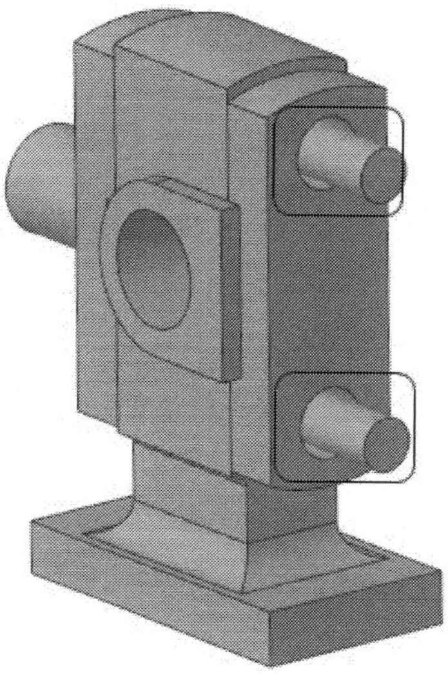

Figure 1-5 *Crosshead assembly illustrating difference in diameter of the hole and the bolt*

The diameter of the hole can be changed easily by opening the file in the **Part** mode and making the necessary modifications in the part. This modification is reflected in the assembly, as shown in Figure 1-6. This is due to the bidirectional associative nature of Creo Parametric.

Since all modes of Creo Parametric are interrelated, it becomes very easy to modify your model at any time.

Parametric Nature

Creo Parametric is parametric in nature, which means that the features of a part become interrelated if they are drawn by taking the reference of each other. You can redefine the dimensions or the attributes of a feature at any time. The changes will propagate automatically throughout the model. Thus, they develop a relationship among themselves. This relationship is known as the parent-child relationship. So if you want to change the placement of the child feature, you can make alterations in the dimensions of the references and hence change the design as per your requirement. The parent-child relationship will be discussed in detail while discussing the datums in later chapters.

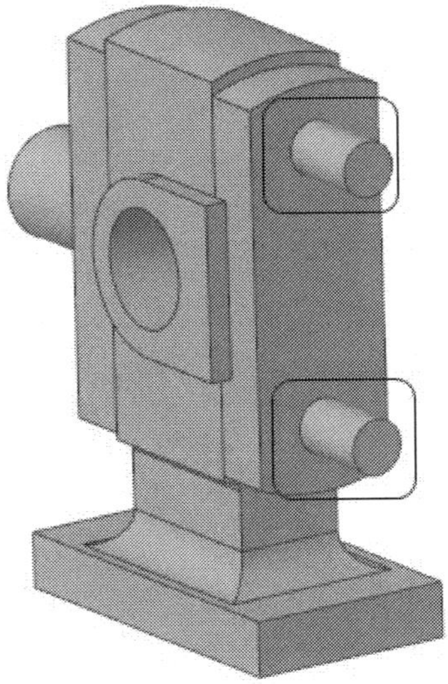

Figure 1-6 *Assembly after modifying the diameter of the hole*

SYSTEM REQUIREMENTS

The system requirements for Creo Parametric are given below.

1. Operating System: Windows 7 64-bit Edition (Professional, Ultimate, and Enterprise), Windows 8.1 64-bit Edition or Windows 8.1 Pro 64-bit Edition, Windows 10 64-bit Edition (Professional, Enterprise).
2. Monitor: 1280 x 1024 (or higher) resolution support with 32-bit color.
3. Processor: 2.5 GHz minimum (Core i5 or higher).
4. Memory: 4 GB or higher.
5. Hard disk space: 5 GB minimum.
6. Microsoft TCP/IP Ethernet Network Adapter.
7. Microsoft approved 3-button mouse.
8. Microsoft Internet Explorer 6.0 or later.
9. A certified and supported graphics card.

IMPORTANT TERMS AND DEFINITIONS

Some important terms that will be used in this book while working with Creo Parametric are discussed next.

Entity

An element of the section geometry is called an entity. The entity can be an arc, line, circle, point, conic, coordinate system, and so on. When one entity is divided at a point, then the total number of entities are said to be two.

Dimension

It is the measurement of one or more entities.

Constraint

Constraints are logical operations that are performed on the selected geometry to make it more accurate in defining its position and size with respect to the other geometry.

Parameter

It is defined as a numeric value or a word that defines a feature. For example, all dimensions in a sketch are parameters. The parameters can be modified at any time.

Relation

A relation is an equation that relates two entities.

Weak Dimensions and Weak Constraints

Weak dimensions and weak constraints are temporary dimensions or constraints that appear in light blue color. These are automatically applied to the sketch. They are removed from the sketch without any confirmation from the user. The weak dimensions or the weak constraints should be changed to strong dimensions or constraints if they seem to be useful for the sketch. This only saves an extra step of dimensioning the sketch or applying constraints to it.

Strong Dimensions and Strong Constraints

Strong dimensions and strong constraints appear in dark blue color. These dimensions and constraints are not removed automatically. All dimensions added manually to a sketch are strong dimensions.

Tip

When several strong dimensions or constraints conflict, Creo Parametric makes the constraints and dimensions appear in blue box, and prompts you to delete one or more of them.

MANAGING FILES

As discussed earlier, a new file is generated whenever you save an object. The number of files generated are directly proportional to the number of times you save that object. So, these files occupy a lot of disk space. The latest version of the file which is currently being used should be stored. Latest version refers to the highest number that is suffixed with the file name of that object. The rest of the files are old versions and should be deleted from the hard disk, if they are not required.

Note

*To save disk space, you should keep deleting the old versions of a file. This can be done by using the **File > Manage File > Delete Old Versions** option from the menu bar.*

MODEL TREE

The **Model Tree** stores and displays all features in a chronicle. You can select any desired feature of a model or an assembly from the **Model Tree** and apply different operations on the selected feature. You can also select the feature by right-clicking on it; a shortcut menu will be displayed. Move the cursor on the shortcut menu and choose the required option from it by using the left mouse button.

Tip

*Looking at the **Model Tree**, you can understand the method and approach used to create the model. Using the **Model Tree**, you can modify the features of a model. Generally, when you import a model in a different file format in Creo Parametric, the features of the model are not displayed in the **Model Tree** and therefore, you will not be able to modify it.*

UNDERSTANDING THE FUNCTIONS OF THE MOUSE BUTTONS

While working with Creo Parametric, it is important to understand the function of the three buttons of the mouse to make an efficient use of this device. The various combinations of the keys and three buttons of the mouse are listed below:

1. Figure 1-7 shows the functions of the left mouse button. The left mouse button is used to make a selection. Using CTRL+left mouse button, you can add or remove items from the selection set.

2. Figure 1-8 shows the functions of the right mouse button. The right mouse button is used to invoke the shortcut menus and to query select the items. When you bring the cursor on an item, it is highlighted in green color. Now, if you hold the right mouse button, a shortcut menu is displayed. Choose the **Pick From List** option from the shortcut menu; the **Pick From List** dialog box will be displayed. You can make selections from this dialog box.

3. Figure 1-9 shows the functions of the middle mouse button in the 3D mode. The middle mouse button is used to spin the model in the drawing area and view it from different directions.

   The CTRL+middle mouse button is used to dynamically zoom in and out the view. When you press and hold the CTRL+middle mouse button and move the cursor up, the view is reduced and you zoom out. When the mouse is moved down, the view is enlarged and you zoom in.

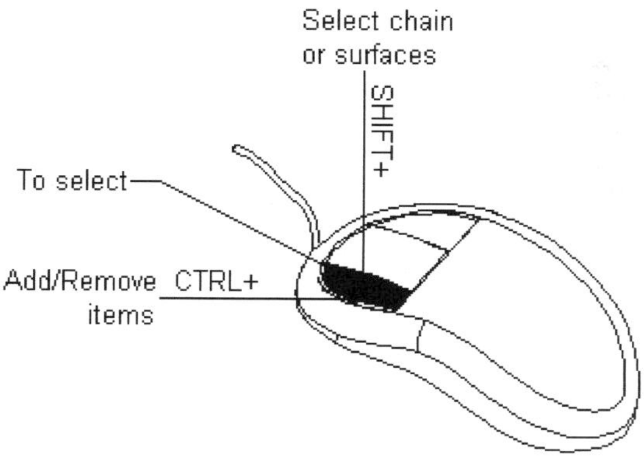

Figure 1-7 Functions of the left mouse button

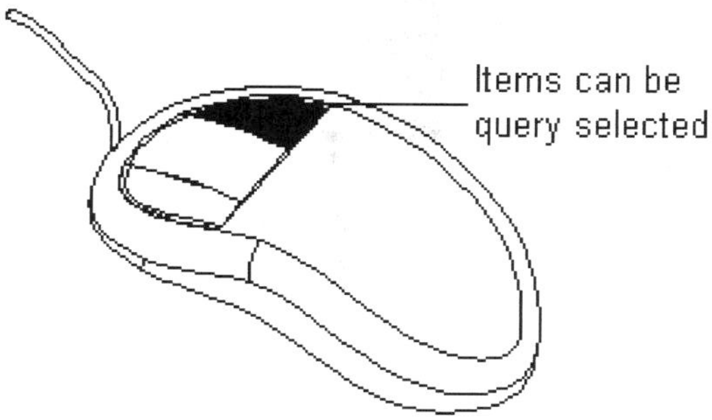

Figure 1-8 Functions of the right mouse button

When you use CTRL+middle mouse button and move the mouse horizontally, the model is rotated about a point that is specified as center.

The SHIFT+middle mouse button is used to pan the object on the screen.

4. Figure 1-10 shows the functions of the middle mouse button in the 2D mode (sketcher environment and **Drawing** mode). It is used to place dimensions in the drawing area. It is also used to confirm an option or to abort the creation of an entity.

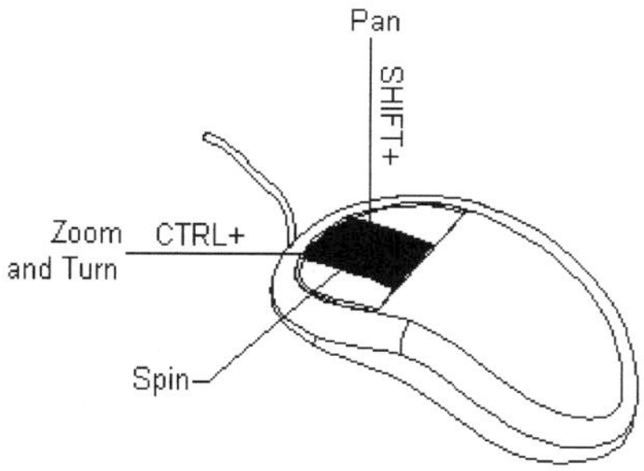

Figure 1-9 Functions of the middle mouse button in the 3D mode

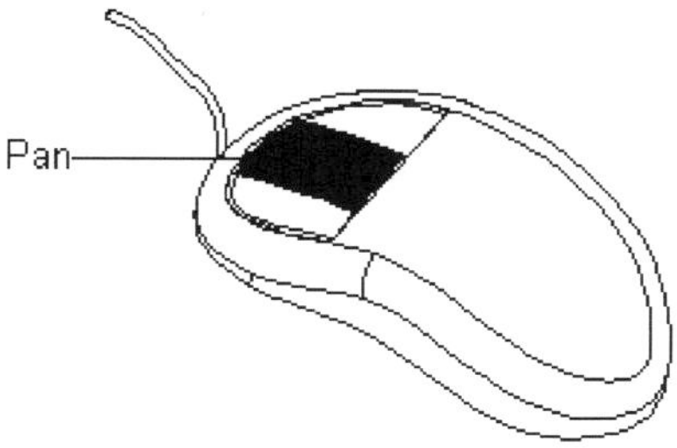

*Figure 1-10 Functions of the middle mouse button in the **Sketch** mode*

The middle mouse button is used to pan view in the **Sketch** mode and the **Drawing** mode.

Note

*When you spin the model with the **Spin Center** button turned on, the model rotates about the spin center origin. If this button is turned off, then the model rotates about the specified point.*

RIBBON

Before you start working on Creo Parametric, it is very important for you to understand the default **Ribbon** and tools in the main window. Figure 1-10 shows various default interface components in Creo Parametric. The **Ribbon** is composed of a series of groups, which are organized into tabs depending on their functionality. The groups in the **Ribbon** that initially appear on the screen are shown in Figure 1-11. You will notice that all the tools in the groups are not enabled.

These tools will be enabled only after you create a part or open an existing file. However, the tools that are required for the current session are already enabled. As you proceed to enter one of the modes provided by Creo Parametric, you will notice that the tools required by that mode are enabled. Additionally, to make the designing easy and user-friendly, this software package provides you with a number of groups. Different modes of Creo Parametric display different groups. Some of the frequently used groups are shown in Figure 1-11.

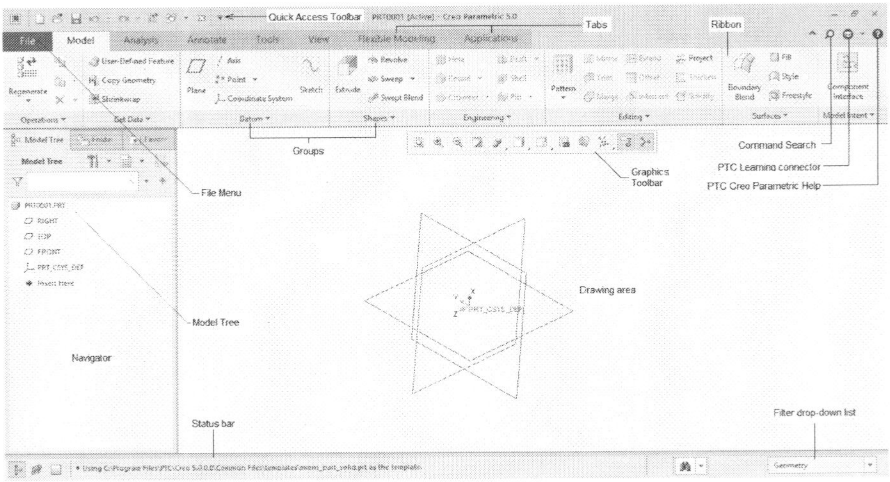

Figure 1-11 Default interface components

Creo Parametric Help

When you choose the **Creo Parametric Help** button, you are directed to **support.ptc.com** and the online Help page is displayed in your default browser. In this page, you can search for any help topic on Creo. The help database is accessible only to the users who have customer account with PTC.

Note

*If you want to change the predefined unit system, choose **File > Prepare > Model Properties** from the menu bar to display the **Model Properties** dialog box. In this dialog box, click on the **change** on the right of the **Units** option in the **Materials** head; the **Units Manager** dialog box will be displayed with the **System of Units** tab chosen. Next, select the desired unit system from the list box and then choose the **Set** button; the **Changing Model Units** message box will be displayed. Choose the **OK** button from the message box; the new unit system will be set and displayed with the red arrow on the left.*

RENDERING IN Creo Parametric

Rendering is a process of generating two-dimensional image of a three-dimensional scene or an object to make it more realistic. A rendered image makes it easier to visualize the shape and size of 3D object as compared to a wireframe or a shaded image. Rendering also helps you express your design intent to other people. You need to alter environments, lights, textures, and so on to get a high quality rendered image.

Chapter 2

Creating Sketches in the Sketch Mode-I

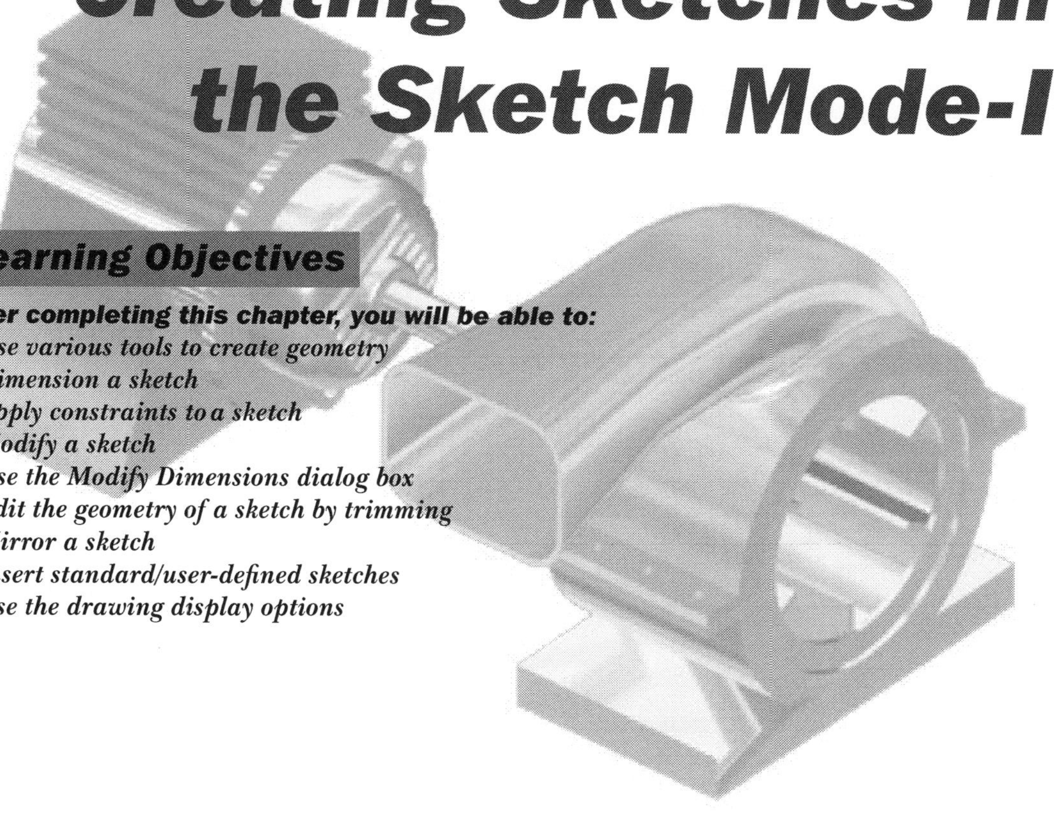

Learning Objectives

After completing this chapter, you will be able to:

- *Use various tools to create geometry*
- *Dimension a sketch*
- *Apply constraints to a sketch*
- *Modify a sketch*
- *Use the Modify Dimensions dialog box*
- *Edit the geometry of a sketch by trimming*
- *Mirror a sketch*
- *Insert standard/user-defined sketches*
- *Use the drawing display options*

THE SKETCH MODE

A sketch is a 2D entity that graphically captures an idea using lines, curves, constraints, and dimensions. To create a three-dimensional (3D) feature, it is necessary to draw or import a two-dimensional (2D) sketch. Almost all models designed in Creo Parametric consist of datums, sketched features, and placed features. When you enter the **Part** mode and select options to create any sketched feature, the system automatically takes you to the sketcher environment. In the sketcher environment, the sketch of the feature is created, dimensioned, and constrained. The sketches created in the **Sketch** mode are stored in the *.sec* format. After creating the sketch, you need to return to the **Part** mode to create the required feature.

In Creo Parametric, a sketch can be drawn using the **Sketch** mode in the sketcher environment or can be imported from other softwares. You can draw a 2D sketch of the product and assign the required dimensions and constraints to capture the design intent. By assigning dimensions and constrains to sketches, you can ensure predictable results when a model is modified.

DRAWING A SKETCH USING THE TOOLS AVAILABLE IN THE SKETCH TAB

In the sketcher environment, the **Sketch** tab is active in the **Ribbon** by default. There are various groups in this tab which contain tools to draw and modify a sketch and its dimensions. In this section, you will draw sketched entities using the tools available in the **Sketch** tab. The draw tools are; line, circle, arc, rectangle, ellipse, etc. The modify tools are; modify, divide, delete segment, corner, etc.

DIMENSIONING THE SKETCH

The basic purpose of dimensioning is to locate and control the size of geometric entities with some reference. Dimensions are created to capture the design intent because these dimensions are displayed when you edit the model and when you create drawings of the model.

In Creo Parametric, sketched entities are dimensioned and constrained automatically while sketching. However, you need to add some additional dimensions and constraints to the sketch to make it fully constrained. The dimensioning tools are available in the **Dimension** group of the **Sketch** tab, refer to Figure 2-1. The **Dimension** tool in this group is used to manually dimension the entities.

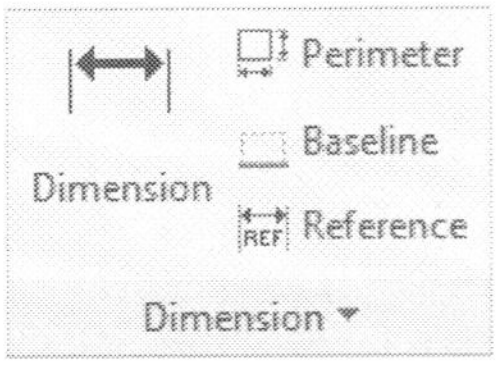

***Figure 2-1** The **Dimension** group*

Note

*If you do not want the weak dimensions to be applied automatically, then clear the **Show weak dimensions** check box from **File > Options > Sketcher > Object display settings**.*

Converting a Weak Dimension into a Strong Dimension

As discussed earlier when you draw a sketch, some weak dimensions are automatically applied to the sketch. These dimensions are displayed in light blue color. As you proceed to manually dimension the sketch, these weak dimensions are automatically deleted without any confirmation.

When you select a weak dimension from the drawing area the selected dimension will be highlighted in green color and a mini popup toolbar will be displayed near the selected dimension. From the mini popup toolbar, choose the **Strong** option, as shown in Figure 2-2, and press the middle mouse button; the select dimension will be converted into strong dimension. Alternatively, press CTRL+T to convert the selected dimension into a strong dimension. Note that a strong dimension is displayed in blue color.

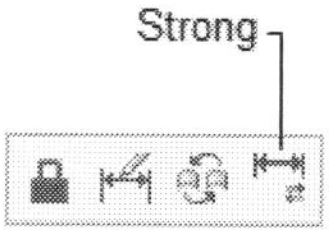

***Figure 2-2** The **Strong** option in the dimension mini popup toolbar*

WORKING WITH CONSTRAINTS

Constraints are rules that are enforced on the sketched geometries. In other words, constraints are the logical operations that are performed on the selected geometries to fully define their size, shape, orientation, and location with respect to other geometries. One of the key benefits of using constraints in a sketch is that it reduces the number of dimensions that are required to fully constraint that sketch.

Creo Parametric automatically applies some constraints to geometries while sketching. For example, if you are creating a line which is nearly parallel to another line, Creo Parametric will automatically place and highlight the parallel constraint symbol on the line being created. If you confirm the line creation, a line will be drawn parallel to the other line. Alternatively, you can also apply constraints manually. The procedure to do so is discussed next.

Types of Constraints

There are two types of constraints in Creo Parametric: **Geometry** and **Assembly**. In this chapter, you will learn about the **Geometry** constraints only and the **Assembly** constraints will be discussed in later chapters.

The geometric constraints are available in the **Constrain** group of the **Sketch** tab. The tools available in this group are shown in Figure 2-3.

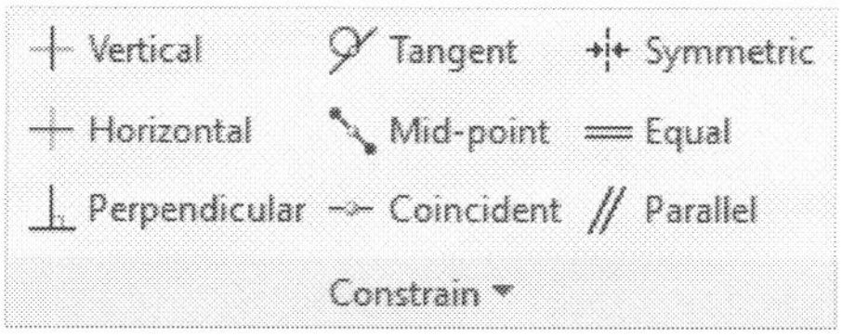

***Figure 2-3** The constraints in the **Constrain** group*

Note

*In Creo Parametric, the name of tools in several workbenches is displayed only in high screen resolution systems, such as 1900*1200 and 2880x1800. If the resolution is low, then only symbols are displayed.*

Note

1. To remove the temporary information, you can repaint the screen by choosing the ***Repaint*** *tool from the* ***Graphics*** *toolbar or pressing the CTRL+R keys.*

2. In the ***Sketch*** *mode, you can pan the sketch using the middle mouse button but in the* ***Part*** *mode, use SHIFT+middle mouse button to pan the model.*

TUTORIALS

Tutorial 1

In this tutorial, you will draw the sketch for model shown in Figure 2-4. The sketch of the model is shown in Figure 2-5. **(Expected time: 30 min)**

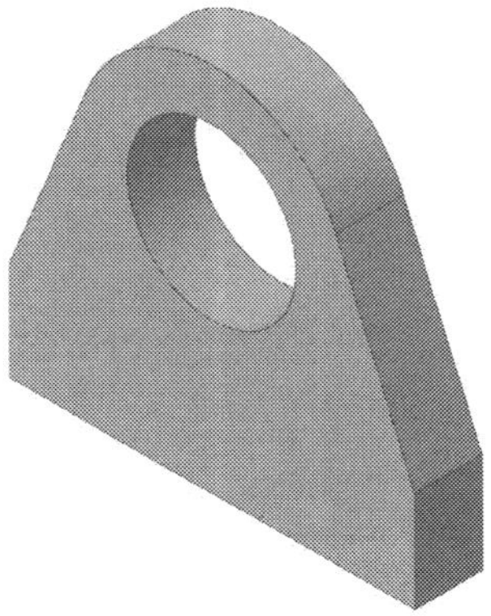

Figure 2-4 *Model for Tutorial 1*

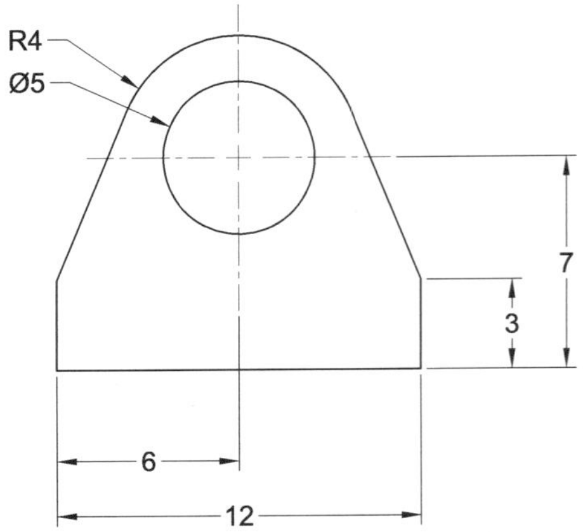

Figure 2-5 *Sketch of the model*

The following steps are required to complete this tutorial:

a. Start Creo Parametric session.
b. Set the working directory and create a new sketch file.
c. Draw lines by using the **Line Chain** tool, refer to Figures 2-6 and 2-7.
d. Draw an arc and a circle, refer to Figures 2-8 through 2-10.
e. Dimension the sketch and then modify the dimensions of the sketch, refer to Figure 2-10.
f. Save the sketch and close the file.

Starting Creo Parametric

1. Start Creo Parametric by double-clicking on the **Creo Parametric** icon on the desktop of your computer.

Setting the Working Directory

After the Creo Parametric session is started, the first task is to set the working directory. A working directory is a directory on your system where you can save the work done in the current session of Creo Parametric. You can set any existing directory on your system as the working directory. Since it is the first tutorial of this chapter, you need to create a folder with the name *c02*.

1. Choose the **Select Working Directory** option from the **Manage Session** flyout of the **File** menu; the **Select Working Directory** dialog box is displayed.

2. In this dialog box, browse to *C:\Creo-5.0* folder. If this folder does not exist, create this folder before setting the working directory.

   For selecting *C:\Creo-5.0,* click on the arrow at the right of the **C:** option in the top address box of the **Select Working Directory** dialog box; the folders in the **C** drive are displayed in a flyout. Now, choose **Creo-5.0** from the flyout, refer to Figure 2-6. Alternatively, you can use the **Folder Tree** node available at the bottom left corner of the screen to set the working directory. To do so, click on the **Folder Tree** node; the **Folder Tree** expands. In the **Folder Tree**, browse to the desired location using the nodes corresponding to the folders. Select the required folder and choose the **OK** button; the selected folder will become the current working directory.

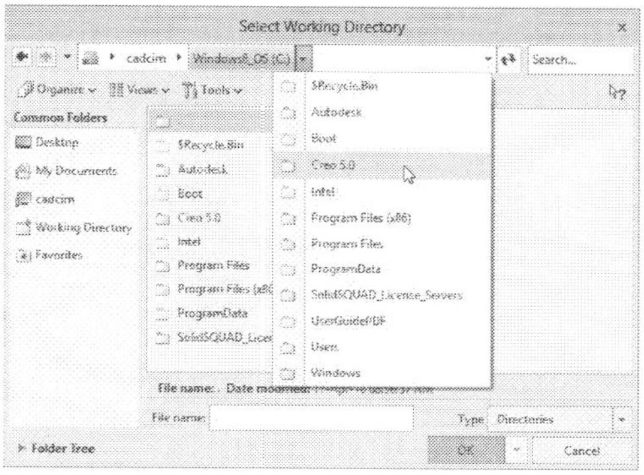

***Figure 2-6** The **Creo5.0** folder chosen from the flyout*

3. Choose the **Organize** button from the top pane of the **Select Working Directory** dialog box to display the flyout. From the flyout, choose the **New Folder** option; the **New Folder** dialog box is displayed.

4. Enter **c02** in the **New Directory** edit box and choose the **OK** button from the **New Folder** dialog box; a new folder named *c02* is created in ***C:\Creo-5.0*** location.

5. Choose the **OK** button from the **Select Working Directory** dialog box to set the working directory to *C:\Creo-5.0\c02*; the message **Successfully changed to C:\Creo-5.0\c02 directory** is displayed in the message area.

Starting a New Object File

Any sketch drawn in the **Sketch** mode is saved with the *.sec* file extension. This file format is one of the file formats available in Creo Parametric.

1. Choose the **New** button from the **Data** group in the **Ribbon** or **Quick Access** toolbar or press CTRL+N; the **New** dialog box is displayed. In this dialog box, select the **Sketch** radio button from the **Type** area; the default name of the sketch appears in the **Name** edit box.

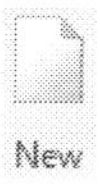

2. Enter **c02tut01** in the **Name** edit box and choose the **OK** button.

   You are in the sketcher environment of the **Sketch** mode. When the **Sketch** mode is invoked, the **Show Navigator** is displayed on the left in the drawing area.

3. Choose the **Show Navigator** button available at the bottom left corner of the screen to close the **Show Navigator**. On closing the tree, the drawing area is increased.

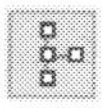

Drawing the Lines of the Sketch

You need to start drawing the sketch with the right vertical line.

1. Choose the **Line Chain** tool from the **Line** drop-down available in the **Sketching** group.

2. Specify the start point by clicking on the right in the drawing area. One end of the line is attached to the cursor. Move the cursor down to an approximate length.

   Notice that when the cursor moves vertically downward, a green colored symbol appears in the drawing area, next to the line. Now, if you draw a line, the vertical constraint will be applied to it.

3. Click to specify the endpoint of the line. The vertical constraint is applied to the line, but it is not visible in the drawing area until the line creation is active.

   Also, another rubber-band line is attached to the cursor with its start point at the endpoint of the last line.

4. Move the cursor horizontally toward the left; a horizontal rubber-band line extends to the left as you move the mouse.

   Notice that when the cursor moves horizontally toward the left, a green colored symbol appears in the drawing area next to the line. Now, if you draw a line, a horizontal constraint will be applied to it.

5. After getting the desired size of the line created, click to end the line. The horizontal constraint is applied to the line, but it is not visible in the drawing area until the line creation is active.

6. Move the cursor upward in the drawing area; a vertical rubber-band line extends as you move the mouse. As you move the cursor upward, you will notice that at a particular point where the length of the left vertical line is equal to the length of the right vertical line, a = symbol is displayed on the vertical line being created and the right line is highlighted in green color. This symbol suggests that the equal length constraint is applied to the two vertical lines.

7. When equal constraint appears on the vertical line, click to specify the endpoint of the vertical line. The rubber-band line is still attached to the cursor.

   You can also apply constraints later. However, to save an extra step of adding the constraints, you will use the constraints that are applied automatically while drawing.

8. Move the cursor to size the line and specify the endpoint of the left inclined line, as shown in Figure 2-7.

9. Press the middle mouse button to end the line creation.

10. The line option is still active. Move the cursor close to the top end of the right vertical line; the cursor snaps to the point that is at equal length of the left vertical line. Select the point by clicking.

11. Size the inclined line and specify the endpoint of the right inclined line. Press the middle mouse button twice; lines are created and all the constraints that you have applied become visible, refer to Figure 2-8. Now, you need to draw the arc and the circle.

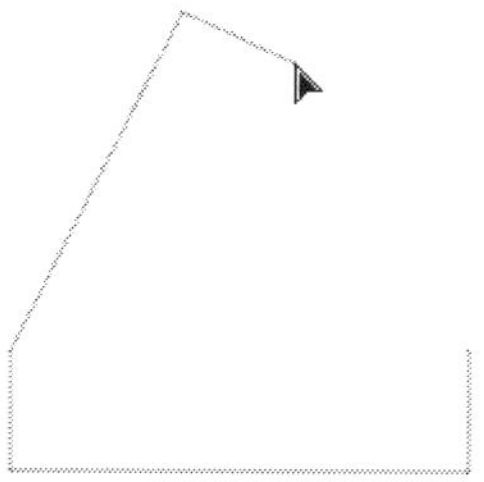

Figure 2-7 *Sketch with left inclined line*

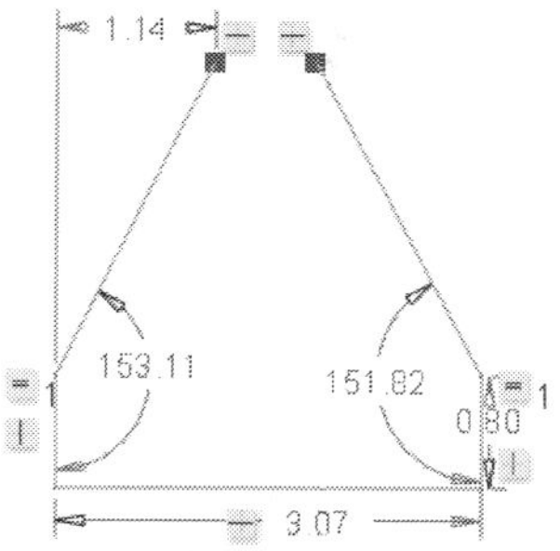

Figure 2-8 *Sketch of the line drawn with weak dimensions*

Note

1. The numbers in blue color with the constraint symbols refer to the label assigned to same type of constrained geometries in the sketch.

2. The horizontal and equal constraint = appear in green color when selected. The label with the constraint appears in blue color which indicates that the constraint is strong. This means you cannot change the orientation of the line until you delete the constraint applied to the line.

Drawing the Arc

1. Choose the **3-Point/Tangent End** tool from the **Arc** drop-down in the **Sketching** group; you are prompted to select the start point of the arc.

2. Select the endpoint of the left inclined line; the Target symbol appears in green color.

3. Move the cursor along the tangent direction through a small distance; a rubber-band arc that is tangent to the inclined line appears. Move the cursor to the endpoint of the right inclined line and click on the end point of the right inclined line to create the arc. As you exit the **Arc** tool, the tangent constraint is applied to the end points of the arc which is indicated by the symbol .

Note

1. The weak dimensions in Figure 2-8 are not displayed in Figure 2-9 because after an arc is drawn, some of the weak dimensions automatically get deleted.

*2. If the tangent constraint symbol is not displayed on any of the inclined lines, apply the constraint manually by choosing the **Tangent** button from the **Constrain** group.*

Drawing the Circle

1. Choose the **Concentric** tool from the **Circle** drop-down; you are prompted to select an arc.

2. Select the arc by clicking on it. Move the mouse; a circle appears.

3. Click to select a point inside the sketch to draw the circle.

4. Press the middle mouse button to end the circle creation. The sketch is completed.

Dimensioning the Sketch

The right vertical line, the bottom horizontal line, the arc, and the circle are dimensioned automatically and the weak dimensions are applied to them. You will use these dimensions. Hence, there is no need to dimension these entities again.

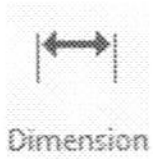

1. Choose the **Dimension** tool from the **Dimension** group.

2. Select the center of the arc and then the bottom horizontal line; the center turns red and the line turns green in color.

3. Place the dimension on the right of the sketch by pressing the middle mouse button.

4. Select the center of the arc and then the left vertical line; the center turns red and the vertical line turns green in color.

5. Press the middle mouse button to place the dimension below the sketch, refer to Figure 2-10.

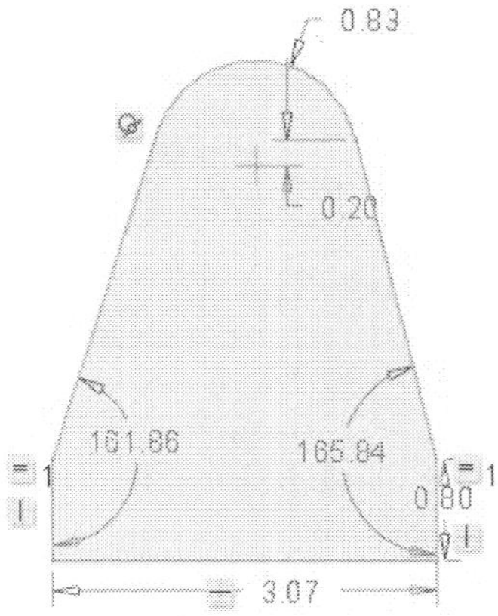

Figure 2-9 Sketch with arc

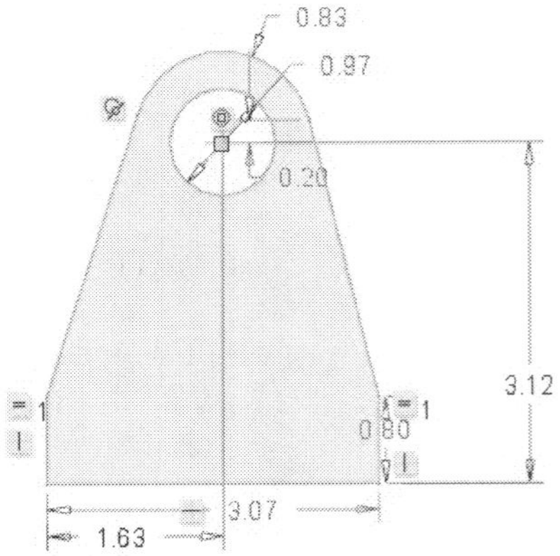

Figure 2-10 Sketch with all the entities, weak dimensions, and weak constraints

Modifying the Dimensions

The sketch is dimensioned with default values. You need to modify these values to the given values.

1. Select all dimensions by specifying a window around them.

Note

You can also use CTRL+ALT+A to select the entire sketch with dimensions.

2. When all dimensions turn green in color, choose the **Modify** tool from the **Editing** group; the **Modify Dimensions** dialog box is displayed.

   All dimensions in the sketch are displayed in this dialog box. Each dimension has a separate thumbwheel and an edit box. You can use the thumbwheel or the edit box to modify the dimensions. It is recommended that you use the edit boxes to modify the dimensions, if the change in the dimension value is large.

3. Clear the **Regenerate** check box and then modify the values of the dimensions.

   Once you clear this check box, any modification in a dimension value is not updated in the sketch. It is recommended that you clear the **Regenerate** check box when more than one dimension has to be modified.

   Notice that the dimensions that you select in the **Modify Dimensions** dialog box get enclosed in a blue box in the drawing area.

4. Modify all dimensions according to the dimensions shown in Figure 2-5. After modifying the dimensions, choose the **OK** button from the **Modify Dimensions** dialog box; the message **Dimension modifications successfully completed** is displayed in the message area.

Tip

You can modify the location of the dimensions as they appear on the screen by selecting and dragging them to a new location.

The completed sketch is shown in Figure 2-11.

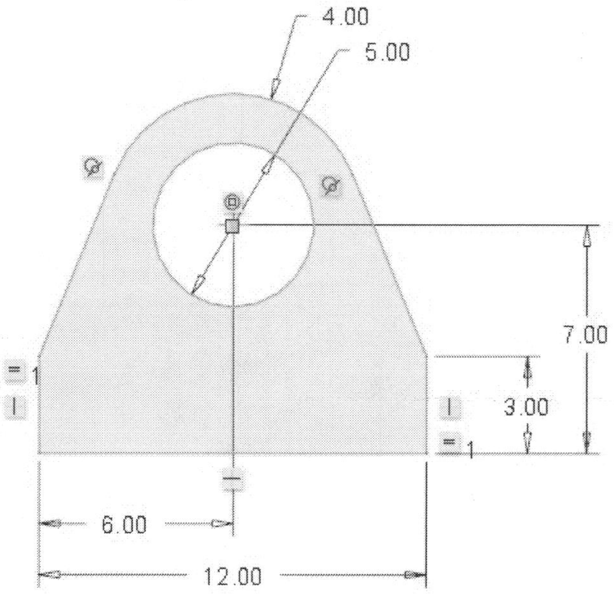

Figure 2-11 *The complete sketch with dimensions and constraints*

Saving the Sketch

Now, the sketch needs to be saved because you may need the sketch later in the **Part** mode to create a 3D model.

1. Choose the **Save** button from the **Quick Access** toolbar; the **Save Object** dialog box is displayed with the name of the sketch that you had entered earlier.

2. Choose the **OK** button; the sketch is saved.
3. After saving the sketch, choose the **Close** button from the **Quick Access** toolbar.

Tutorial 2

In this tutorial, you will draw the sketch of the model shown in Figure 2-12. The sketch of the model is shown in Figure 2-13. For your reference, all entities in the sketch are labeled alphabetically. Also, you have to print the sketch. **(Expected time: 30 min)**

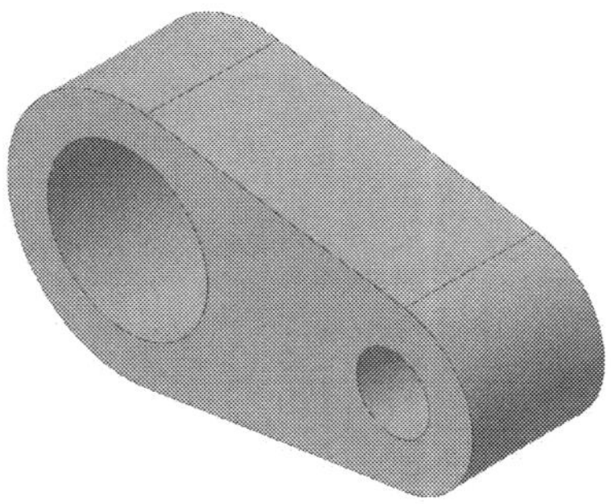

Figure 2-12 *Model for Tutorial 2*

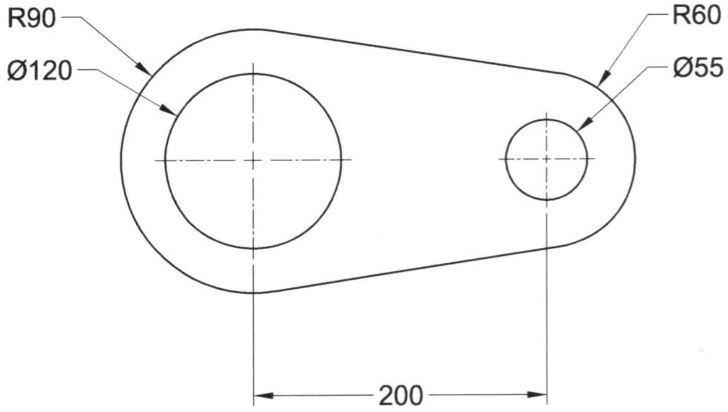

Figure 2-13 *Sketch of the model*

The following steps are required to complete this tutorial:

a. Set the working directory and create a new sketch file.
b. Draw the sketch by using the sketcher tools, refer to Figures 2-14 through 2-17.
c. Dimension the sketch and then modify the dimensions of the sketch, refer to Figure 2-18.
d. Save the sketch and print it.

Setting the Working Directory

The working directory was selected in Tutorial 1, therefore, you do not need to select the working directory again. But if a new session of Creo Parametric is started, then you have to set the working directory again by following the steps given next.

1. Open the Navigator by sliding it out. In the Navigator, the **Folder Tree** is displayed at the bottom. Click on the black arrow available on the right of the **Folder Tree**; the **Folder Tree** expands. Click on the arrow adjacent to the *Creo-5.0* folder in the Navigator; the contents of the *Creo-5.0* folder are displayed.

2. Now, right-click on the *c02* folder to display a shortcut menu. From the shortcut menu, choose the **Set Working Directory** option; the working directory is set to *c02*. Close the Navigator.

Starting a New Object File

1. Choose the **New** button from the **Data** group; the **New** dialog box is displayed. Select the **Sketch** radio button from the **Type** area of the **New** dialog box; the default name of the sketch appears in the **Name** edit box.

2. Enter **c02tut02** in the **Name** edit box. Choose the **OK** button to enter the sketcher environment of the **Sketch** mode.

Drawing the Circles

1. Choose the **Center and Point** tool from the **Circle** drop-down in the **Sketching** group and specify the center of the circle.

2. Move the cursor to size the circle and then click to complete the circle.

3. Draw another circle whose center is collinear with the center of the previous circle.

Figure 2-14 shows the two collinear circles drawn by using the **Center and Point** tool.

Drawing the Tangent Lines

1. Choose the **Line Tangent** tool from the **Line** drop-down in the **Sketching** group; you are prompted to select the start location on the arc or the circle.

2. Select the left circle at the top; a rubber-band line appears whose one end is attached to the circle and the other end is attached to the cursor.

3. Click on the top of the right circle; a tangent connecting the two circles is drawn.

4. Similarly, draw a tangent by selecting the two circles at the bottom.

Figure 2-15 shows the sketch after drawing the tangent lines.

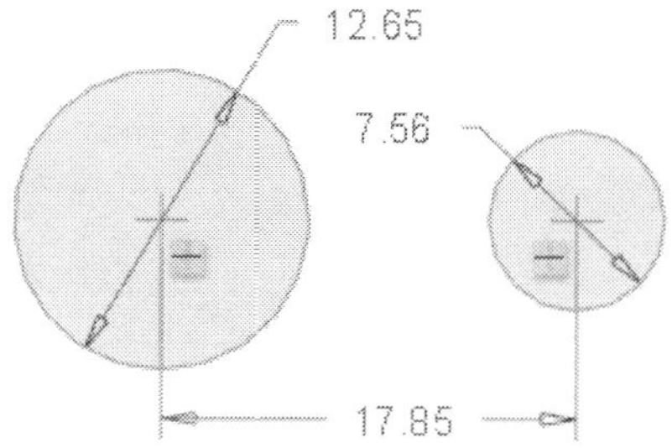

Figure 2-14 *Two collinear circles drawn using the* ***Centre and Point*** *tool*

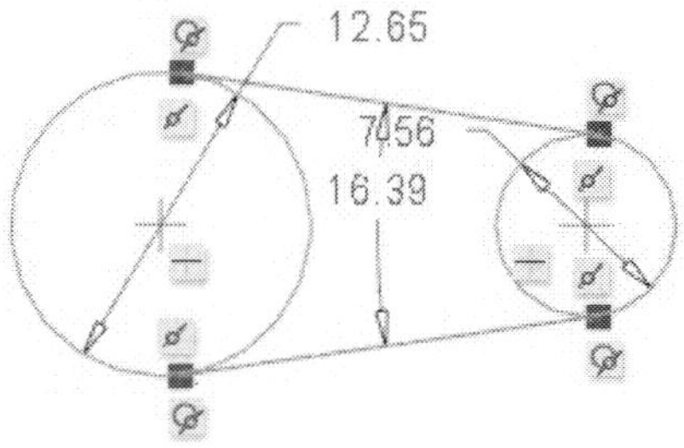

Figure 2-15 *Circles joined by tangent lines*

Trimming the Circles

As evident from Figure 2-14, the tangents that are drawn intersect the circles at the point where they meet the circle. Therefore, the part of the circle that is not required can be dynamically trimmed.

1. Choose the **Delete Segment** tool from the **Editing** group.

2. Select the two circles individually to trim them at the locations shown in Figure 2-16.

   Figure 2-17 shows the two circles after deleting the unwanted portions of the circle.

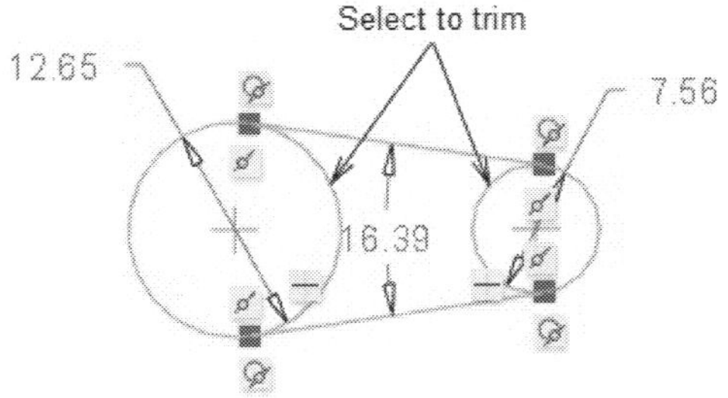

***Figure 2-16** Locations to trim*

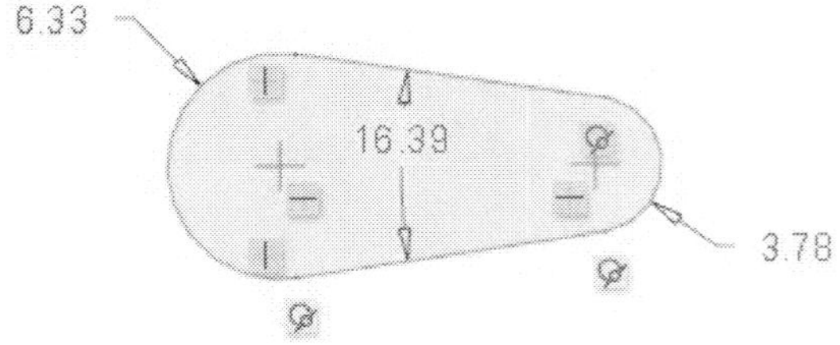

***Figure 2-17** Sketch after trimming*

Drawing the Circles

1. Choose the down arrow on the right of the **Center and Point** tool to display the flyout. Choose the **Concentric** tool from the flyout; you are prompted to select an arc.

2. Select the left arc and create a circle concentric to the arc. Similarly, select the right arc to create a concentric circle (refer to Figure 2-13).

   Notice that the radius dimension is applied to the two arcs, whereas the diameter dimension is applied to the circles. It is so because the arcs are applied with radius dimension and circles are applied with diameter dimension by default.

Dimensioning the Sketch

In order to fully define a sketch, you need to dimension it.

1. Choose the **Dimension** tool.

   Dimension

2. Select the centers of the two circles and place the dimension at the bottom of the sketch.

Modifying the Dimensions

1. Select all dimensions by defining a window.

Note
You can also use CTRL+ALT+A from the keyboard to select all entities and items in the sketch.

2. When all dimensions turn green in color, choose the **Modify** tool; the **Modify Dimensions** dialog box is displayed.

3. Clear the **Regenerate** check box and then modify the values of the dimensions. You will notice that the dimension that you edit in the **Modify Dimensions** dialog box is enclosed by a blue box in the drawing area.

4. When all dimensions are modified, choose the **OK** button from the **Modify Dimensions** dialog box; the message **Dimension modifications successfully completed** is displayed in the message area.

   The completed sketch is shown in Figure 2-18.

5. Save the sketch as discussed earlier. Next, you need to print the sketch.

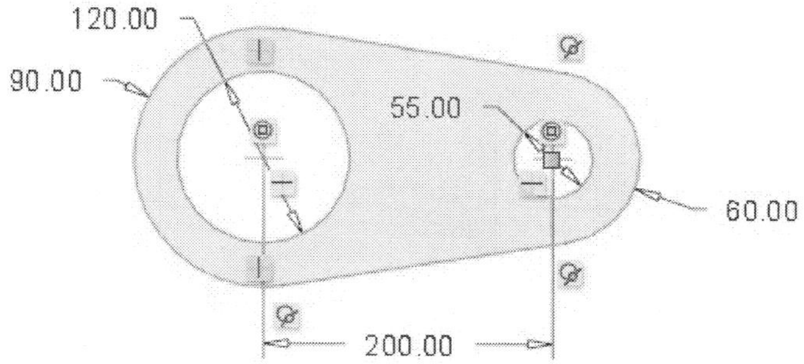

Figure 2-18 *Complete sketch with dimensions and constraints*

Printing the Sketch Using the Plot Option

1. Choose the **Print** option from the **File** menu or press CTRL+P; the **Printer Configuration** dialog box is displayed, as shown in Figure 2-19. 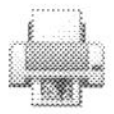

2. Choose the **Commands and Settings** button from this dialog box; a shortcut menu is displayed, as shown in Figure 2-20. 

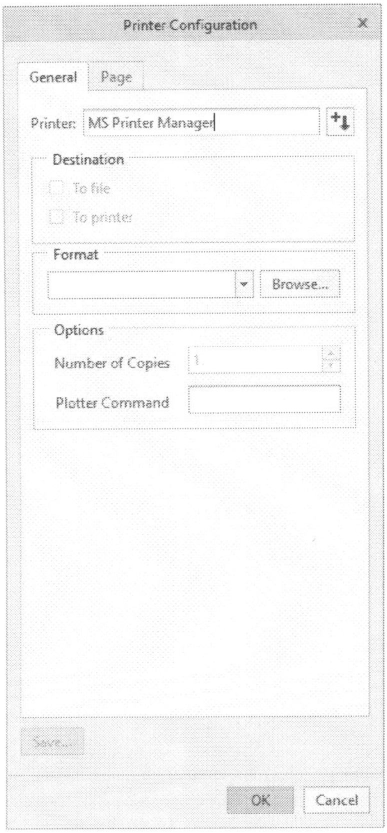

Figure 2-19 The ***Printer Configuration*** *dialog box*

Figure 2-20 Commands and Settings *shortcut menu*

3. Choose the **Add Printer Type** option from the shortcut menu; the **Add Printer Type** dialog box is displayed.

4. From the printers listed in the **Add Printer Type** dialog box, select the printer that is installed on your system and choose the **OK** button.

5. From the **Printer Configuration** dialog box, choose the **Page** tab and select the **A** option from the **Size** drop-down list, if not already selected; the dimensions of the sheet are set, by default.

6. Also, select the desired image resolution and the image depth from the dialog box under the **Resolution** area of the **Printer Configuration** dialog box.

7. Next, choose the **OK** button from the **Printer Configuration** dialog box to complete the printing.

EXERCISES

Exercise 1

In this exercise, you will draw the sketch of the model shown in Figure 2-21. The dimensions of the model are shown in Figure 2-22. **(Expected time: 30 min)**

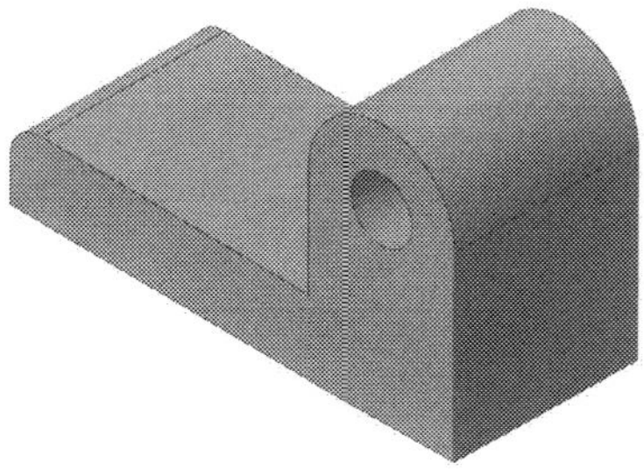

Figure 2-21 Solid model for Exercise 1

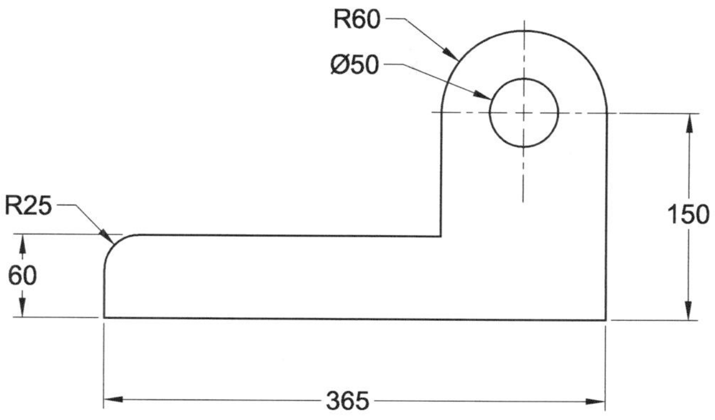

Figure 2-22 Dimensions of the model

Exercise 2

In this exercise, you will draw the sketch of the model shown in Figure 2-23. The dimensions of the model are shown in Figure 2-24. **(Expected time: 30 min)**

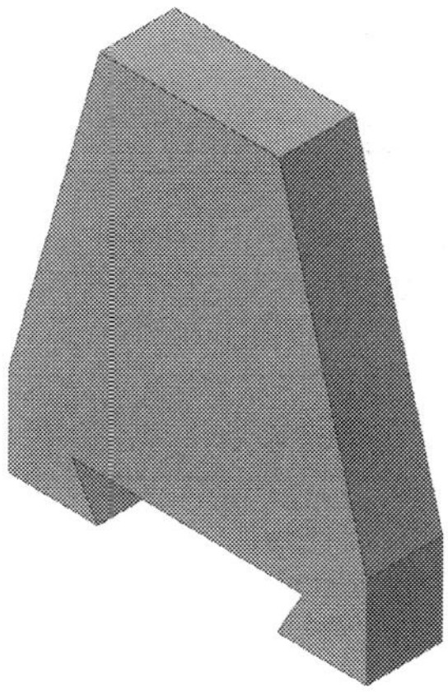

Figure 2-23 Solid model for Exercise 2

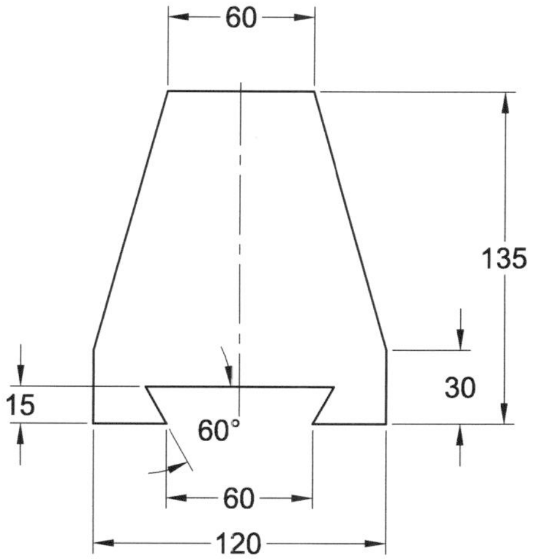

Figure 2-24 Dimensions of the model

This page is intentionally left blank

Chapter 3

Creating Sketches in the Sketch Mode-II

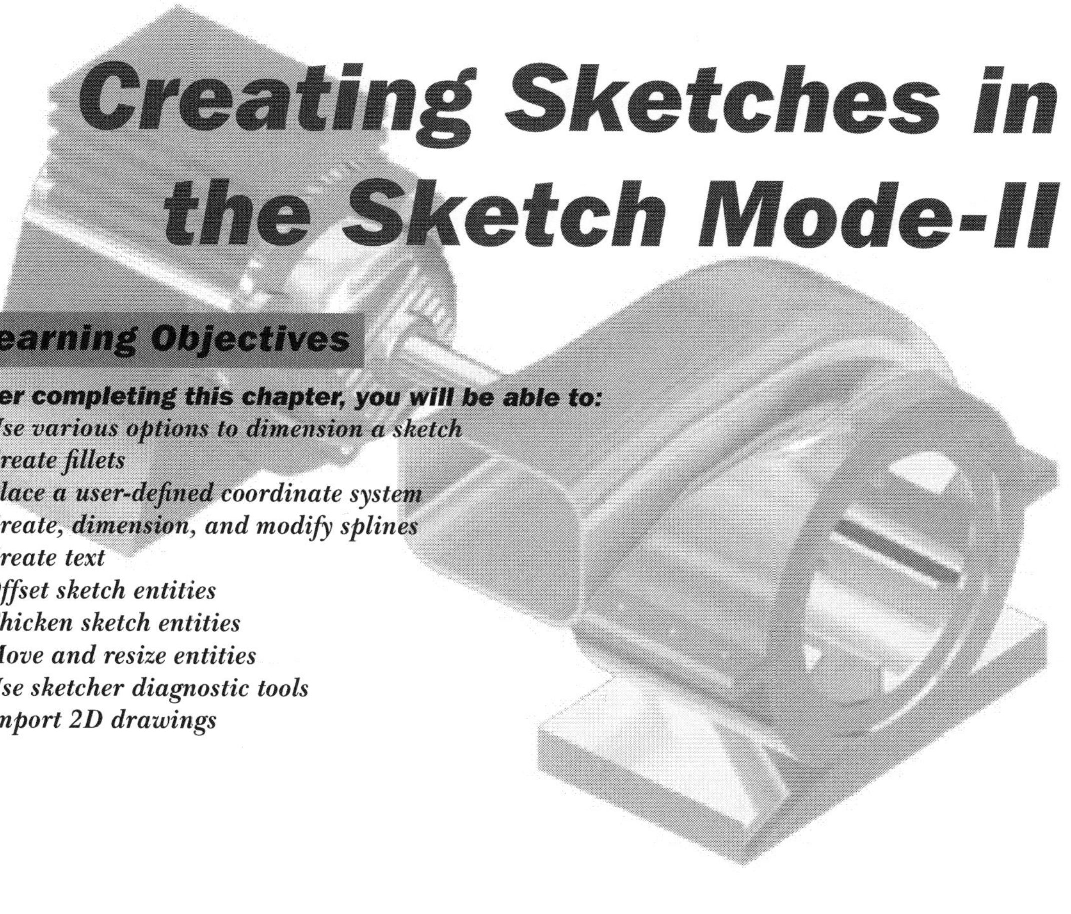

Learning Objectives

After completing this chapter, you will be able to:

- *Use various options to dimension a sketch*
- *Create fillets*
- *Place a user-defined coordinate system*
- *Create, dimension, and modify splines*
- *Create text*
- *Offset sketch entities*
- *Thicken sketch entities*
- *Move and resize entities*
- *Use sketcher diagnostic tools*
- *Import 2D drawings*

DIMENSIONING THE SKETCH

In Chapter 2, you learned dimensioning a sketch using the **Dimension** tool from the **Dimension** group. In this chapter, you will learn the use of the **Baseline** tool for dimensioning a sketch.

Dimensioning a Sketch Using the Baseline Tool

Ribbon: Sketch > Dimension > Baseline

In Creo Parametric, the **Baseline** tool is used to create dimensions in terms of horizontal and vertical distance values of an entity with respect to a specified baseline. This type of dimensioning in a drawing makes writing a CNC program for manufacturing a component easy.

The orientation of the dimension will depend upon the baseline dimension and the entity selected. Figure 3-1 shows a sketch dimensioned using the above-mentioned method. In this figure, the two baselines are dimensioned using the **Baseline** tool. Therefore, the dimensions of these lines are displayed as 0.00. The remaining lines are dimensioned by selecting the baseline dimension and then the required entity by using the **Dimension** tool.

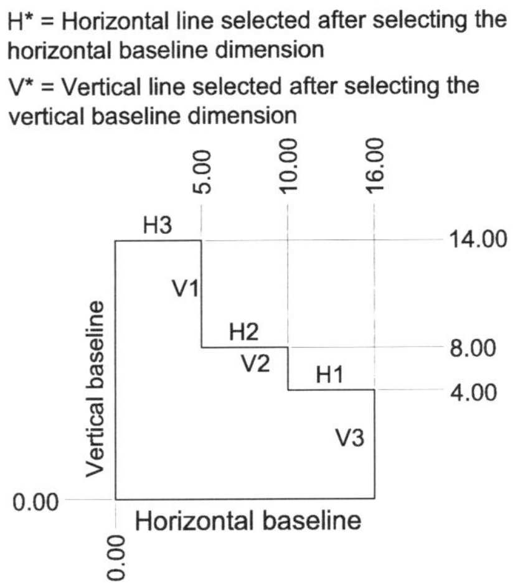

Figure 3-1 *Baseline dimensioning of a sketch*

Replacing the Dimensions of a Sketch Using the Replace Tool

Ribbon: Sketch > Operations > Replace

The **Replace** tool is used to replace a dimension with a new dimension in a sketch. To use this option, you must have a dimensioned sketch.

CREATING FILLETS

In the sketcher environment, you can create the following two types of fillets:

Creating Circular Fillets

Ribbon: Sketch > Sketching > Fillet drop-down > Circular

A circular fillet is the arc formed at the intersection of two lines, a line and an arc, or two arcs. This type of fillet is controlled by the radius or diameter dimension of the fillet. The resulting fillet will depend on the location where the elements are selected. Figure 3-2 shows two non-parallel lines and Figure 3-3 shows the circular fillet created between them and construction lines are added that extend to the intersection of both lines. The circular fillet thus created is an arc with its endpoints tangent to the two lines.

Figure 3-2 *Two lines that do not join*

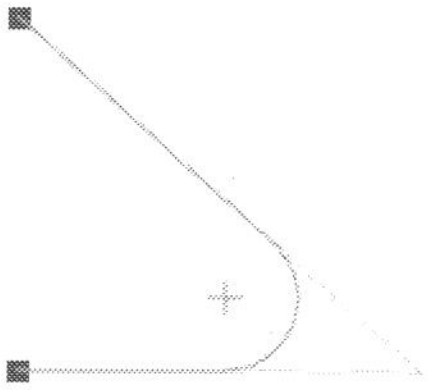

Figure 3-3 *Fillet created between the two lines*

Creating Elliptical Fillets

Ribbon: Sketch > Sketching > Fillet drop-down > Elliptical

An elliptical fillet is the arc in the form of an ellipse that joins two lines, two arcs, or a line and an arc. The advantage of elliptical fillets over circular fillets is that the geometry of elliptical fillets can be controlled by dimensions in two directions. Figures 3-4 and 3-5 illustrate the elliptical fillet. Notice that a tangent constraint is automatically applied when you create a fillet.

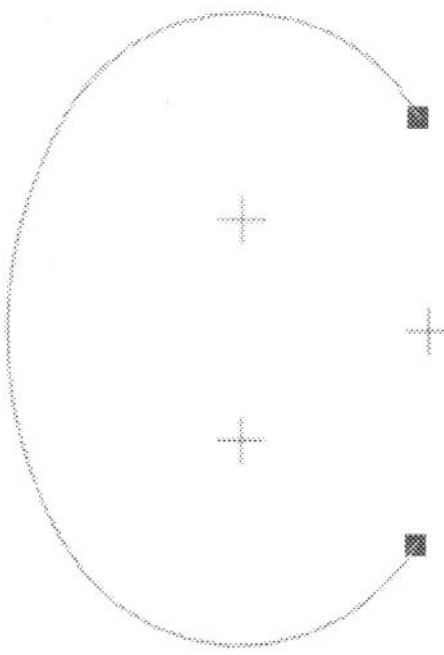

Figure 3-4 *Arcs to be filleted*

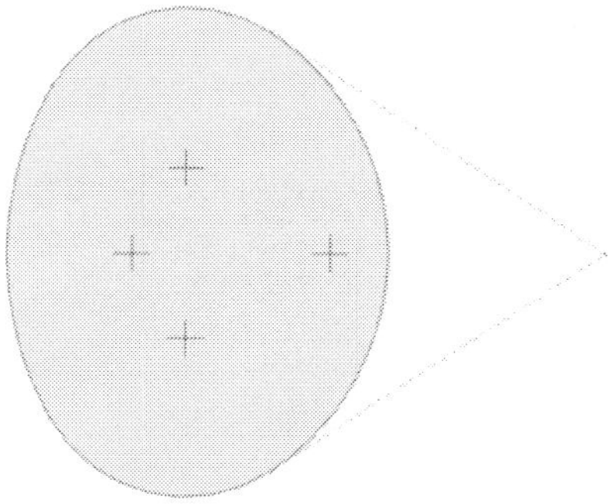

Figure 3-5 *Elliptical fillet created*

CREATING A REFERENCE COORDINATE SYSTEM

There are two types of coordinate systems, construction and geometric. As discussed earlier, the construction entities cannot be referenced outside the Sketcher environment whereas, the geometric entities can be. The **Coordinate System** tool in the **Datum** group is used to create a geometric coordinate system that will act as a reference for dimensioning. You can dimension the splines using the coordinate system. Thus, it provides you the flexibility to modify the spline points by specifying different coordinates with respect to the coordinate system.

Note

If you add a coordinate system to a sketch, it must be dimensioned. But if the coordinate system is placed at the endpoints of a line, an arc, a spline, or at the center of an arc or a circle, it need not be dimensioned. In other words, a coordinate system must be referenced to an entity in a sketch.

WORKING WITH SPLINES

Splines are curved entities that pass through a number of intermediate points. Generally, splines are used to define the outer surface of a model. This is because the splines can provide different shape to curves and the flexibility to modify the surfaces that result from the splines. The application of splines is widely found in vehicle body designing.

Tip

A dimension can be moved by pressing and holding the left mouse button on the dimension and moving it. The dimension text is replaced by a green colored box. You can drag the dimension to the desired location in the graphics window and release the left mouse button to place the dimension at that point.

WRITING TEXT IN THE SKETCHER ENVIRONMENT

Ribbon: Sketch > Sketching > Text

In the sketcher environment, the text is written using the **Text** tool from the **Sketching** group.

Tip

You can also change the height and width of the text by choosing the Width and Height options available in the mini popup toolbar.

OFFSETTING SKETCHED ENTITIES

Ribbon: Sketch > Sketching > Offset

In the **Sketch** mode, the entities in a sketch can be offset by a specified distance. You can create offset entities of lines, arcs, or splines by using the **Offset** tool available in the **Sketching** group. The offset can be a positive or a negative value. The options available in this tool can be used to create offset of a single entity, a chain of two or more entities, or a loop of two or more entities.

Note

If you offset entities by a large dimension, it is possible that Creo Parametric creates an offset chain that might have a different number of entities. For example, offsetting a spline by a large value causes the resulting spline to be broken into several pieces. If the offset value is changed, the system can piece together the broken spline so it becomes a single entity again.

THICKENING SKETCHED ENTITIES

Ribbon: Sketch > Sketching > Thicken

A thickened entity is a bidirectional offset entity whose components are separated by a user-defined distance. You can also add flat or circular end caps to connect the two offset entities. Depending on the offset and thickness values you enter, the resulting thickened entities can be on the either side, or both sides of the parent geometry.

ROTATING AND RESIZING ENTITIES

Ribbon: Sketch > Editing > Rotate Resize

The sketches can be translated, scaled, or rotated by using the **Rotate Resize** tool available in the **Editing** group. Select a sketch and then choose the **Rotate Resize** tool from the **Editing** group. On choosing this tool, the sketch, which consists of various entities, will act as a single entity. Also, the sketch appears green in color and is enclosed within a boundary box, as shown in Figure 3-6.

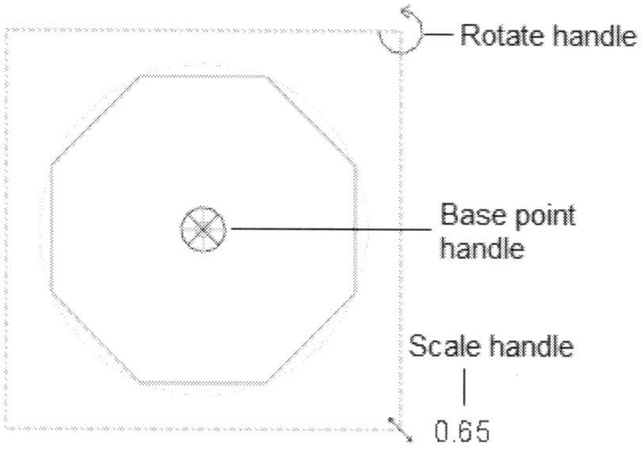

Figure 3-6 *Selected entities with move, rotate, and resize handles*

SKETCH DIAGNOSTIC TOOLS

As discussed earlier that a sketch-based feature uses a sketch to create desired geometry. Hence it is very necessary that the sketch should fulfill the requirements of a sketch-based feature. Creo Parametric provides you tools to diagnose and analyze the requirements of the sketch in the **Sketch** mode. The tools available in the **Inspect** group of the **Sketch** mode are used to diagnose the sketches.

IMPORTING 2D DRAWINGS IN THE SKETCH MODE

Ribbon: Sketch > Get Data > File System

The two-dimensional (2D) drawings when opened in the sketcher environment can be saved in the *.sec* format. The *.sec* file can be used to create a solid model. You can use a prestored sketch by importing it in the modeling environment. The **File System** button in the **Get Data** group is used to import 2D sketches from other type of files stored in your computer. Using this button, you can save time in drawing the same or similar section again. The file formats from which the data can be imported are shown in Figure 3-7.

When you choose the **File System** button from the **Get Data** group; the **Open** dialog box will be displayed, as shown in Figure 3-8. You can use this dialog box to select and open the file.

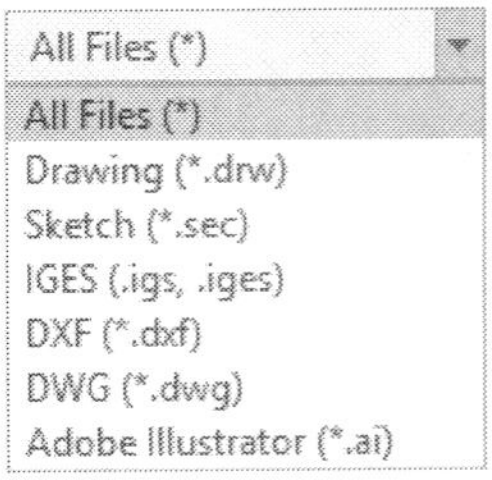

Figure 3-7 *File formats*

When you select a drawing file created in the **Drawing** mode of Creo Parametric, the draft entities of that file will be imported and selected drawing will be opened in a window. Also, you will be prompted to select the entities to copy from the window. Select the draft entities and then press the middle mouse button; the window disappears and a plus sign gets attached to the cursor indicating that you need to select a point on the Creo Parametric screen to insert the file. Select a point on the screen; the selected entities get inserted and are displayed within an enclosed boundary. Also, the **Import/Drawing** tab will be invoked. You can use this tab to set the position, scale, and orientation of the imported sketch. Note that if the *.drw* file does not consist of draft entities, no data will be imported.

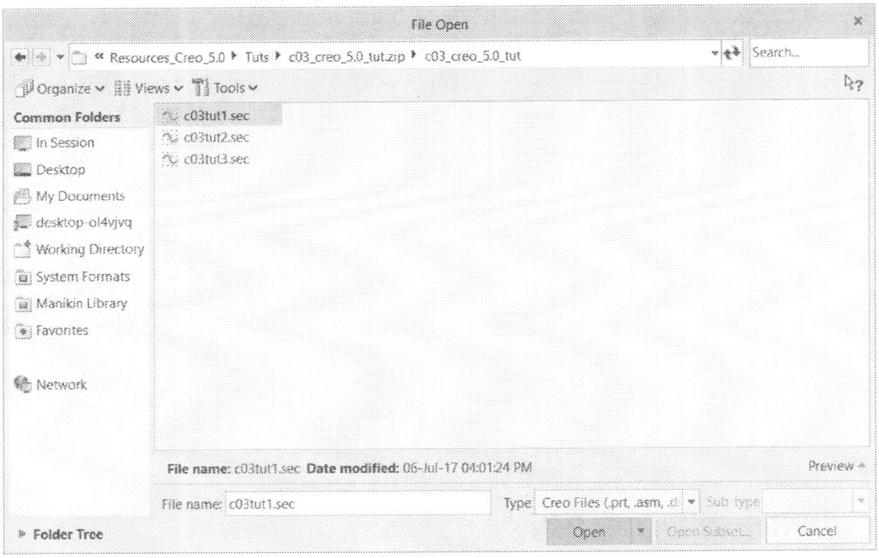

Figure 3-8 *The* ***Open*** *dialog box*

The section imported using the **File System** button in the current sketch is an independent copy. The imported section will be no longer associated with the source section. The units, dimensions, grid parameters, and accuracy are acquired from the current sketch.

TUTORIALS

Tutorial 1

In this tutorial, you will import an existing sketch that you had drawn in Tutorial 2 of Chapter 2. After placing the sketch, draw the keyway, as shown in Figure 3-9.

(Expected time: 15 min)

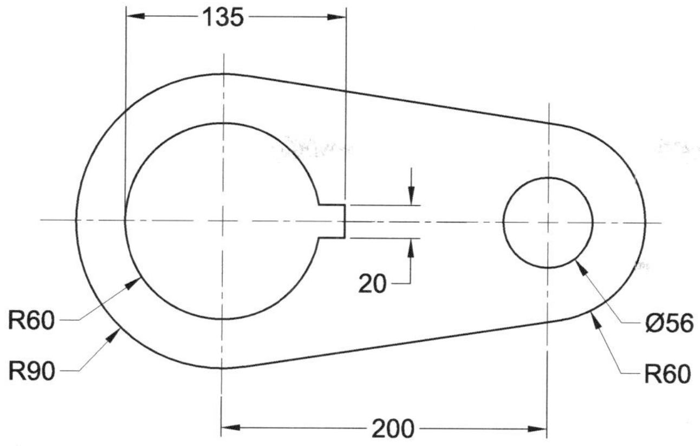

Figure 3-9 *Sketch for Tutorial 1*

The following steps are required to complete this tutorial:

a. Start Creo Parametric.
b. Set the working directory and create a new object file.
c. Import the section by using the **File System** button, refer to Figure 3-10.
d. Draw the keyway and dimension it, refer to Figures 3-11 and 3-12.
e. Modify the dimensions, refer to Figure 3-13.
f. Save the sketch and exit the sketcher environment.

Starting Creo Parametric

1. Start Creo Parametric by double-clicking on the Creo Parametric icon on the desktop of your computer.

Setting the Working Directory

When the Creo Parametric session starts, the first task is to set the working directory. As mentioned earlier, working directory is a directory on your system where you can save the work done in the current session of Creo Parametric. You can set any existing directory on your system as the working directory. Since this is the first tutorial of this chapter, you need to create a folder named *c03* in the *C:\Creo-5.0* folder.

1. Choose **Manage Session > Select Working Directory** option from the **File** menu; the **Select Working Directory** dialog box is displayed.

2. Select *C:>Creo-5.0*. If this folder does not exist, then first create it and then set the working directory.

   Alternatively, you can use the **Folder Tree** available on the bottom left corner of the screen to set the working directory. To do so, click on the **Folder Tree** node; the Folder Tree will expand. In the Folder Tree, browse to the desired location using the nodes corresponding to the folders and select the required folder. After selecting the folder, the **Folder Content** window will be displayed. Close this window and the selected folder will become your current working directory.

3. Choose the **Organize** button from the **Select Working Directory** dialog box or right-click in this dialog box to display a shortcut menu. From the shortcut menu, choose the **New Folder** option; the **New Folder** dialog box is displayed.

4. Enter **c03** in the **New Directory** edit box of the **New Folder** dialog box and then choose the **OK** button; a folder with the name *c03* is created at *C:\Creo-5.0*.

5. Choose the **OK** button from the **Select Working Directory** dialog box; *C:\Creo-5.0\c03* is set as the working directory.

Starting a New Object File

1. Choose the **New** button from the **Data** group; the **New** dialog box is displayed. Select the **Sketch** radio button from the **Type** area of the **New** dialog box; the default name of the sketch appears in the **Name** edit box.

2. Enter *c03tut1* in the **Name** edit box and choose the **OK** button.

   You are in the sketcher environment of the **Sketch** mode. When you enter the sketcher environment, the Navigator is displayed on the left in the graphics window. Slide-in the Navigator by clicking on the **Show Navigator** button present on the bottom left corner of the drawing area. Now, the drawing area is increased.

Importing the Section

1. Choose **Get Data > File System** from the **Ribbon**; the **Open** dialog box is displayed with the working directory as the current directory.

2. Click on the black arrow beside the **Creo-5.0** option in the address bar and choose **c02** from the flyout displayed. Make sure the **Sketch (*.sec)** option is selected in the **Type** drop-down list. Select *c02tut3.sec* and choose the **Open** button from the **Open** dialog box.

3. Move the cursor in the drawing area. Notice that the cursor is attached with a plus sign. Now, click anywhere in the drawing area to place the sketch. The sketch is displayed in the drawing area and the **Import Section** tab is displayed in the **Ribbon**.

4. Enter **1** as a scale factor in the scale edit box and choose the **OK** button to complete importing the sketch.

5. Choose the **Refit** button from the **Graphics** toolbar. The sketch, similar to the one shown in Figure 3-10, is displayed in the drawing area.

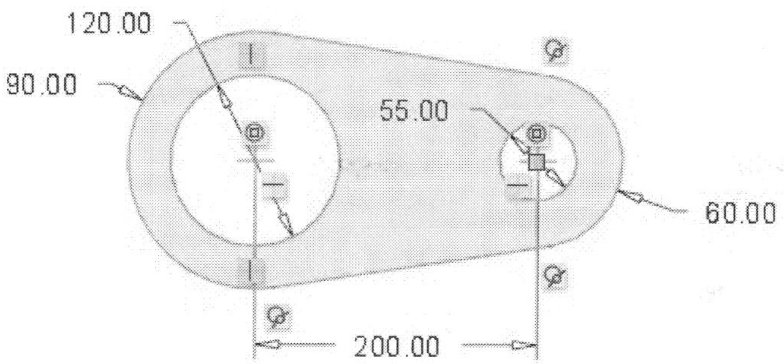

Figure 3-10 Sketch imported and placed in the current file

Drawing the Keyway

To create the keyway, you need to sketch a small rectangle and then remove the portion of the circle that lies between the rectangle.

1. Choose the **Line Chain** tool from the **Line** drop-down in the **Sketching** group.

2. Draw the keyway, as shown in Figure 3-11; the weak dimensions and constraints are automatically applied to the sketch of the keyway.

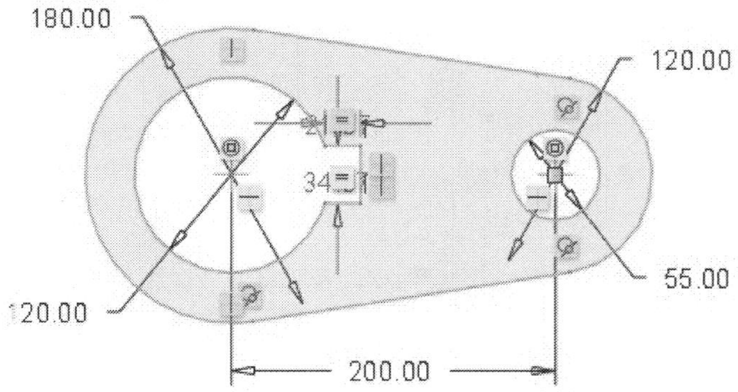

Figure 3-11 Sketch of the keyway with weak dimensions and constraints

The horizontal lines of the keyway and the circle intersect at the points where the lines meet the circle. The portion of the circle that lies between the two horizontal lines of the keyway needs to be deleted from the circle.

3. Choose the **Zoom In** button from the **Orientation** group of the **View** tab; the cursor is converted into a magnifying glass symbol.

4. Draw a window around the keyway to zoom in it. Now, the display of the keyway is enlarged.

5. Choose the **Delete Segment** tool from the **Editing** group.

6. Click to select the part of the circle that lies between the two horizontal lines; the selected part is deleted.

7. Choose the **Refit** button from the **Graphics** toolbar to view the full sketch.

Dimensioning the Keyway

Now, you need to apply dimensions to the keyway.

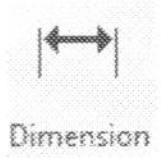

1. Choose the **Dimension** tool from the **Dimension** group.

2. Dimension the keyway, as shown in Figure 3-12.

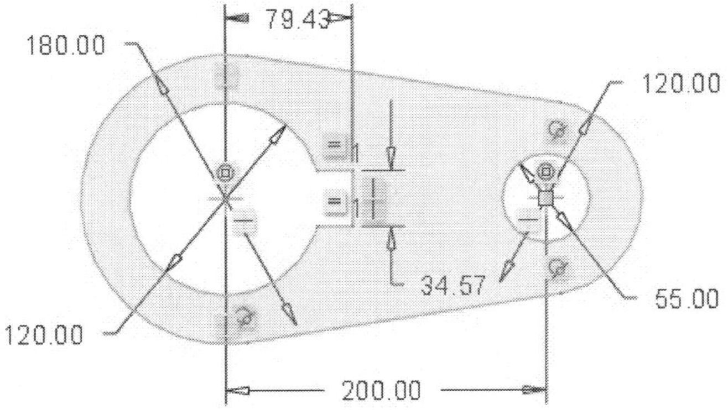

Figure 3-12 *Sketch after dimensioning the keyway*

Modifying the Dimensions

The dimensions of the keyway need to be modified as per the given dimension values.

1. Select the three dimensions of the keyway by pressing CTRL+left mouse button.

2. Choose the **Modify** tool from the **Editing** group; the **Modify Dimensions** dialog box is displayed.

3. Clear the **Regenerate** check box and then modify the dimensions of the keyway. When you clear the check box, the sketch does not regenerate as you modify the dimensions.

   The dimension that you edit in the **Modify Dimensions** dialog box gets enclosed in a blue box in the sketch.

4. Modify all dimensions. Refer to Figure 3-9 for dimension values.

5. After the dimensions are modified, choose the **OK** button from the **Modify Dimensions** dialog box; the message **Dimension modifications successfully completed** is displayed in the message area.

   The sketch after modifying the dimension values of the sketch is shown in Figure 3-13.

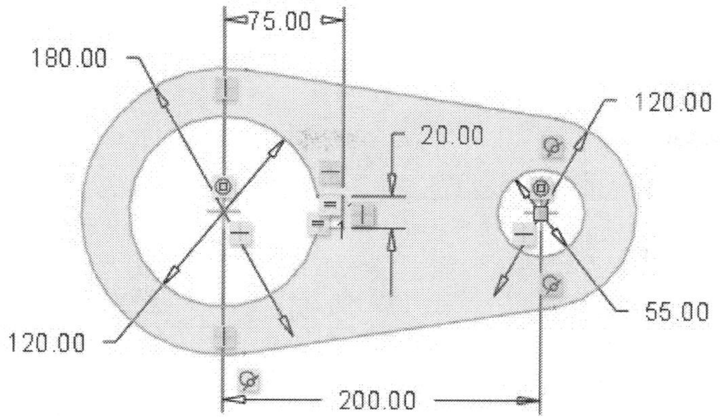

Figure 3-13 Sketch after modifying the dimensions

Saving the Sketch

As you may need the sketch later, you must save it.

1. Choose the **Save** button from the **File** menu; the **Save Object** dialog box is displayed with the name of the sketch entered earlier.

2. Choose the **OK** button; the sketch is saved.

3. After saving the sketch, choose the **Close** button from the **Quick Access** toolbar to exit the **Sketch** mode.

Tutorial 2

In this tutorial, you will draw the sketch of the model shown in Figure 3-14. The sketch is shown in Figure 3-15. **(Expected time: 30 min)**

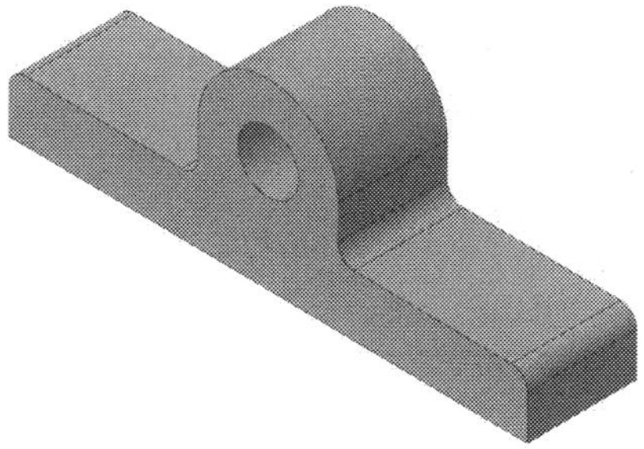

Figure 3-14 Model for Tutorial 2

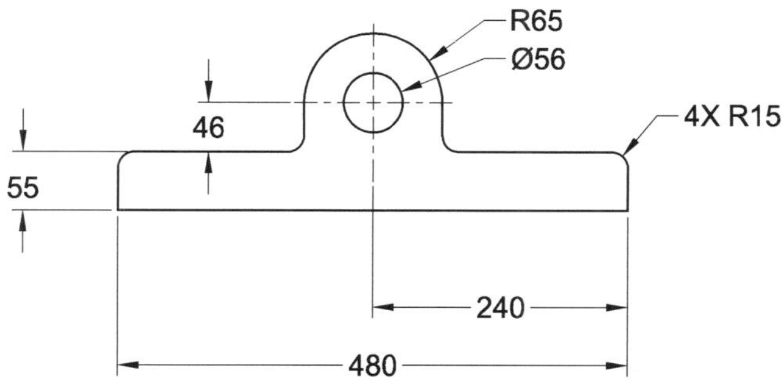

Figure 3-15 Sketch of the model

The following steps are required to complete this tutorial:

a. Set the working directory and create a new object file.
b. Draw the sketch using sketcher tools, refer to Figures 3-16 and 3-17.

c. Apply fillets at two corners of the sketch, refer to Figures 3-18 and 3-19.
d. Dimension the sketch, refer to Figure 3-20.
e. Modify dimensions of the sketch, refer to Figure 3-21.
f. Save the sketch and exit the **Sketch** mode.

Setting the Working Directory

The working directory was selected in Tutorial 1, and therefore there is no need to select the working directory again. But if a new session of Creo Parametric start, then you have to set the working directory again by following the steps given next.

1. Open the Navigator by clicking on the **Show Navigator** button in the left edge of the Creo Parametric window; the Navigator slides out. In the Navigator, the Folder Tree is displayed at the bottom. Click on the black arrow that is available at the right-side of the Folder Tree to expand it.

2. Click on the node adjacent to the *Creo-5.0* folder in the Navigator to display the content of this folder.

3. Now, right-click on the *c03* folder to display a shortcut menu. From this shortcut menu, choose the **Set Working Directory** option; *c03* is set as the working directory.

4. Close the Navigator by clicking on the **Show Navigator** on the right edge of the Navigator; the Navigator slides in.

Starting a New Object File and Drawing the Sketch

1. Start a new object file in the **Sketch** mode. Name the file as *c03tut2*.

2. Choose the **Line Chain** tool from the **Line** drop-down of the **Sketching** group.

3. Draw the lines with constraints, as shown in Figure 3-16.

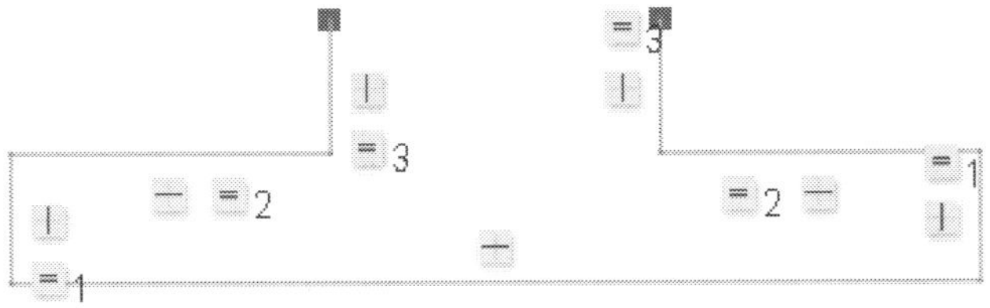

Figure 3-16 *Lines in the sketch with the dimensions turned off for clarity*

4. Choose the **3-Point / Tangent End** tool from the **Arc** drop-down available in the **Sketching** group. 

5. Select the endpoint of the left vertical line as the start point of the arc. Complete the arc at the endpoint of the right vertical line.

6. Choose the **Concentric** tool from the **Circle** drop-down available in the **Sketching** group; you are prompted to select an arc. 

7. Click on the arc; a rubber-band circle appears. Size the circle by moving the cursor and click to complete it. The sketch after drawing the circle is shown in Figure 3-17.

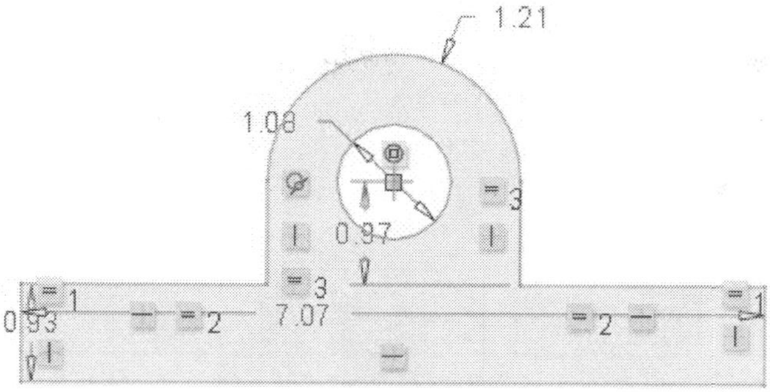

Figure 3-17 *Sketch after drawing the arc and the circle*

Note

*1. Choose the **Disp Dims** button from the **Display** group in the **View** tab to turn the dimensions on or off.*

2. Creo Parametric does not have the options like midpoint, endpoint, or center of an arc or a circle. However, while drawing a sketch, these options are applied in the form of weak constraints. For example, while drawing an entity, the endpoint of the entity snaps to the cursor. The middle point constraint appears when you bring the cursor near to the middle point of the line to draw another line.

Filleting the Corners

1. Choose the **Circular Trim** option from the **Fillet** drop-down in the **Sketching** group; you are prompted to select the two entities to be filleted. The corners that you need to fillet are shown in Figure 3-18.

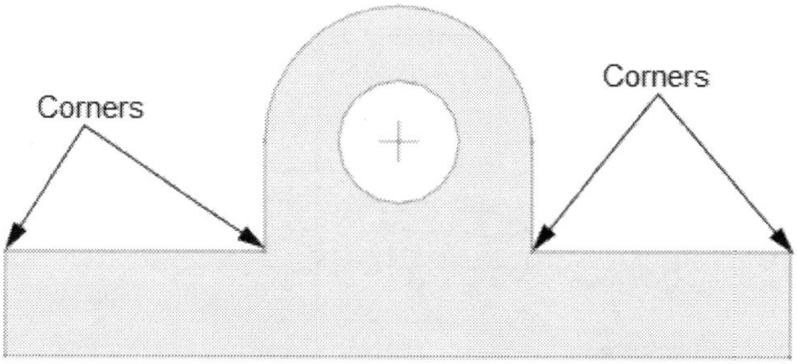

Figure 3-18 *Corners to be filleted*

2. Select the two entities one by one using the left mouse button to fillet the corners of these entities.

The sketch after creating the fillets is shown in Figure 3-19.

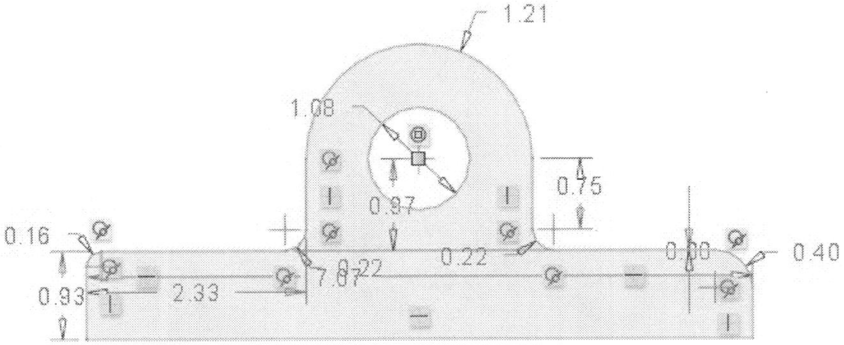

Figure 3-19 Sketch after creating fillets

Applying the Constraints

1. Choose the **Equal** tool from the **Constrain** group.
2. Click to select the fillets and apply the equal constraint to all fillets.

Dimensioning the Sketch

The weak dimensions are applied to the sketch automatically. These are not the required dimensions and therefore, you need to dimension the sketch manually.

1. Choose the **Dimens**ion tool from the **Dimension** group.
2. Dimension the sketch, as shown in Figure 3-20.

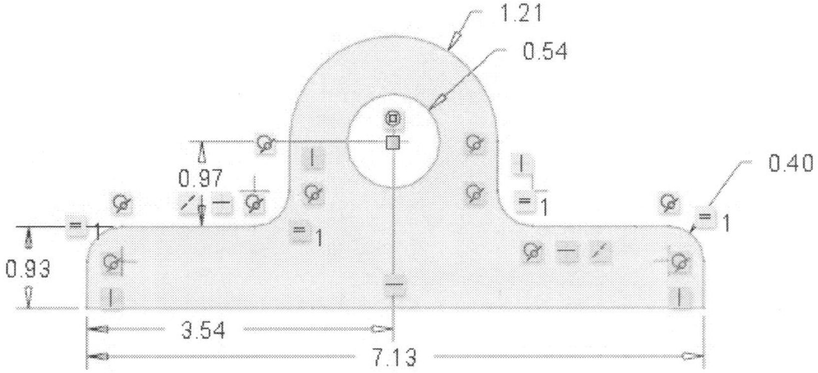

Figure 3-20 Sketch after dimensioning

Modifying the Dimensions

You need to modify the dimension values that are assigned to the sketch.

1. Select all dimensions using CTRL+ALT+A.
2. Choose the **Modify** tool from the **Editing** group; the **Modify Dimensions** dialog box is displayed.

3. Clear the **Regenerate** check box and then modify the values of the dimensions. If this check box is cleared, the sketch does not regenerate while modifying the dimensions.

   The dimension that you edit in the **Modify Dimensions** dialog box is enclosed in a blue box in the sketch.

4. Modify all dimensions. Refer to Figure 3-15 for dimension values.

5. Choose the **OK (Regenerate the section and close the dialog)** button from the **Modify Dimensions** dialog box.

   The sketch after modifying the dimension values is shown in Figure 3-21.

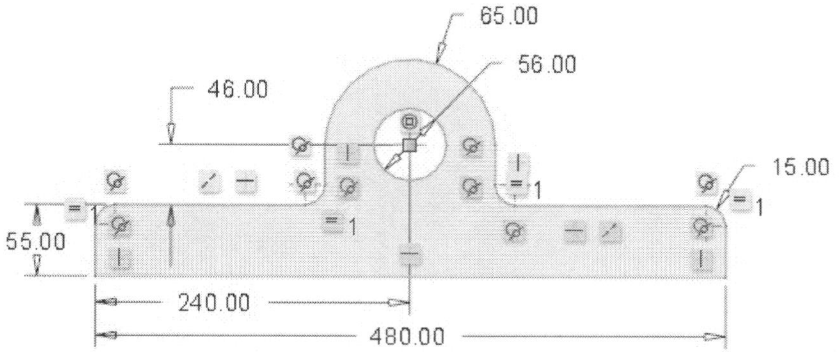

Figure 3-21 *Sketch after modifying the dimensions*

Saving the Sketch and Exiting the Sketch Mode

1. Choose the **Save** button from the **File** menu and save the sketch.

2. Choose the **Close** button from the **Quick Access** toolbar to exit the **Sketch** mode.

EXERCISES

Exercise 1

In this exercise, you will draw the sketch of the model shown in Figure 3-22. The sketch to be drawn is shown in Figure 3-23. **(Expected time: 30 min)**

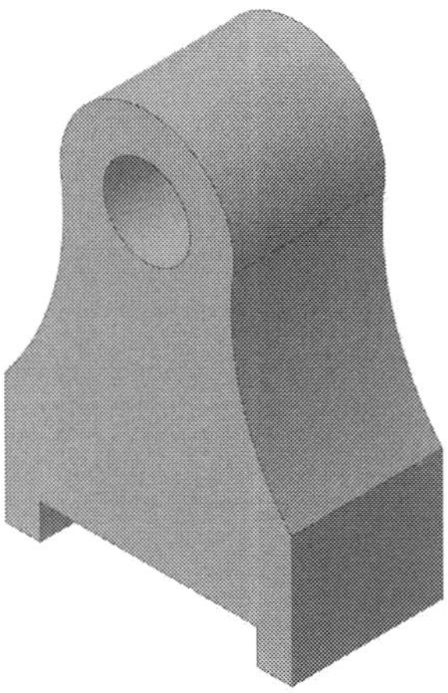

Figure 3-22 *Solid model for Exercise 1*

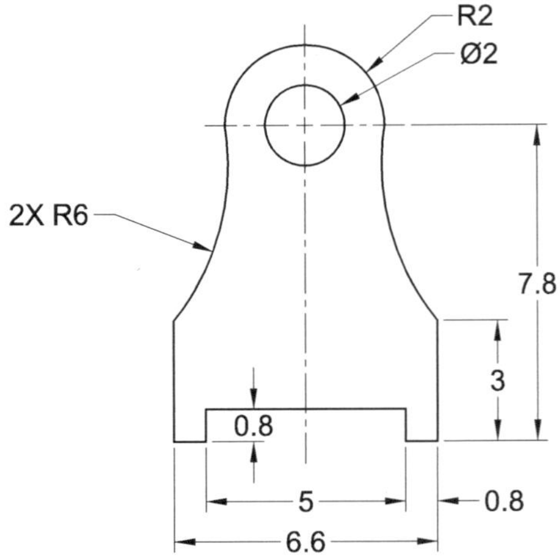

Figure 3-23 *Sketch of the model*

Exercise 2

In this exercise, you will draw the sketch of the model shown in Figure 3-24. The sketch to be drawn is shown in Figure 3-25. **(Expected time: 15 min)**

Figure 3-24 *Solid model for Exercise 2*

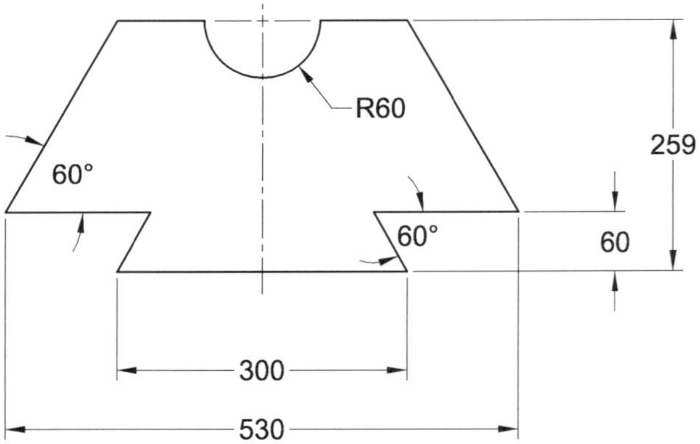

Figure 3-25 *Sketch of the model*

Chapter 4

Creating Base Features

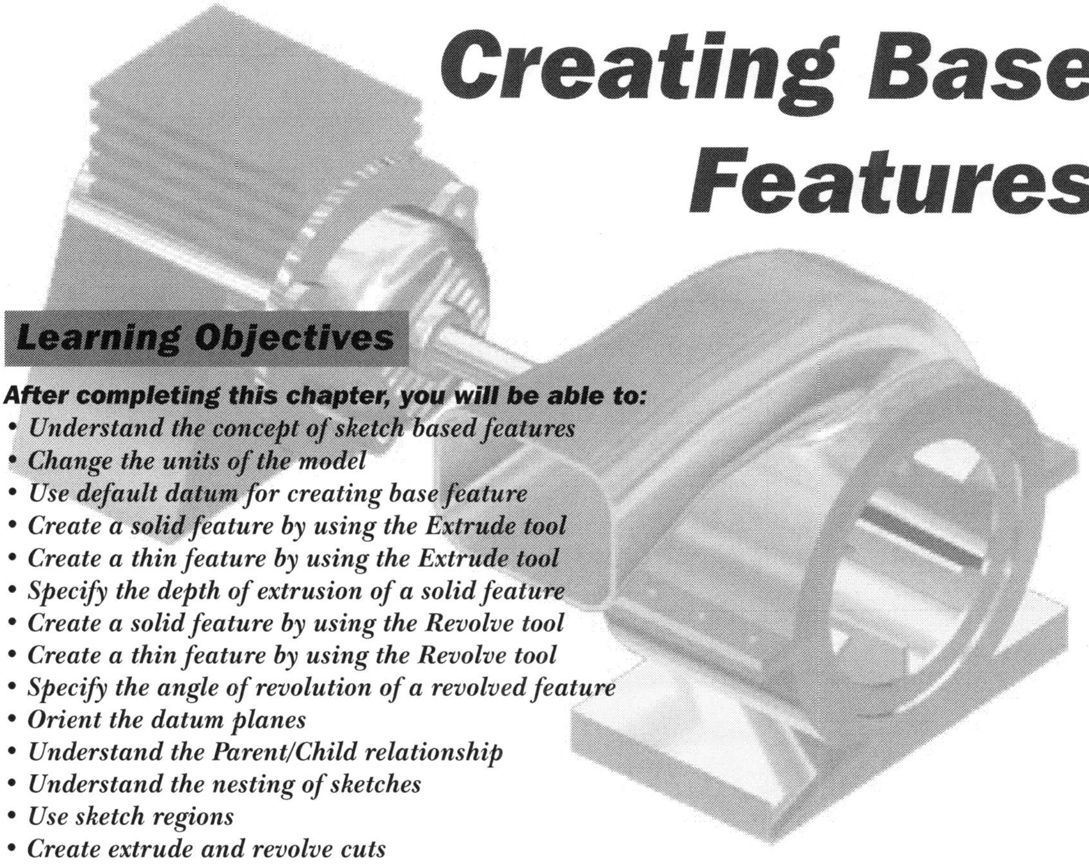

Learning Objectives

After completing this chapter, you will be able to:

- *Understand the concept of sketch based features*
- *Change the units of the model*
- *Use default datum for creating base feature*
- *Create a solid feature by using the Extrude tool*
- *Create a thin feature by using the Extrude tool*
- *Specify the depth of extrusion of a solid feature*
- *Create a solid feature by using the Revolve tool*
- *Create a thin feature by using the Revolve tool*
- *Specify the angle of revolution of a revolved feature*
- *Orient the datum planes*
- *Understand the Parent/Child relationship*
- *Understand the nesting of sketches*
- *Use sketch regions*
- *Create extrude and revolve cuts*

CONCEPT OF SKETCH BASED FEATURES

Sketch-based feature uses a sketch to define its shape, dimensions, and general placements. All the features that are used to create solid models are known as sketch based features. A sketch based feature uses a sketch in the following two ways:

1. As an internal section, the sketch is created within a feature.

2. As an external sketch, the existing sketch is used to define the feature.

Note

1. In this book, the sketch based features are defined using the internal sections.

2. Sketch-based features may have requirements such as a closed loop section (sketch), reference of a horizontal or a vertical axis, or a coordinate system. If the sketch is invalid, Creo Parametric displays a warning dialog box. If regeneration of the feature is failed, a notification is displayed in the notification area.

CREATING BASE FEATURES

The base feature is the first feature created while creating a model in the **Part** mode. The base feature is created using the datum planes. Although you can create a base feature without using the datum planes, but in that case, you will not have proper control over the orientation of feature and direction of feature creation.

Tip

It is recommended that you set the working directory before starting a new file.

Invoking the Part Mode

To invoke the **Part** mode, choose the **New** option from the **File** menu or choose the **New** button from the **Data** group; the **New** dialog box will be displayed with various modes of Creo Parametric. The **Part** radio button in the **Type** area and the **Solid** radio button in the **Sub-type** area of the **New** dialog box are selected by default. The default name of the part file also appears in the **File name** edit box. You can change the part name as desired and then choose the **OK** button to enter the **Part** mode. Note that you cannot use spaces in the name of a file.

Figure 4-1 shows the initial screen appearance in the **Part** mode. It displays **Model Tree**, three default datum planes, **Ribbon**, and toolbars.

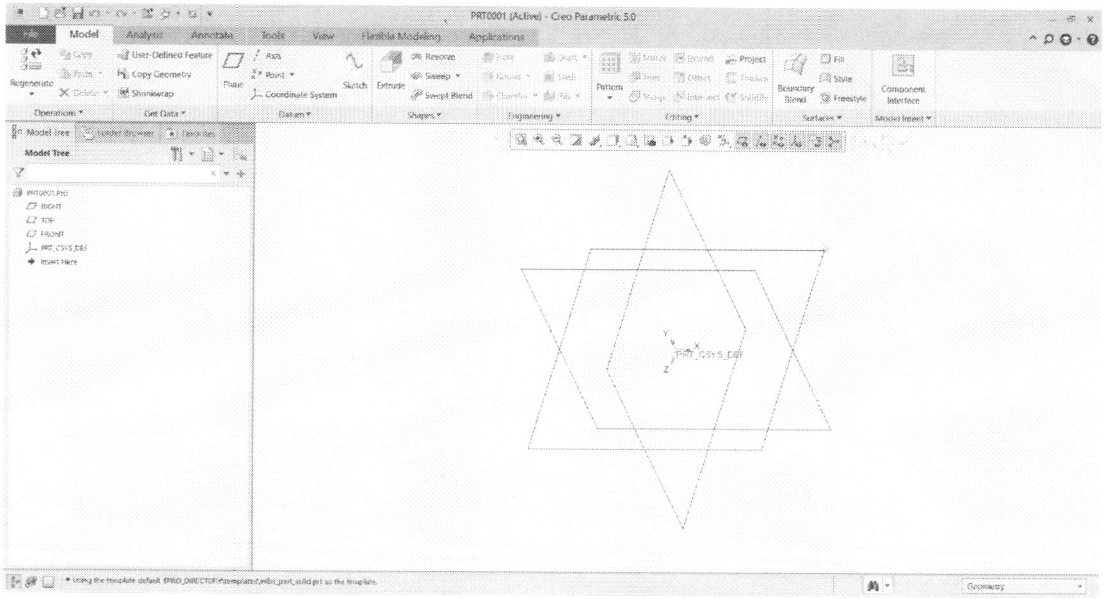

Figure 4-1 *The initial screen appearance in the* ***Part*** *mode*

Tip

It is recommended that you always use the datum planes to create a base feature. This is because the model created using the datum planes can be easily oriented. The uses of datum planes are discussed in Chapter 5.

Note

Although, the three default datum planes, ***RIGHT****,* ***TOP****, and* ***FRONT****, are the primary features in the* ***Part*** *mode. In the* ***Model Tree****, datum planes appear as three separate features. If you delete any one of them, only that datum plane will be deleted.*

PARENT-CHILD RELATIONSHIP

Every model created in Creo Parametric is composed of features that in some way or the other are related to other features in the model. The feature that occurs first in the **Model Tree** is called the parent feature and any feature(s) that is related to this feature is called the child feature(s).

SKETCH REGION

You can use the **Sketch Region** option to make quick selection to create geometry with the selected sketch-based features. A sketch region is a closed contour defined by sketched entities and their intersection with coplanar 3D edges in the part geometry. Sketch-based feature geometry creation is faster and easier. The **Sketch Region** option reduces the need to perform project and trim operations within the sketch. It also offers a flexible way to use portions of a single sketch as the basis for several sketch-based features.

NESTING OF SKETCHES

The process of creating more than one closed loops in a single sketch for a single feature is known as nesting of sketches. These sketches are drawn in the sketcher environment.

TUTORIALS

Tutorial 1

In this tutorial, you will create the model shown in Figure 4-2. The dimensions of the model are shown in Figure 4-3. **(Expected time: 30 min)**

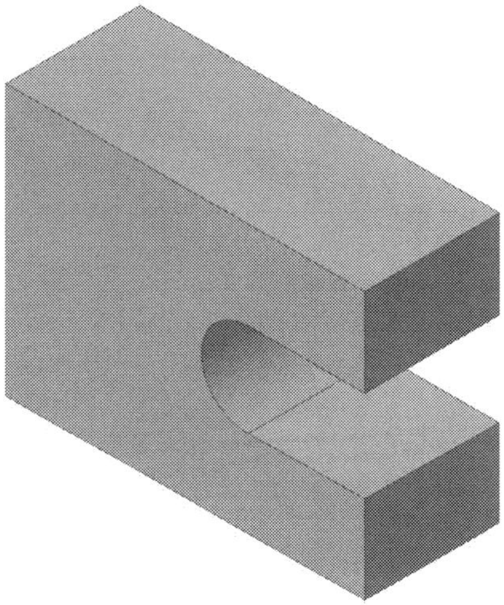

Figure 4-2 *The isometric view of the model*

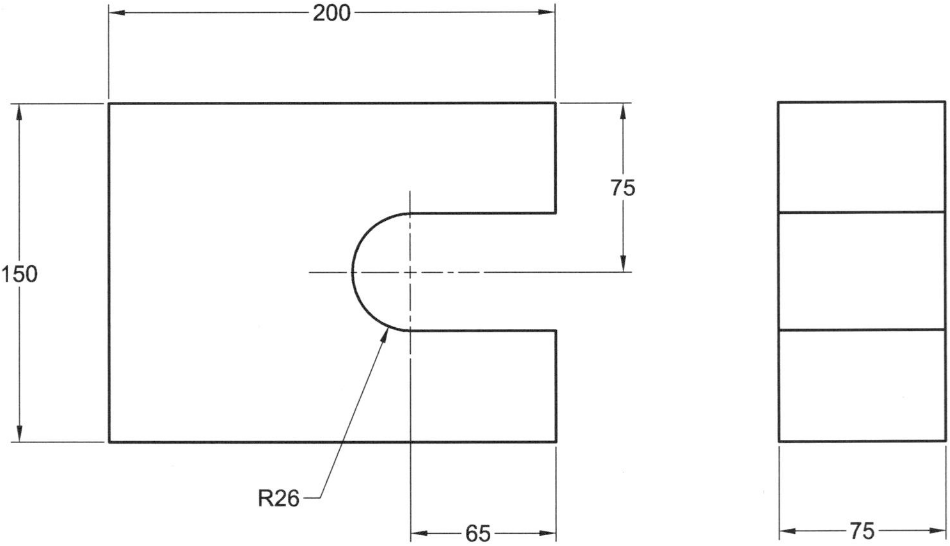

Figure 4-3 *The front and right views of the model with dimensions*

The following steps are required to complete this tutorial:

a. Set the working directory and create a new object file in the **Part** mode.
b. First examine the model and then determine the type of protrusion for the model. Select the sketching plane for the model and orient it parallel to the screen.
c. Draw the sketch by using the sketching tools and apply constraints and dimensions, refer to Figures 4-3 through 4-8.
d. Exit the sketcher environment and define the model attributes, refer to Figures 4-9 and 4-10.

Setting the Working Directory

After starting the Creo Parametric session, the first task is to set the working directory. A working directory is a directory on your system where you can save the work done in the current session of Creo Parametric. You can set any existing directory on your system as the working directory.

1. Choose the **Manage Session > Select Working Directory** from the **File** menu; the **Select Working Directory** dialog box is displayed. Browse to the *C:\Creo-5.0* folder.

2. Choose the **Organize** button from the **Select Working Directory** dialog box to display the flyout. Next, choose the **New Folder** option from the flyout; the **New Folder** dialog box is displayed.

3. Enter **c04** in the **New Directory** edit box and choose the **OK** button from the **New Folder** dialog box. Now, you have created a folder named *c04* in *C:\Creo-5.0*.

4. Next, choose the **OK** button from the **Select Working Directory** dialog box; the working directory is set to *C:\Creo-5.0\c04* and a message **Successfully changed to C:\Creo-5.0\c04 directory** is displayed in the message area.

Starting a New Object File

Solid models are created in the **Part** mode of Creo Parametric. The file extension for the files created in this mode is *.prt*.

1. Choose the **New** button from the **File** menu; the **New** dialog box is displayed. The **Part** radio button is selected by default in the **Type** area and the **Solid** radio button is selected by default in the **Sub-type** area of the **New** dialog box.

2. Enter the file name as *c04tut1* in the **Name** edit box and choose the **OK** button. The three default datum planes are displayed in the drawing area. Also, the **Model Tree** appears on the left of the drawing area in the Navigator.

> **Tip**
> *By default, a Creo file opens in* ***inlbs_part_solid*** *unit. However, you can change it by choosing* ***Prepare > Model Properties > Units*** *option from the* ***File*** *menu.*

Selecting the Extrude Option

The given solid model is created by extruding the sketch to a distance of 75 units. Therefore, the sketch will be extruded as a solid to create the model.

1. Choose **Extrude** from the **Shapes** group in the **Model** tab; the **Extrude** dashboard is displayed on the top of the drawing area. All the attributes needed to create the model will be defined after the sketch is drawn.

Selecting the Sketching Plane

To create a sketch for the extruded feature, first you need to select the sketching plane for the model. The **FRONT** datum plane will be selected as the sketching plane. The sketching plane is selected such that the direction of extrusion of the solid model is perpendicular to it. From the isometric view of the model shown in Figure 4-2, it is evident that the direction of extrusion of the model is perpendicular to the **FRONT** datum plane.

1. Choose the **Placement** tab from the **Extrude** dashboard to display the slide-down panel. Next, choose the **Define** button from the slide-down panel; the **Sketch** dialog box is displayed.

2. Select the **FRONT** datum plane from the drawing area as the sketching plane. As you select the sketching plane, the reference plane and its orientation are set automatically. The reference plane is selected to orient the sketching plane.

   The arrow appearing on the sketching plane indicates the direction of viewing the sketch.

   In the **Sketch** dialog box, the **Reference** collector displays **RIGHT:F1(DATUM PLANE)**. This indicates that the **RIGHT** datum plane is selected as the reference plane. In the **Orientation** drop-down list, the **Right** option is selected by default. This means while drawing the sketch, the **RIGHT** datum plane will be on the right. The **RIGHT** datum plane will be perpendicular to the sketching plane and the sketching plane will be parallel to the screen.

3. Choose the **Sketch** button in the **Sketch** dialog box to enter the Sketcher environment.

Drawing the Sketch

You need to draw the sketch of the solid model that will be extruded later to create the 3D model.

1. Choose the **Corner Rectangle** tool from the **Sketching** group.

2. Draw a rectangle by defining its lower left corner and upper right corner. The rectangle is created and weak dimensions are applied to it.

   You will notice that some of the constraints are applied to the lines composing the rectangle. This is because while drawing the rectangle some constraints are applied automatically to the lines composing the rectangle.

3. Select the right vertical line; the line turns green. Press the DELETE key to delete it. The sketch after deleting the vertical line is shown in Figure 4-4.

4. Now, draw the lines and arc. Some weak dimensions and constraints are applied to the sketch, as shown in Figure 4-5.

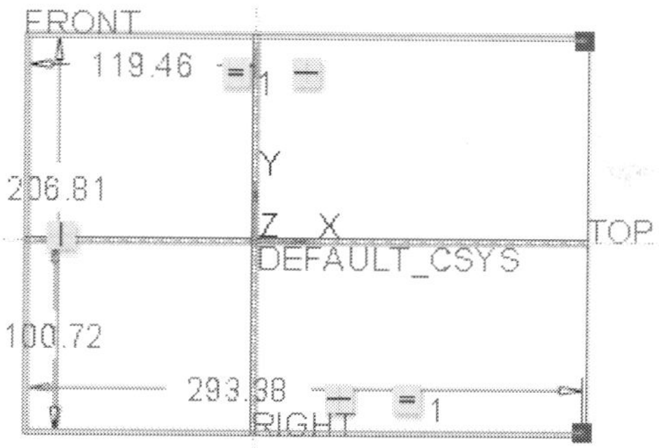

Figure 4-4 The sketch after deleting the vertical line

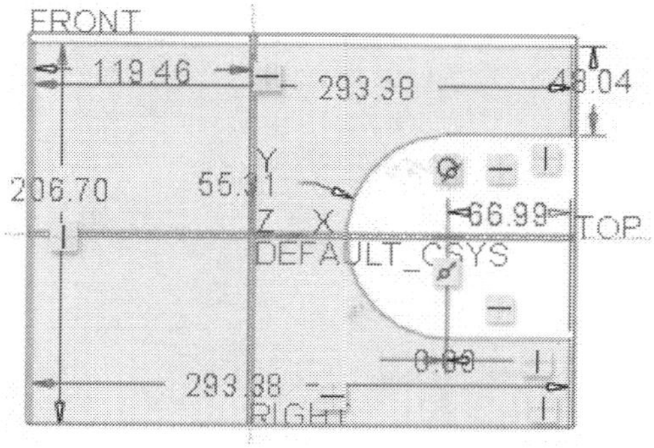

Figure 4-5 The sketch with weak dimensions

Note

The center of the arc and the ***TOP*** *datum plane are aligned by default. In case, they are not aligned, you need to align them. To do so, choose the* ***Coincident*** *button from the* ***Constrain*** *group. Select the center of the arc and then select the* ***TOP*** *datum plane. Now, the center and the datum plane are aligned.*

Tip

Since you are creating a solid sketch based feature, it is suggested that you do not turn off the ***Shade Closed Loops*** *and* ***Highlight Open Ends*** *buttons available in the* ***Inspect*** *group of the* ***Sketch*** *tab.*

Applying Constraints to the Sketch

You need to apply equal length constraints to the sketch in order to maintain the design intent of the model.

1. Choose the **Equal** tool from the **Constrain** group in the **Sketch** tab and select the two vertical lines on the right of the sketch. Now, the equal length constraint = is applied to both the lines. Next, double-click on the middle mouse button to exit the tool.

2. Select the two horizontal lines that are connected to the arc to apply the equal length constraint. As you apply the constraint to the lines, some of the weak dimensions will disappear from the drawing area and a label will be assigned to the equal constraint. The sketch after applying the equal length constraints is shown in Figure 4-6.

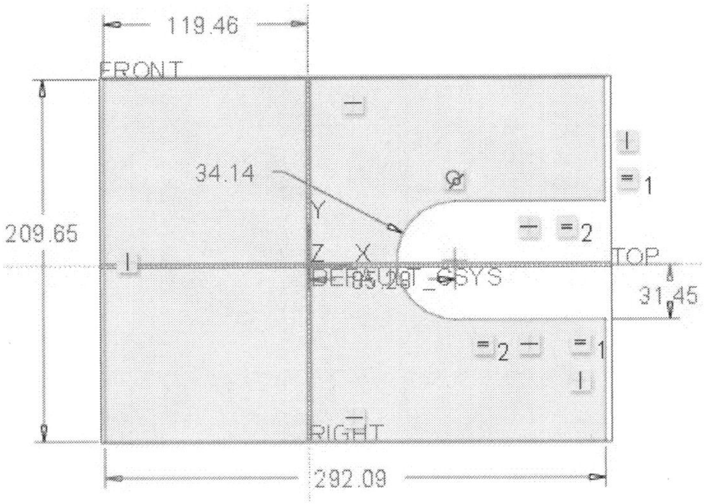

Figure 4-6 *Equal length constraints applied to the sketch*

Dimensioning the Sketch

Although some weak dimensions are applied to the sketch, you need to add a dimension to the sketch.

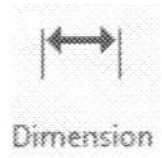

1. Choose the **Dimension** tool from the **Dimension** group.
2. Select the center of the arc and the upper right vertical line and then place the dimension by using the middle mouse button, as shown in Figure 4-7.

   You need to dimension only these entities because the rest of the weak dimensions are useful dimensions and can be modified directly.

Note

If the dimensions in your sketch are different from those shown in Figure 4-3, add the missing dimensions and delete the dimensions that are not required.

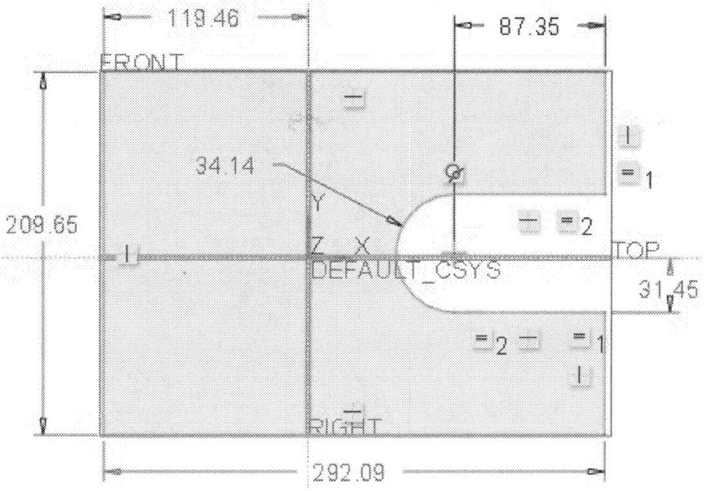

Figure 4-7 *Dimension added to the sketch*

Modifying the Dimensions

You need to modify the dimension values of the sketch. The default dimensions shown in Figure 4-7 also include the length and width of the rectangle and the distance of the sketched section from the selected references. It is recommended that you draw the base feature symmetrical with the other two datum planes. As a result, the distance from the **RIGHT** datum plane is 100 (200 divided by 2 is equal to 100).

1. Select the sketch and dimensions using the CTRL+ALT+A keys.

2. Choose the **Modify** tool from the **Editing** group; the **Modify Dimensions** dialog box is displayed. All dimensions in the sketch are displayed in this dialog box, and each dimension has a separate thumbwheel and an edit box. You can use the thumbwheel or the edit box to modify dimensions. It is recommended to use the edit boxes to modify dimensions, if the desired value is not obtained using the thumbwheel. 

3. Clear the **Regenerate** check box; any modification done in a dimension value does not update the sketch during modification. The dimensions get modified after you exit the **Modify Dimensions** dialog box. It is recommended to clear the **Regenerate** check box when more than one dimension has to be modified.

4. Modify all dimensions one by one, as shown in Figure 4-8. You will notice that the dimension you select in the **Modify Dimensions** dialog box is enclosed in a blue box in the drawing area.

5. After modifying the dimensions, choose the **OK** button from the **Modify Dimensions** dialog box; the message **Dimension modifications successfully completed** is displayed in the message area.

6. Choose the **OK** button from the **Close** group to exit the sketcher environment.

Specifying the Model Attributes

Next, you need to specify the attributes to create the model.

1. Choose the **Standard Orientation** option from the **Saved Orientations** flyout in the **Graphics** toolbar or press CTRL+D; the default trimetric view of the model is displayed in the drawing area.

   This display gives you a better view of the sketch in the 3D space. The model is displayed in orange. Also, the pink colored arrow is displayed on the model indicating the direction of extrusion. The model may not fit fully in the drawing area.

2. Press CTRL+middle mouse button and drag the mouse downward in the drawing area. Notice that a red rubber band line is attached to the cursor. The length of the rubber band line gives an idea about the extent you need to zoom the model. After the model is visible in the drawing area, release the middle mouse button and the CTRL key.

Note

*You can also fit the model into the drawing area by choosing the **Refit** button from the **Graphics** toolbar.*

The model appears, as shown in Figure 4-9. All attributes selected by default in the **Extrude** tab will be used to create the model. You need to change only the depth of extrusion.

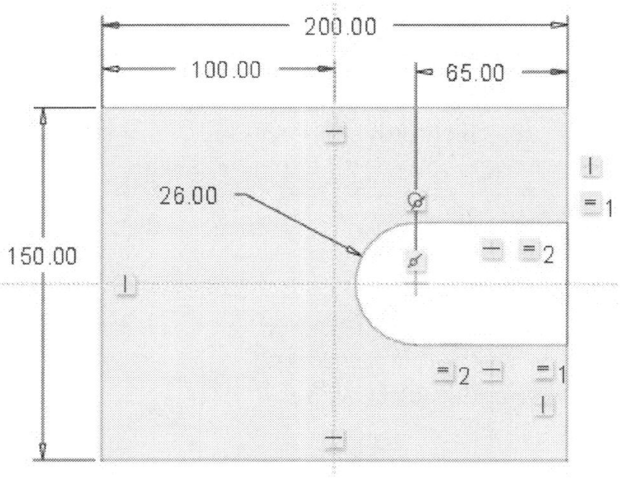

Figure 4-8 *Sketch after modifying the dimensions*

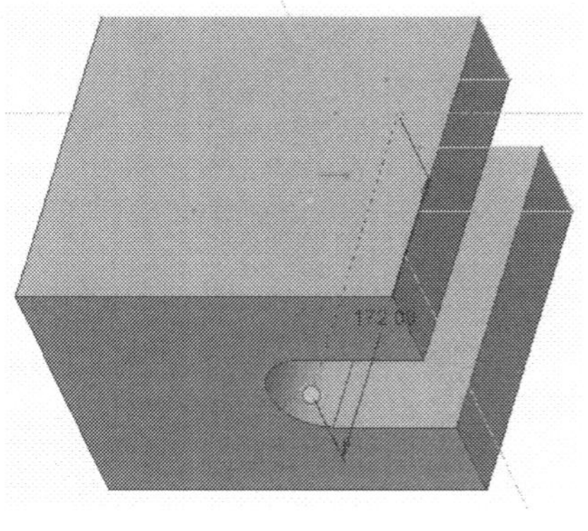

Figure 4-9 *Arrow showing the direction of feature creation*

3. Enter **75** in the edit box present on the **Extrude** tab, and press ENTER; the model in the drawing area is displayed with the specified depth of extrusion.

4. Choose the **OK** button from the **Extrude** dashboard to confirm the feature creation and exit the current feature creation tool. The trimetric view of the model is shown in Figure 4-10.

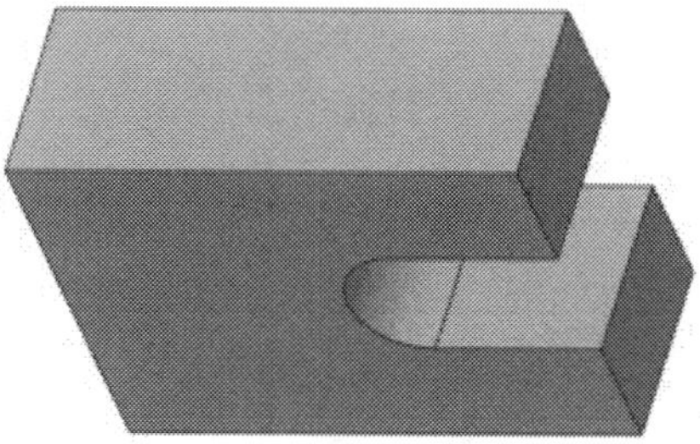

Figure 4-10 *Default trimetric view of the model*

Note

In Figure 4-10, the display of datum planes and the coordinate system are turned off by clearing the ***Plane Display*** *and* ***Csys Display*** *check boxes, respectively from the* ***Datum Display Filter*** *drop-down list in the* ***Graphics*** *toolbar.*

Saving the Model and Closing the File

1. Choose the **Save** option from the **File** menu or choose the **Save** button from the **Quick Access** toolbar; the **Save Object** dialog box is displayed with the name of the object file that you have specified earlier.

2. Choose the **OK** button from the **Save Object** dialog box to save the file.

Note

You can also change the current name of the file before saving it by choosing the ***Rename*** *option from the* ***Manage File*** *flyout in the* ***File*** *menu. If you do so, the* ***Rename*** *dialog box is displayed. Enter the required name of the file in the* ***New file name*** *edit box and then choose the* ***OK*** *button to accept the new name.*

3. Choose **File > Close** from the menu bar or choose the **Close** button from the **Window** group in the **View** tab to close the current window.

Tutorial 2

In this tutorial, you will create the model shown in Figure 4-11. The dimensions of the model are shown in Figure 4-12. **(Expected time: 30 min)**

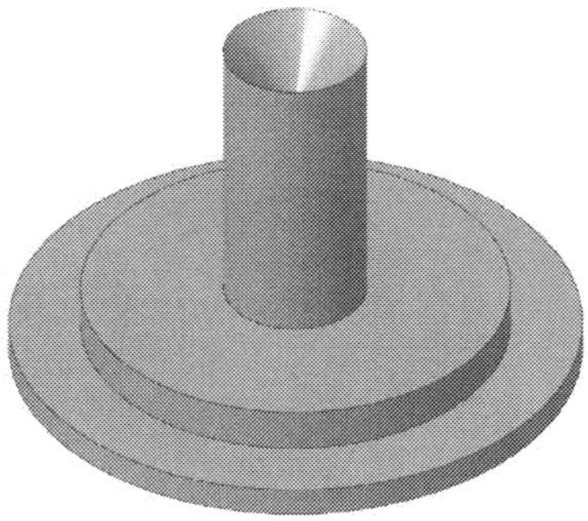

Figure 4-11 *The isometric view of the solid model*

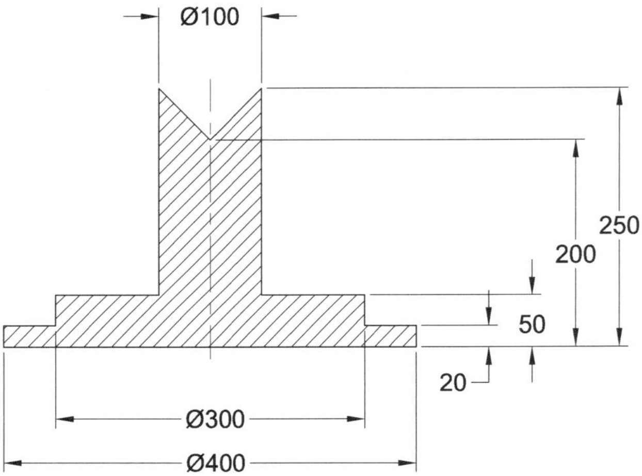

Figure 4-12 *The front section view of the solid model*

The following steps are required to complete this tutorial:

a. Create a new object file in the **Part** mode.
b. First examine the model and then determine the type of protrusion required for the model. Next, select the sketching plane for the model.
c. Draw the sketch for the revolved feature and a centerline to revolve it using the sketching tools. Next, apply dimensions to the model, refer to Figures 4-14 through 4-16.
d. Exit the sketcher environment and define the model attributes, refer to Figures 4-17 and 4-18.

Setting the Working Directory

The working directory was selected in Tutorial 1; therefore, there is no need to select it again. But if you are starting a new session of Creo Parametric, you need to set the working directory again by following the steps discussed earlier.

Starting a New Object File

1. Open a new object file in the **Part** mode and then name the file as *c04tut2*.

The three default datum planes are displayed in the drawing area. However, if the default datum planes were turned off in the previous tutorial, then they will not appear in the drawing area.

2. Turn on the display of datum planes by selecting the **Plane Display** check box available in the **Datum Display Filter** drop-down list in the **Graphics** toolbar.

Selecting the Revolve Tool

The given solid model is a revolved feature that will be created by revolving the sketch through an angle of 360-degree about an axis. Therefore, the **Revolve** tool will be used to create the model. The procedure to create the solid model using the revolve tool is given next.

1. Choose the **Revolve** tool from the **Shapes** group in the **Model** tab; the **Revolve** dashboard is displayed in the **Ribbon**.

   Some of the attributes like rotation angle and direction of rotation of the sketch will be defined after the sketch has been created.

Selecting the Sketching Plane

To create the sketch of the model, first you need to select a sketching plane for the model. Note that the axis of revolution of the revolved feature is normal to the **TOP** datum plane in the model. Therefore, any of the other two datum planes, other than the **TOP** datum plane, can be selected as the sketching plane. Here, you will select the **FRONT** datum plane as the sketching plane. To do so, you need to follow the steps given next.

1. Choose the **Placement** tab from the dashboard to display the slide-down panel. Choose the **Define** button from the side-down group; the **Sketch** dialog box is displayed.

2. Select the **FRONT** datum plane as the sketching plane. As you select the sketching plane, the reference plane and its orientation are set automatically. The reference plane is selected in order to orient the sketching plane.

   In the **Sketch** dialog box, the **Reference** collector displays **RIGHT:F2(DATUM PLANE)**. This indicates that the **RIGHT** datum plane is selected as the reference plane. In the **Orientation** drop-down list, the **Right** option is selected by default. As a result, while drawing the sketch, the **RIGHT** datum plane will be on the right.

   The **RIGHT** datum plane will be perpendicular to the sketching plane and the sketching plane will be parallel to the screen.

3. Choose the **Sketch** button from the **Sketch** dialog box.

Drawing the Sketch

Now, you need to draw the sketch of the revolved feature. The sketch to be drawn is the cross-section of the revolved feature, which will be revolved about the centerline.

1. Choose the **Line Chain** tool from the **Sketching** group and then draw the sketch, refer to Figure 4-11. The sketch should be a closed loop and the bottom horizontal line should be aligned to the **TOP** datum plane.

   As you draw the sketch, weak dimensions and strong constraints are applied to the sketch.

2. Choose the **Centerline** tool from the **Datum** group or from the **Sketching** group. Draw a centerline for the axis of revolution. The centerline should be drawn such that it is aligned with the **RIGHT** datum plane, refer to Figure 4-13.

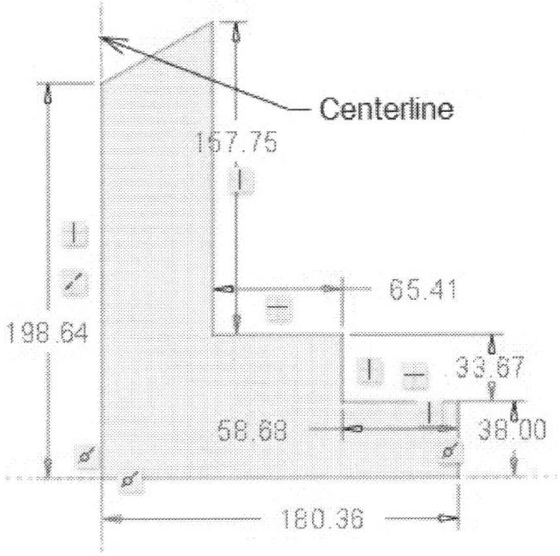

Figure 4-13 *Sketch with weak dimensions*

> **Tip**
> *If you have drawn a centerline on both axes, then press and hold right mouse button on the desired centerline to display a shortcut menu. Choose the* ***Designate Axis of Revolution*** *option from the shortcut menu to revolve the sketch around that particular centerline.*

Dimensioning the Sketch

Weak dimensions are automatically applied to the sketch. Since the model is a revolved feature, you need to manually apply the linear diameter dimensions to the sketch. The linear diameter dimensions are applied by using the centerline that was drawn in the sketch.

> **Tip**
> *The linear diameter dimensioning is necessary for all revolved features because mostly all revolved models are machined on a lathe. Therefore, while machining a revolved model, it is necessary that the operator of the machine has a drawing of the model that is diametrically dimensioned.*

1. Choose the **Dimension** tool from the **Dimension** group.

2. Select the centerline and the first right vertical line.

3. Now, use the middle mouse button to place the dimension on the top of the sketch; the diameter dimension is placed. Alternatively, you can also convert a linear dimension into

a diameter dimension. To do so, exit the **Dimension** tool, select the linear dimension and choose the **Diameter** option from the mini popup toolbar; the linear dimension is converted into the diameter dimension.

4. Select the centerline, the second right vertical line, and then again the centerline.

5. Now, press the middle mouse button to place the dimension below the sketch.

6. Select the centerline, the third right vertical line, and then again the centerline. Now, press the middle mouse button to place the dimension below the previous dimension.

   Dimension the remaining entities in the sketch, as shown in Figure 4-14.

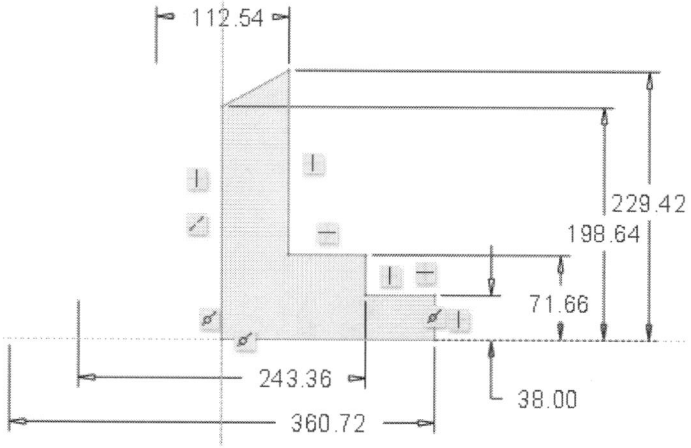

Figure 4-14 *Sketch after dimensioning*

Modifying the Dimensions

When you dimension a sketch, default dimension values are applied to the sketch. You need to modify the dimension values of the sketch.

1. Select the sketch and dimensions by using the CTRL+ALT+A keys.

2. Choose the **Modify** tool from the **Editing** group; the **Modify Dimensions** dialog box is displayed.

3. In this dialog box, clear the **Regenerate** check box and then modify the values in the dimensions, as shown in Figure 4-15. If you clear this check box, any modification of the dimension value will not update the sketch. It is recommended that you clear the **Regenerate** check box if more than one dimension has to be modified.

   You will notice that the dimension that you have selected in the **Modify Dimensions** dialog box is enclosed in a blue box in the drawing area.

4. After modifying all dimensions, choose the **OK** button from the **Modify Dimensions** dialog box; the message **Dimension modifications successfully completed** is displayed in the message area.

5. Choose the **OK** button to exit the sketcher environment.

Specifying the Model Attributes

When you exit the sketcher environment, the **Revolve** dashboard above the drawing area is enabled again. Using this dashboard, you can specify the angle of revolution for the revolved feature.

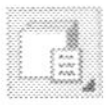

1. Choose the **Saved Orientations** button from the **Orientation** group in the **View** tab to display a flyout. Choose the **Default Orientation** option from the flyout or press CTRL+D; the model is oriented in its default orientation, that is, trimetric view, as shown in Figure 4-16.

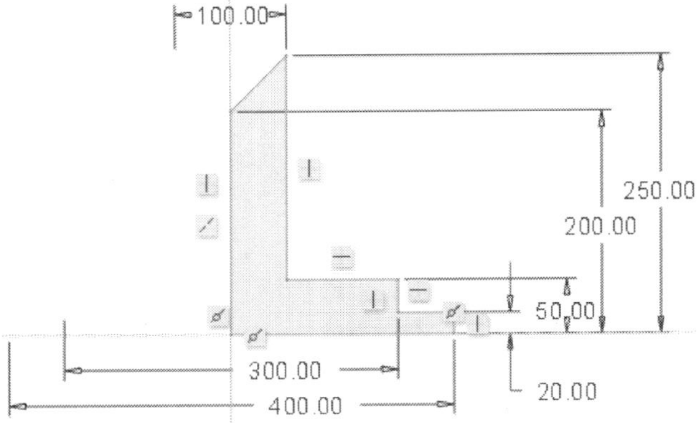

Figure 4-15 *Sketch after modifying the dimensions*

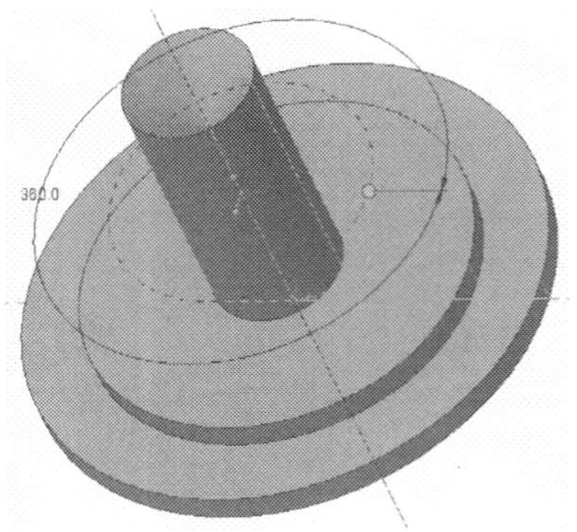

Figure 4-16 *Preview of the model in the default trimetric view*

This display gives you a better view of the sketch in the 3D space and the model appears in orange color. The drag handle is also available on the model that can be used to modify the angle of revolution dynamically.

All attributes needed to create a solid model are selected by default in the **Revolve** dashboard.

Note

*If the model is not fully displayed in the drawing area, you can choose the **Refit** option from the graphics toolbar to fit the model in the drawing area.*

2. Choose the **OK** button in the **Revolve** dashboard. The model appears as the one shown in Figure 4-17.

Figure 4-17 *The default trimetric view of the model*

Saving the Model

1. Choose the **Save** option from the **File** menu or choose the **Save** button from the **Quick Access** toolbar; the **Save Object** dialog box is displayed with the name of the object file specified earlier.

2. Choose the **OK** button to save the file.

Closing the Current Window

The given model is completed and is also saved. Now, you can close the current window.

1. Choose **File > Close** from the menu bar.

EXERCISES

Exercise 1

Create the model shown in Figure 4-18. The dimensions of the model are shown in Figure 4-19. **(Expected time: 20 min)**

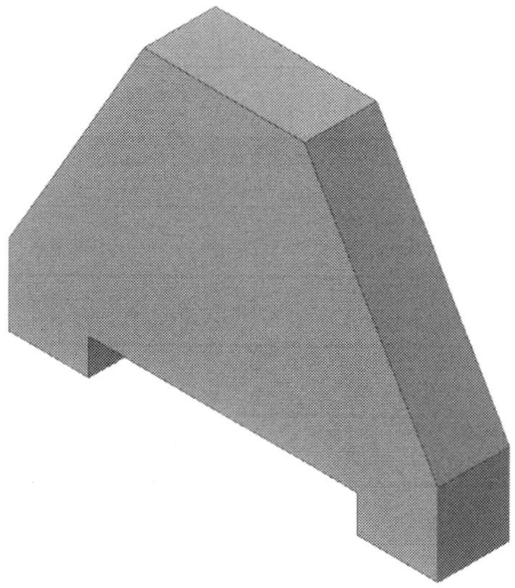

Figure 4-18 *The isometric view of the model*

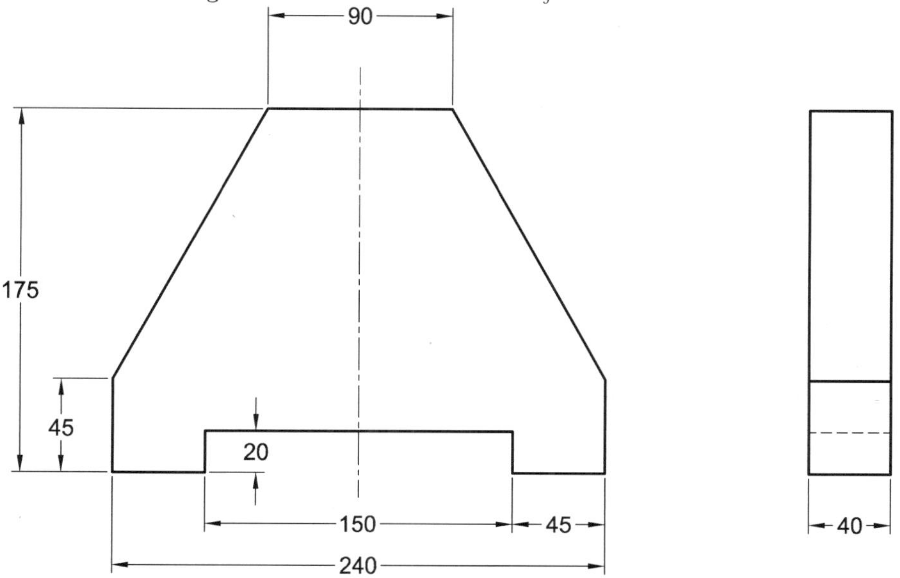

Figure 4-19 *The front and right views of the model*

Exercise 2

Create the model shown in Figure 4-20 The dimensions of the model are shown in Figure 4-21. **(Expected time: 30 min)**

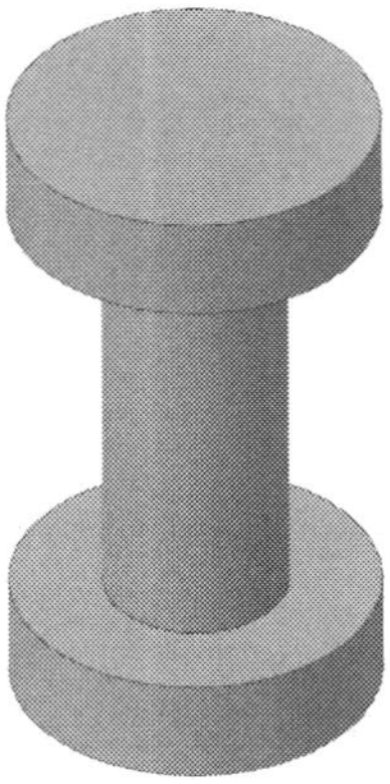

Figure 4-20 *The isometric view of the model*

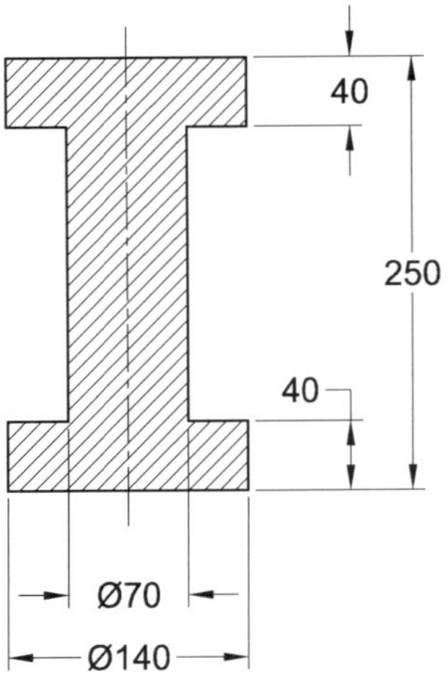

Figure 4-21 *The front section view of the solid model*

Chapter 5

Datums

Learning Objectives

After completing this chapter, you will be able to:

- *Understand the need of datums in modelling*
- *Work with three default datum planes*
- *Understand selection methods in Creo Parametric*
- *Create datum planes using different constraints*
- *Create datum axes using different constraints*
- *Create datum points*
- *Create datum coordinate system*
- *Create datum curves*
- *Understand asynchronous datum features*

DATUMS

Datums are references used to describe the position of a feature(s). They act as reference for sketching a feature, orienting a model, assembling components, and so on. Datums are imaginary features with no mass or volume. Remember that datums play a very important role in creating complex models in Creo Parametric; therefore, you must have a good understanding of datums. Datums are considered to be features but not model geometry. In Creo Parametric, datums exist as datum plane, datum curve, datum point, datum coordinate system, datum graph, and so on.

DATUM OPTIONS

Datums are also considered as features having no geometry. Figure 5-1 shows the **Datum** group and various datum types options from the group.

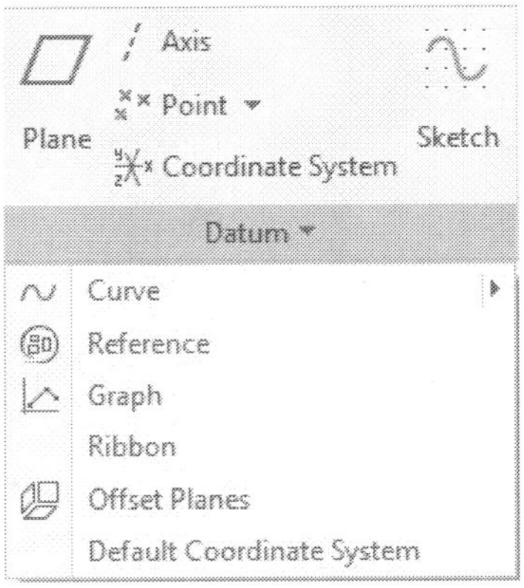

Figure 5-1 *The* ***Datum*** *group*

Datum Planes

Ribbon: Model > Datum > Plane

You can create datum planes other than the three default datum planes by using the **Plane** tool available in the **Datum** group. You can create a datum plane even when any other tool is invoked. You can turn on or off the display of the datum planes by using the **Plane Display** button from the **Show** group of the **View** tab or from the **Graphics** toolbar.

Datum Axes

Ribbon: Model > Datum > Axis

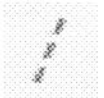

Similar to datum planes, datum axes can also be used as references for feature creation. Datum axes are created automatically when any cylindrical feature is created but you can also create Datum axes manually by using the **Axis** tool available in the **Datum** group. The display of the datum axis can be turned on or off by selecting the **Axis Display** check box from the **Datum Display Filters** drop-down in the **Graphics** toolbar.

Datum Points

Ribbon: Model > Datum > Point > Point

Datum points are imaginary points created to help creating models and drawings, analyzing models, and so on. Creo Parametric creates datum point as a feature. A datum point feature can contain multiple datum points that are created during the same operation. Datum points created in one operation appear under one feature in the **Model Tree**.

Datum Coordinate System

Ribbon: Model > Datum > Coordinate System

Coordinate system is used to determine the position of a point, planes or other geometric element. A coordinate system helps you to create datum planes and points, calculate mass properties, set modeling and assembly references, create manufacturing operation references for toolpaths, and define specific location in space.

Datum Curves

Ribbon: Model > Datum > Curve

Datum curves are 2D curves that can be used to create features such as extrusion or revolve feature. Datum curves can also be used to create trajectories for swept features. Swept features are very useful when you are working with complex surface design, piping design, or wiring and harness design. To access the tools for creating datum curves, click the down arrow in the **Datum** group of the **Model** tab and select the desired tool from the **Curve** flyout.

Asynchronous Datum Features

Asynchronous datum features are those datum features which are created while feature-creation tools are active. To create an asynchronous feature, choose any tool from the **Datum** drop-down of the feature dashboard. The asynchronous feature thus created is embedded with the feature that you are creating. However, you can convert the embedded feature into a stand-alone feature by dragging and dropping it above the **Insert Here** prompt on the **Model Tree**. When you create an asynchronous datum feature, it is neither visible in the drawing area nor it is displayed under the default display of the **Model Tree**. But, it can be seen in the **Model Tree** when you click the arrow sign (▶) that appears on the left of the feature created. In case, you have converted an asynchronous feature into a stand-alone datum feature, it will automatically become visible in the drawing area and also in the model tree.

TUTORIALS

Tutorial 1

In this tutorial, you will create the model shown in Figure 5-2. The front and right views of the solid model are shown in Figure 5-3. **(Expected time: 30 min)**

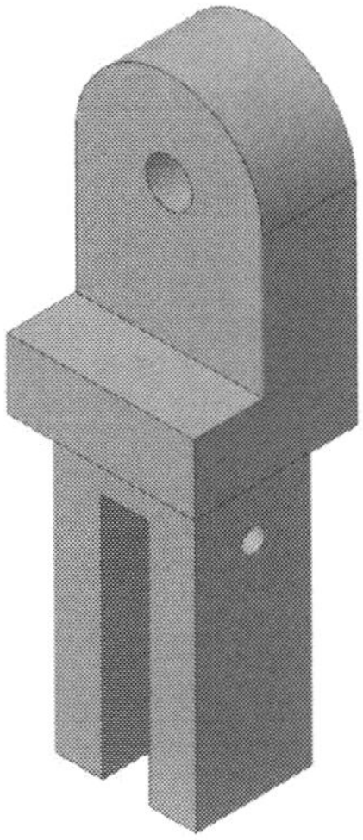

Figure 5-2 *Model for Tutorial 1*

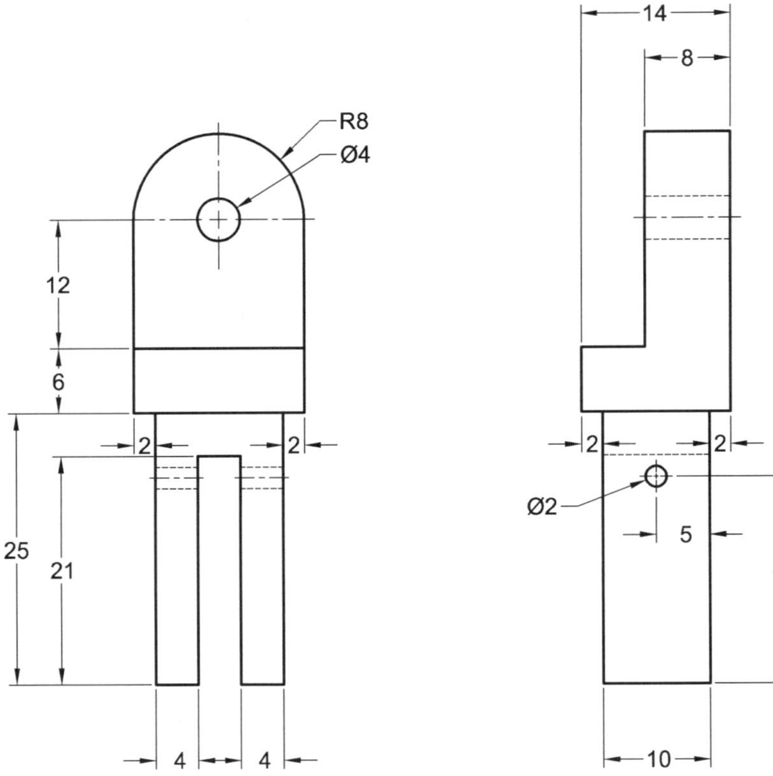

Figure 5-3 *Front and right views of the model*

The following steps are required to complete this tutorial:

Examine the model and determine the number of features in it, refer to Figure 5-2.

a. Create the base feature, refer to Figures 5-4 and 5-5.
b. Create the second extrude feature, refer to Figures 5-6 through 5-8.
c. Create the third feature on an offset plane, refer to Figures 5-11 through 5-14.
d. Create the circular cut feature, refer to Figures 5-15 through 5-17.

Setting the Working Directory

After starting the Creo Parametric session, the first task is to set the working directory. A working directory is a directory on your system where you can save the work done in the current session of Creo Parametric. You can set any existing directory on your system as the working directory.

1. Choose **Manage Session > Select Working Directory** from the **File** menu; the **Select Working Directory** dialog box is displayed. Select the *C:\Creo-5.0* folder from this dialog box.

2. Choose the **Organize** tab from the **Select Working Directory** dialog box to display the flyout. Next, choose the **New Folder** option from the flyout; the **New Folder** dialog box is displayed.

3. Enter **c05** in the **New Directory** edit box and choose the **OK** button from the dialog box; a folder with the name *c05* is created at *C:\Creo-5.0*.

4. Next, choose the **OK** button from the **Select Working Directory** dialog box. The working directory is set to *C:\Creo-5.0\c05*. Also, a message **Successfully changed to C:\Creo5.0\c05 directory** is displayed in the message area.

Starting a New Object File

1. Choose the **New** button to invoke the **New** dialog box. Next, enter *c05tut1* in the **Name** edit box.

   The three default datum planes are displayed in the drawing area. If required, close the **Model Tree** by choosing the **Show Navigator** button on the bottom left corner of the program window so that the drawing area is increased.

Selecting the Sketching Plane for the Base Feature

To create the sketch for the base feature, you first need to select the sketching plane. In this model, you need to draw the base feature on the **FRONT** datum plane because it is evident from the isometric view of this model that the direction of extrusion for this feature is perpendicular to the **FRONT** datum plane.

Note

You can select any plane as the sketching plane for creating the base feature. The base feature thus created may not have proper orientation. So, you need to be careful while defining the sketching plane for creating the base feature. The desired orientation of the model is shown in Figure 5-2.

1. Choose the **Extrude** tool from the **Shapes** group; the **Extrude** dashboard is displayed above the graphics window.

2. Choose the **Placement** tab from the dashboard; a slide-down panel is displayed. Next, choose the **Define** button from the slide-down panel; the **Sketch** dialog box is displayed.

3. Select the **FRONT** datum plane as the sketching plane; an arrow pointing in the direction of the view is displayed on the **FRONT** datum plane.

4. Select the **TOP** datum plane from the drawing area and then select the **Top** option from the **Orientation** drop-down list.

   The **TOP** datum plane is selected in order to orient the sketching plane.

5. Choose the **Sketch** button from the **Sketch** dialog box to enter into the sketcher environment.

Note

If the sketching plane does not orient parallel to the screen then you need to choose the ***Sketch View*** *button from the* ***Graphics*** *toolbar. You can make the sketching plane parallel to the screen by default. To do so, choose* ***File > Options*** *from the menu bar. Then, choose* ***Sketcher*** *from the list shown in the left area of the* ***Creo Parametric Options*** *dialog box. Now, select the* ***Make the sketching plane parallel to the screen*** *check box from the* ***Sketcher startup*** *area and then choose the* ***OK*** *button from the window; the* ***Creo Parametric Options*** *message window will be displayed. Choose the* ***Yes*** *button from this window; the* ***Save As*** *dialog box will be displayed. Choose the* ***OK*** *button to save and exit.*

Creating and Dimensioning the Sketch for the Base Feature

The base feature can be created by drawing the sketch and then extruding it to the given distance.

1. Draw the sketch using various sketcher tools and then add required constraints and dimensions to the sketch, as shown in Figure 5-4. When you initially draw the sketch, it is dimensioned automatically and some weak dimensions are assigned to it.

Tip

It is recommended that you use the ***Modify*** *button to modify weak dimensions. In the* ***Modify Dimensions*** *dialog box that appears on choosing this button, clear the* ***Regenerate*** *check box and then modify dimensions by using the thumbwheel or the edit boxes. This way the sketch will not be regenerated while editing.*

2. Modify the dimension values, as shown in Figure 5-4.

3. After the sketch is completed, choose the **OK** button. Now, you are out of the sketcher environment.

4. Choose the **Saved Orientations** button from the **Graphics** toolbar; a flyout is displayed. Choose the **Default Orientation** option from the flyout; the model orients itself in default trimetric view, refer to Figure 5-5. The pink colored arrow is displayed on the model indicating the direction of extrusion.

5. In the dimension box of the **Extrude** dashboard, enter **8**; the model looks like the one shown in Figure 5-5.

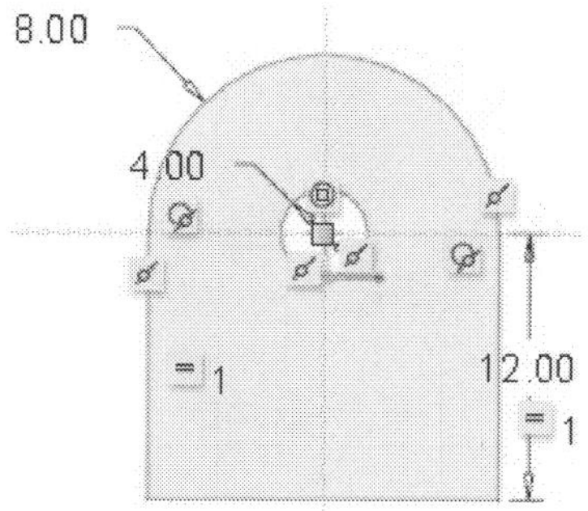

Figure 5-4 Sketch for the base feature with dimensions and constraints

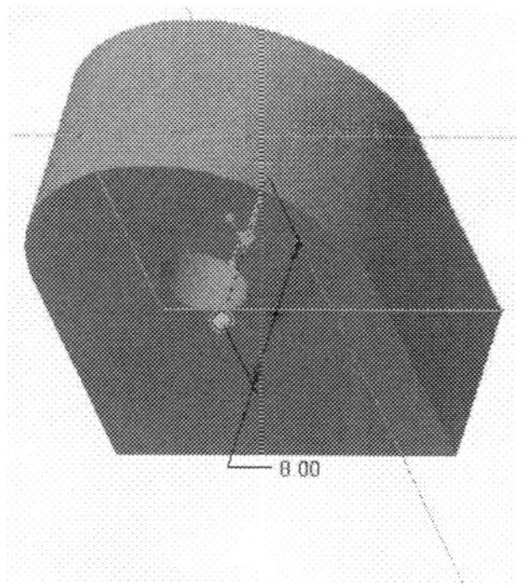

Figure 5-5 Default orientation of the model and the arrow showing the direction of feature creation

6. Choose the **OK** button from the **Extrude** dashboard; the base feature is created. You can use the middle mouse button to spin the model in order to view it from various directions.

Note

*When you choose the **Default Orientation** option from the **Saved Orientations** drop-down, the resulting orientation of the model is trimetric, not isometric. If you want the model to be displayed in the isometric view whenever you choose the **Default Orientation** option, use the **Creo Parametric Options** dialog box. To display this dialog box, choose **Files > Options** from the menu bar. In this dialog box, choose the **Model Display** from the left area and select the **Isometric** option from the **Default model orientation** drop-down list. Next, choose the **OK** button and save these settings. Now, the default orientation will be set to isometric.*

Tip

*It is recommended to check the orientation of the base feature of a model when it is completed. To check whether the plane you specified for sketching was correct, choose the **Saved Orientations** button from the **View** tab; a flyout is displayed. Next, choose the **FRONT** option from the flyout; the base feature will reorient in the drawing area such that you can view the front view of the base feature.*

Selecting the Sketching Plane for the Second Feature

The second feature is an extrude feature. It will be created on the plane that was used to create the base feature.

1. Choose the **Extrude** tool again from the **Shapes** group; the **Extrude** dashboard is displayed above the drawing area.

2. Choose the **Placement** tab from the dashboard. Next, choose the **Define** button from the slide-down panel; the **Sketch** dialog box is displayed.

3. Choose the **Use Previous** button from the **Sketch Plane** area in the **Sketch** dialog box; the **Top** datum plane with default orientation is selected automatically.

   When you choose the **Use Previous** button, the system selects the sketching plane that was used previously to create the base feature. You need to choose this button because the base feature and the second feature are on the same plane, but have different depths of extrusion. In case they had the same depth of extrusion, you could have drawn them on the same plane at the same time as a single feature. The **TOP** datum plane and its orientation are set automatically.

Drawing the Sketch for the Second Feature

The second feature has a rectangular section that will be extruded to a depth of 14 units.

1. Draw the sketch, as shown in Figure 5-6.

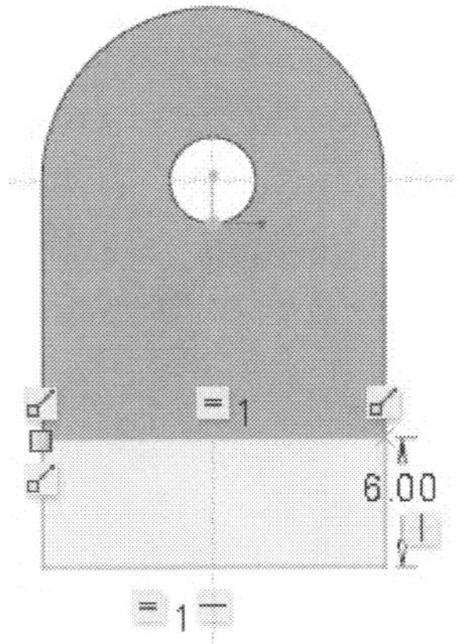

Figure 5-6 *Sketch for the second feature*

2. The sketch is automatically constrained and some weak dimensions are assigned to it. Add required constraints and modify weak dimensions, refer to Figure 5-6.

3. Choose the **OK** button to exit the sketcher environment; the **Extrude** dashboard is enabled above the drawing area.

> **Tip**
> *You can use the **Project** button from the **Sketching** group to use the bottom edge of the base feature for sketching. Else, you need to draw an aligned line on the edge.*

4. Use the middle mouse button to orient the model, as shown in Figure 5-7. This orientation of the model gives you a better view of the sketch in three-dimensional (3D) space. The colored arrow is also displayed on the model indicating the direction of extrusion.

5. Enter **14** in the dimension box present on the **Extrude** dashboard and using ENTER. The second extruded feature is completed and its preview is displayed in the drawing area.

6. Choose the **Saved Orientations** button from the **Graphics** toolbar; a flyout is displayed. Choose the **Default Orientation** option from the flyout; the model orients itself in default trimetric view.

7. Now, choose the **OK** button from the **Extrude** dashboard; the feature is created and oriented, as shown in Figure 5-8. You can also use the middle mouse button to spin the model.

Figure 5-7 Arrow showing the direction of feature depth

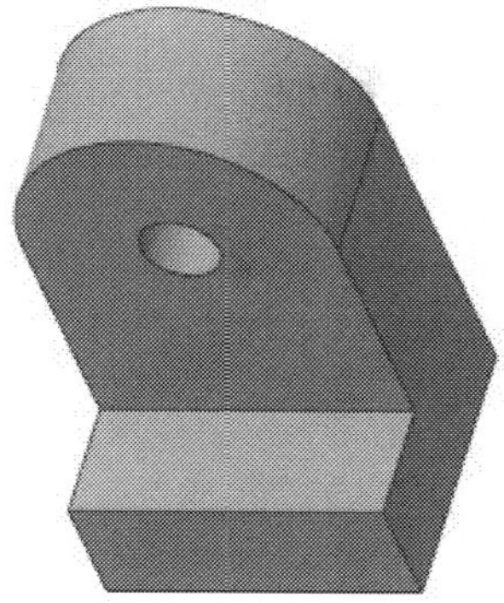

Figure 5-8 Second extruded feature with the base feature

Creating the Datum Plane for the Third Feature

A new datum plane is required to create the next feature. The datum plane will be created at an offset distance of 2 units from the front face of the second feature. You need to turn on the display of datum planes, if it is off.

1. Select the front face of the second feature; a mini popup toolbar appears in the drawing ares, as shown in Figure 5-9. Also, the selected face is highlighted in green.

2. Choose the **Plane** button from the mini popup toolbar; the **Datum Plane** dialog box appears. In the **Datum Plane** dialog box, the **Offset** option automatically becomes available under the **References** collector.

   Now, you need to specify the offset distance. If you enter a positive value, the datum plane will be created along the direction of arrow and if you enter a negative value then the datum plane will be created in the direction opposite to that shown by the arrow.

3. In the **Translation** edit box under the **Offset** area of the **Datum Plane** dialog box, enter **-2** and press ENTER.

   The negative value is entered because the datum plane has to be created in the direction opposite to that shown by the arrow.

4. Choose the **OK** button; the datum plane named **DTM1** is created, as shown in Figure 5-10.

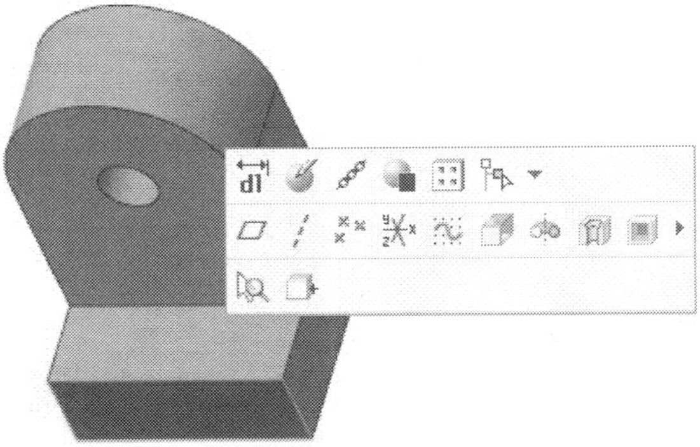

Figure 5-9 The mini popup toolbar displayed on selecting the front face of the second feature

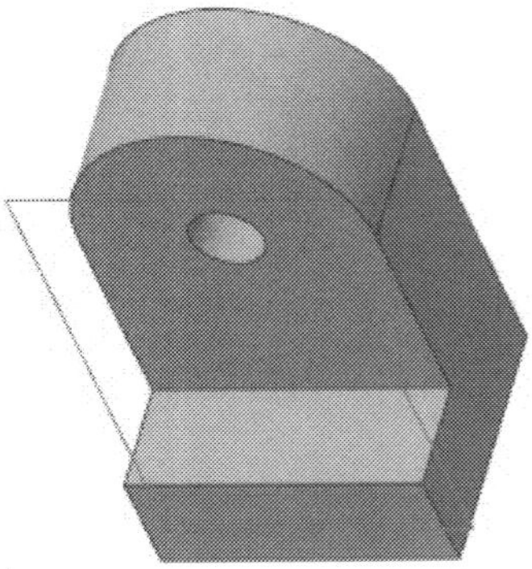

Figure 5-10 The datum plane created

Creating the Third Feature on DTM1

The datum plane **DTM1** is created and can be seen in the **Model Tree** as well as in the drawing area. The sketch of the next feature that will be extruded has to be created on the datum plane **DTM1**.

1. Choose the **Extrude** tool from the **Shapes** group of the **Model** tab; the **Extrude** dashboard is displayed above the drawing area.

2. Choose the **Placement** tab; a slide-down panel is displayed. Now, choose the **Define** button from the panel; the **Sketch** dialog box is displayed.

3. Select **DTM1** as the sketching plane for the third feature; a pink arrow is displayed on the selected datum plane, as shown in Figure 5-11. This arrow shows the direction of viewing the sketching plane.

   The **TOP** datum plane and its orientation are selected by default.

4. Choose the **Sketch** button from the **Sketch** dialog box to enter the sketcher environment. If required, choose the **No hidden** button from the **Display Style** drop-down in the **Graphics** toolbar to view the visible edges of the model.

5. Sketch the section of the third feature of the model and add constraints and dimensions to the sketch, as shown in Figure 5-12.

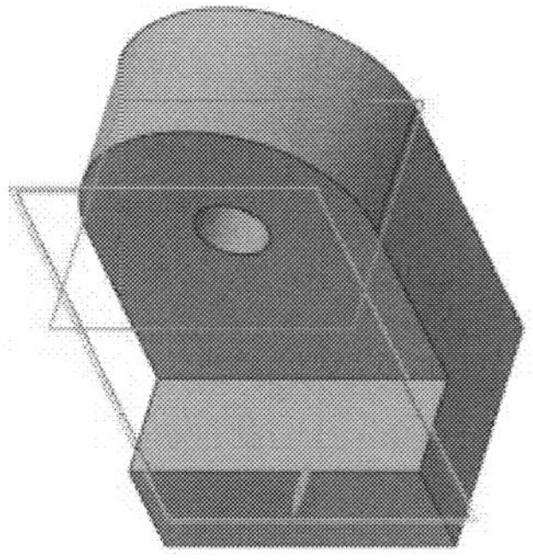

Figure 5-11 Arrow on DTM1 showing the direction of viewing the sketching plane

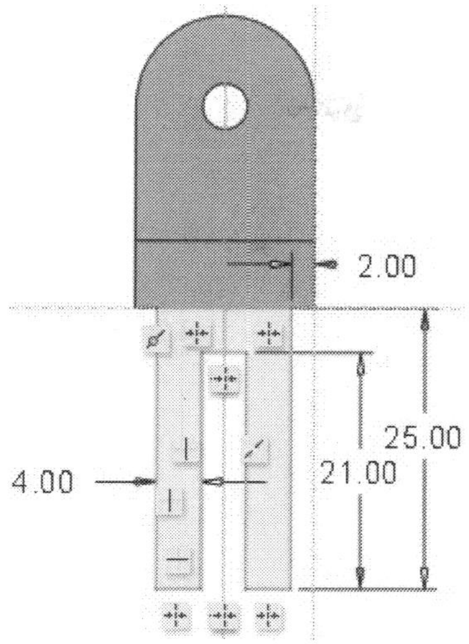

Figure 5-12 Sketch of the third feature with dimensions and constraints

6. Choose the **OK** button to exit the sketcher environment; the **Extrude** dashboard is enabled above the drawing area.

7. Turn the model display to **Shading With Edges**. Use the middle mouse button to orient the model, as shown in Figure 5-13. This orientation gives a better view of the model.

   Notice that the arrow is pointing toward the direction of feature creation. But, you need to extrude the sketch in the opposite direction.

8. Choose the **Change depth direction of extrude to other side of sketch** button from the **Extrude** dashboard to change the direction of arrow.

9. Enter **10** in the dimension box present on the **Extrude** dashboard; preview of the third feature is displayed in the drawing area.

10. Choose the **OK** button from the **Extrude** dashboard to confirm the feature creation.

    Choose the **Saved Orientations** button from the **Graphics** toolbar; a flyout is displayed. Choose the **Default Orientation** option from the flyout; the model orients itself in default trimetric view, as shown in Figure 5-14.

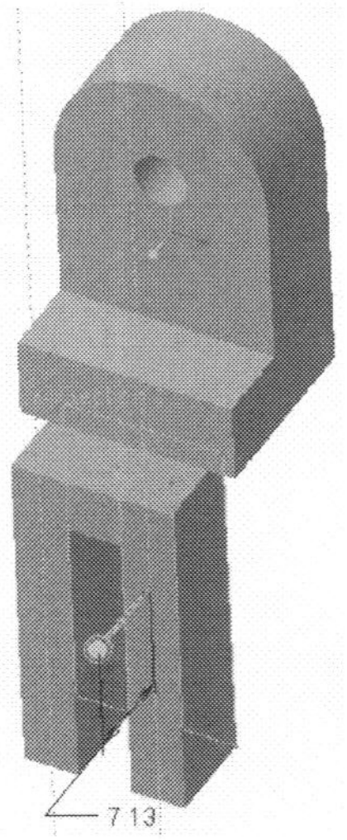

Figure 5-13 Arrow showing the direction of material addition

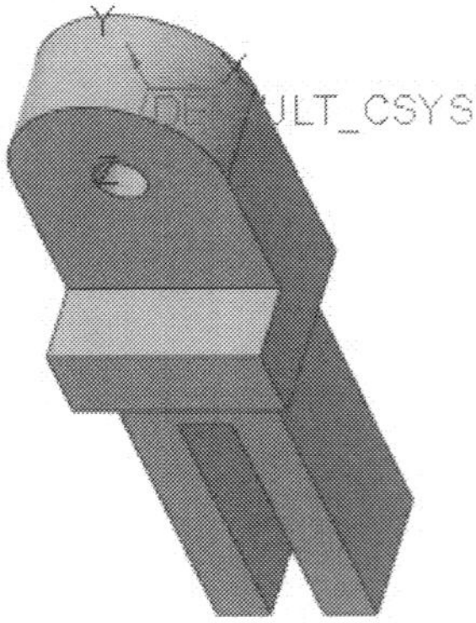

Figure 5-14 Model after creating the third feature

Selecting the Sketching Plane for the Cut Feature

You need to create a circular section for the circular cut feature. The sketching plane for the cut feature is shown in Figure 5-15.

Note

*The circular cut feature can also be created using the **Hole** tool which will be discussed in Chapter 6.*

1. Choose the **Extrude** tool from the **Shapes** group; the **Extrude** dashboard is displayed above the drawing area.

2. Choose the **Remove Material** button from the **Extrude** dashboard.

3. Choose the **Placement** tab from the dashboard; a slide-down panel is displayed. Next, choose the **Define** button from it; the **Sketch** dialog box is displayed.

4. Select the face shown in Figure 5-15 for sketching.

5. Using the left mouse button, select the **TOP** datum plane and then select the **Top** option from the **Orientation** drop-down list.

6. Choose the **Sketch** button; the system takes you to the sketcher environment.

Creating the Sketch for the Cut Feature

1. Draw the sketch of the cut feature and add dimensions to it, as shown in Figure 5-16.

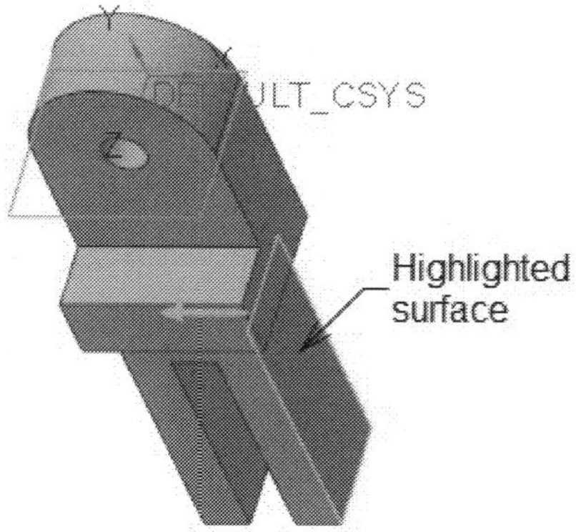

Figure 5-15 *Sketching plane for the cut feature*

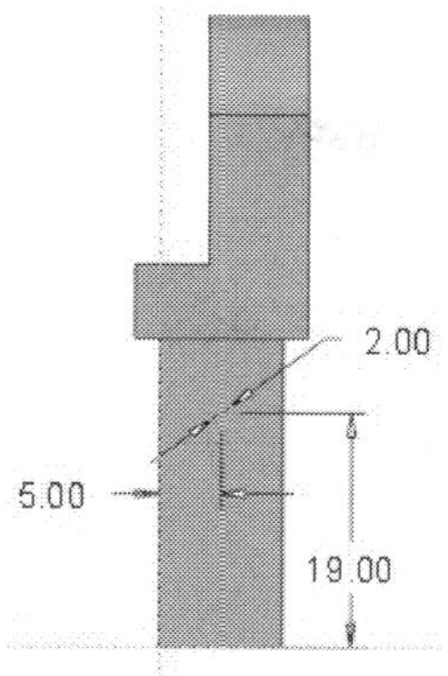

Figure 5-16 *Sketch and dimensions for the cut feature*

2. Choose the **OK** button.

3. Press the CTRL+D keys orient the model to its default orientation, as shown in Figure 5-17; two arrows appear.

   One arrow indicates the direction of feature creation and the other arrow indicates the direction along which the material will be removed.

4. Choose the **Options** tab in the **Extrude** dashboard; the **Depth** slide-down panel is displayed.

5. From the **Side 1** drop-down list, choose the **Through All** option; cut feature is created and can now be previewed in the drawing area.

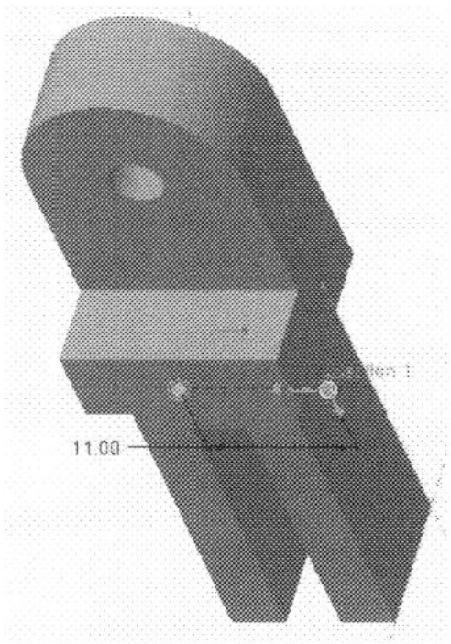

Figure 5-17 *The two arrows on the cut feature*

6. Choose the **OK** button from the **Extrude** dashboard to accept the feature creation. The trimetric view of the model is shown in Figure 5-18.

Saving the Model

1. Choose the **Save** button from the **File** menu and save the model.

The order of feature creation can be seen in the **Model Tree** shown in Figure 5-19.

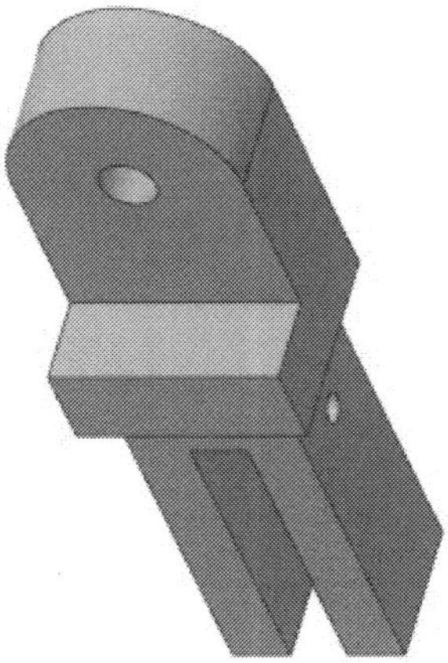

Figure 5-18 Completed model for Tutorial 1

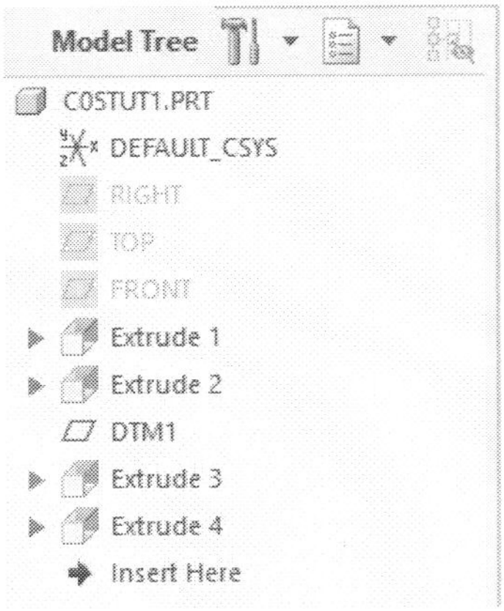

Figure 5-19 The Model Tree for Tutorial 1

Tutorial 2

In this tutorial, you will create the model shown in Figure 5-20. The front and top views of the solid model are shown in Figure 5-21. **(Expected time: 30 min)**

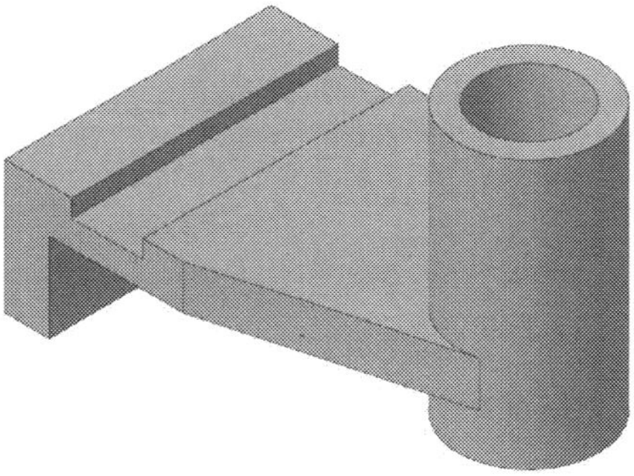

Figure 5-20 Isometric view of the model

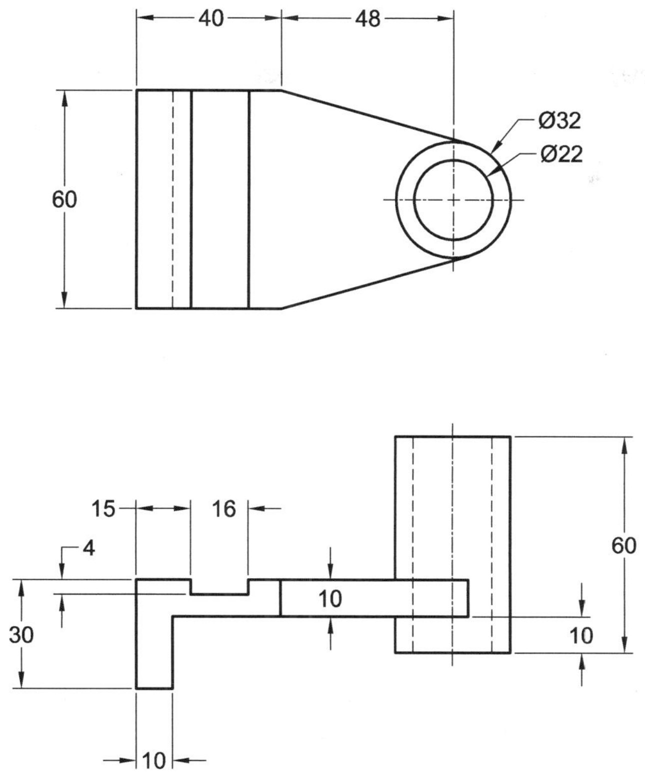

Figure 5-21 *Orthographic views of the model*

Examine the model and then determine the number of features in it, refer to Figure 5-20.

The following steps are required to complete this tutorial:

a. Start a new file and create the base feature, refer to Figures 5-22 and 5-23.
b. Create the second feature, refer to Figures 5-24 through 5-27.
c. Create the hollow cylindrical feature on an offset datum plane, refer to Figures 5-28 through 5-33.

Starting a New Object File

1. Set the working directory, if required, and then start a new part file with the name *c05tut2*.

   The three default datum planes are displayed in the drawing area.

Selecting the Sketching Plane for the Base Feature

To create the sketch for the base feature, you first need to select the sketching plane. In this model, you need to draw the base feature on the **FRONT** datum plane because the direction of extrusion is perpendicular to the **FRONT** datum plane.

1. Choose the **Extrude** tool from the **Shapes** group.

2. Choose the **Placement** tab from the dashboard to display a slide-down panel. Next, choose the **Define** button; the **Sketch** dialog box is displayed.

3. Select the **FRONT** datum plane as the sketching plane.

   A pink arrow is displayed on the **FRONT** datum plane and it points in the direction of viewing the sketch plane. The **RIGHT** datum plane and its orientation are selected automatically.

4. Choose the **Sketch** button to enter the sketcher environment.

Creating and Dimensioning the Sketch for the Base Feature

The section to be extruded for the base feature is evident from the model. The section sketch is shown in Figure 5-22. When this sketch is extruded, it will create the base feature.

1. Draw the sketch using various sketcher tools, as shown in Figure 5-22.

   The sketch is dimensioned automatically and some weak dimensions are assigned to it.

2. Add required constraints and modify weak dimensions, as shown in Figure 5-22.

3. Choose the **OK** button; the **Extrude** dashboard is displayed.

4. Choose the **Saved Orientations** button from the **Graphics** toolbar; a flyout is displayed. Choose the **Default Orientation** option from the flyout; the model orients itself in default trimetric view, and an arrow is also displayed on it indicating the direction of extrusion.

5. Enter **60** in the dimension box that is present on the **Extrude** dashboard and then using ENTER.

6. Choose the **OK** button from the **Extrude** dashboard.

   The base feature is completed, as shown in Figure 5-23. You can use the middle mouse button to spin the model to view it from various directions.

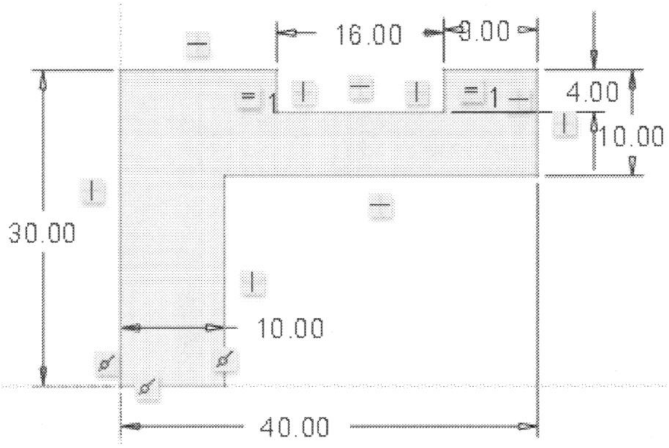

Figure 5-22 Sketch with dimensions and constraints for the base feature

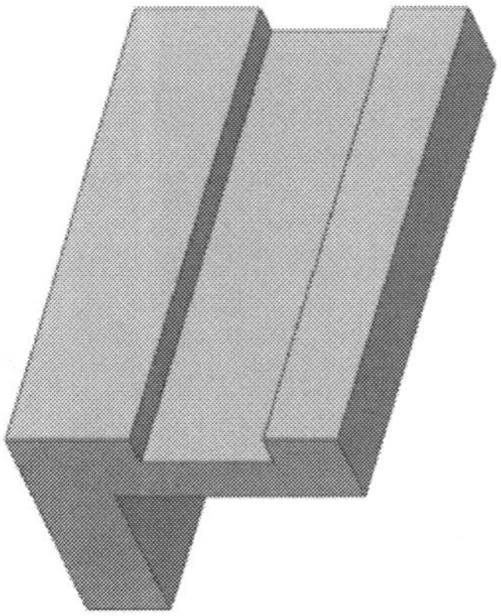

Figure 5-23 Base feature of the model

Selecting the Sketching Plane for the Second Feature

The next feature is an extruded feature. The sketching plane for this feature is the top face of the base feature.

1. Choose the **Extrude** tool from the **Shapes** group.

2. Choose the **Placement** tab from the dashboard; the slide-down panel is displayed. Choose the **Define** button from the slide-down panel; the **Sketch** dialog box is displayed.

3. Select the top face of the base feature as the sketching plane; an arrow pointing in the direction of viewing the sketch is displayed on the top face, refer to Figure 5-24.

4. Choose the **Flip** button to flip the arrow to point in the direction.

5. Select the **RIGHT** datum plane and then select the **Right** option from the **Orientation** drop-down list.

6. Choose the **Sketch** button to enter the sketcher environment.

Creating and Dimensioning the Sketch for the Second Feature

The next feature to be created is an extrude feature. The section for the extrude feature is shown in Figure 5-25.

1. Draw the sketch using the sketcher tools. In the sketch, draw a center line passing through the center of the arc, refer to Figure 5-25. This center line helps in constraining the sketch. Add required constraints and dimensions to the sketch, refer to Figure 5-25.

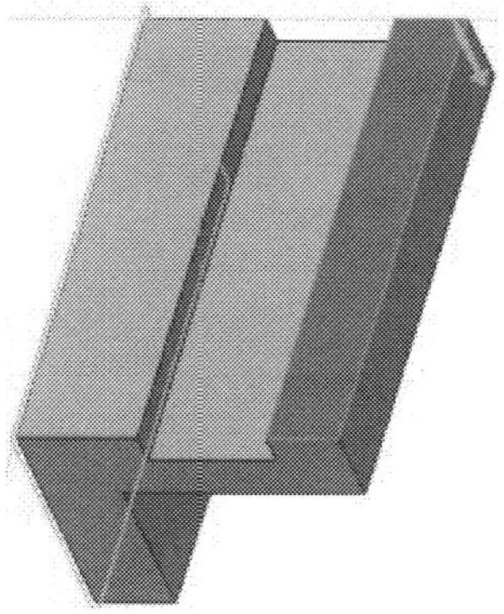

Figure 5-24 *Arrow pointing from the sketching plane in the direction of viewing the sketch*

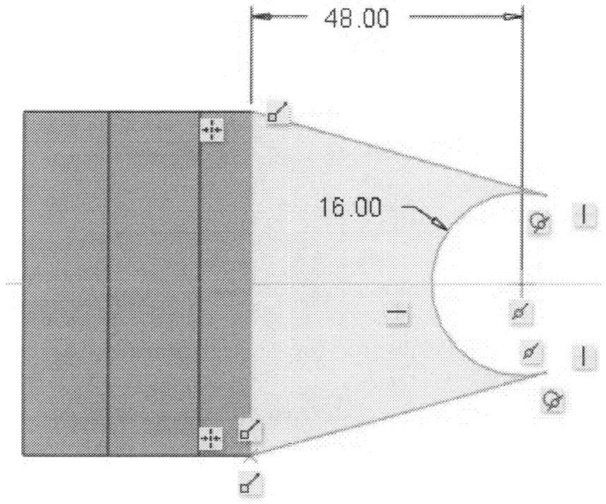

Figure 5-25 *Sketch with dimensions and constraints*

2. Choose the **OK** button; the **Extrude** dashboard is enabled. Use the middle mouse button to orient the model, refer to Figure 5-26.

3. Enter **10** in the dimension box present on the **Extrude** dashboard and using ENTER.

4. Now, choose the **OK** button to confirm the feature creation. The default trimetric view of the extruded feature is shown in Figure 5-27.

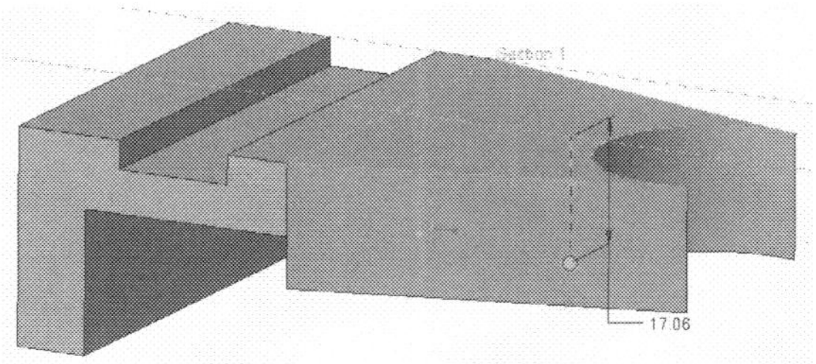

Figure 5-26 *Arrow showing the direction of material addition*

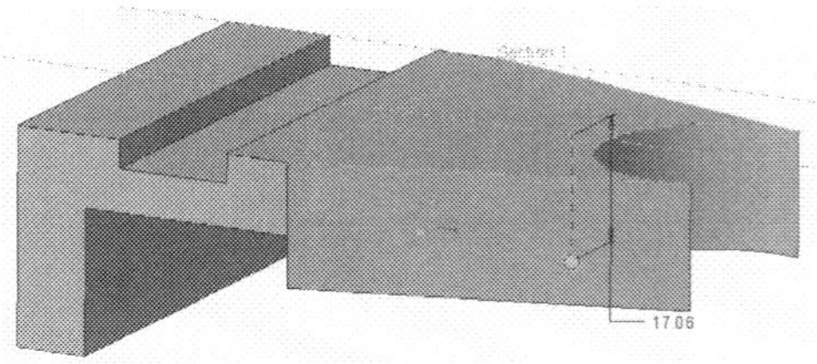

Figure 5-27 *Model with the second extruded feature*

Creating a Datum Plane for the Last Feature

To create the hollow cylindrical feature, you need a datum plane. This datum plane will be created at an offset distance of 10 units from the bottom face of the second feature shown in Figure 5-28.

Note

The other method to create this feature is to select the top planar surface of the second feature as the sketching plane and extrude the sketch on both sides of the sketching plane. The depth of extrusion will be different on both the sides. If you use this method to create this cylindrical feature, you do not need to create a datum plane.

1. Choose the **Plane** tool from the **Datum** group; the **Datum Plane** dialog box is displayed.

2. Spin the model using the middle mouse button and then select the bottom face of the second feature.

   As you select the face of the second feature, the **Offset** constraint is displayed in the **References** collector of the dialog box.

3. In the **Translation** dimension box, enter **10**.

4. Choose the **OK** button from the **Datum Plane** dialog box; the datum plane **DTM1** is created, as shown in Figure 5-29, and is selected as the sketching plane for creating the sketch.

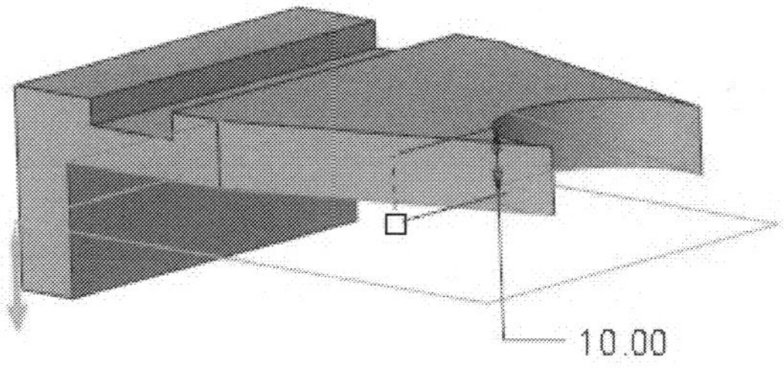

Figure 5-28 Creating the datum plane

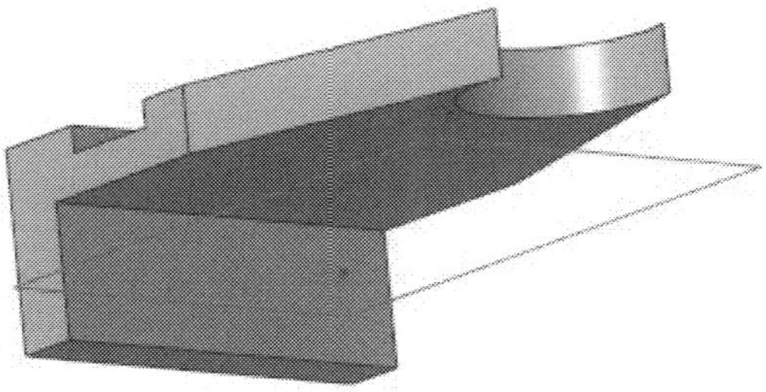

Figure 5-29 Model after creating the datum plane

Selecting the Sketching Plane for the Last Feature

The plane **DTM1** will be selected as the sketching plane and the depth of extrusion will be defined from this plane.

1. Choose the **Extrude** tool from the **Shapes** group.
2. Choose the **Placement** tab from the dashboard; the slide-down panel is displayed. Choose the **Define** button from the slide-down panel; the **Sketch** dialog box is displayed.
3. Select **DTM1** as the sketching plane; an arrow appears on the datum plane.
4. Choose the **Flip** button to reverse the direction of viewing the sketch. Figure 5-30 shows the plane with the direction of the arrow reversed.

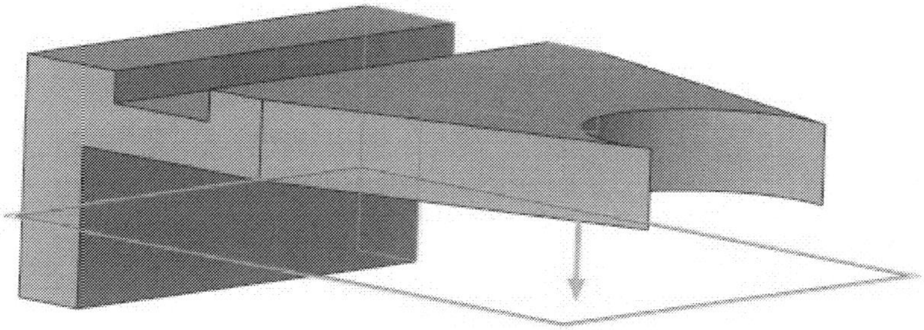

Figure 5-30 Direction of viewing the sketching plane

5. Select the **FRONT** datum plane and select the **Bottom** option from the **Orientation** drop-down list.
6. Choose the **Sketch** button to enter into the sketcher environment.

Creating and Dimensioning the Sketch for the Last Feature

The sketch for the hollow cylindrical feature will be drawn on the datum plane **DTM1**. The sketch for the hollow cylindrical feature consists of two concentric circles.

1. Draw the sketch using the sketcher tools and add required constraints and dimensions to it, as shown in Figure 5-31.

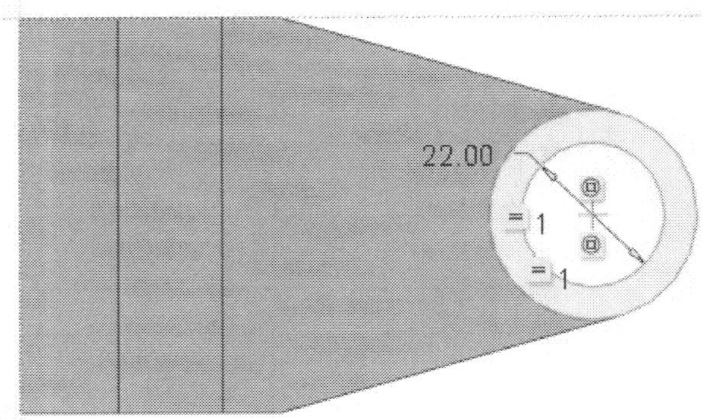

Figure 5-31 Sketch with dimensions and constraints

2. Choose the **OK** button; the **Extrude** dashboard is displayed. Now, turn the model display to **Shading With Edges**.

   Use the middle mouse button to orient the model, as shown in Figure 5-32. This orientation gives you a better view of the sketch in the 3D space.

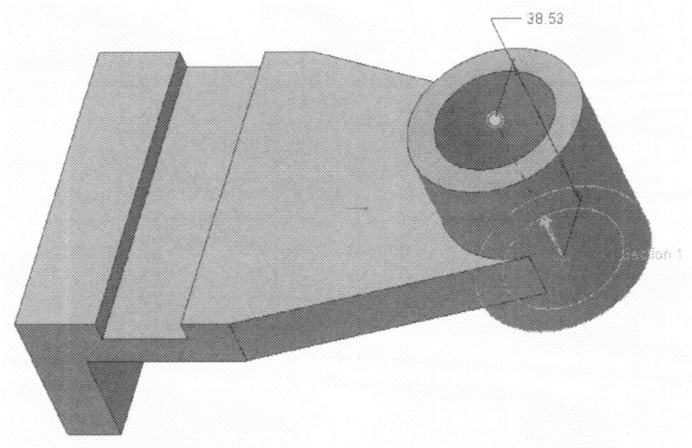

Figure 5-32 Preview of the feature

3. Enter **60** in the dimension box present on the **Extrude** dashboard and then press ENTER.

4. Choose the **OK** button to confirm the feature creation. The default trimetric view of the complete model is shown in Figure 5-33.

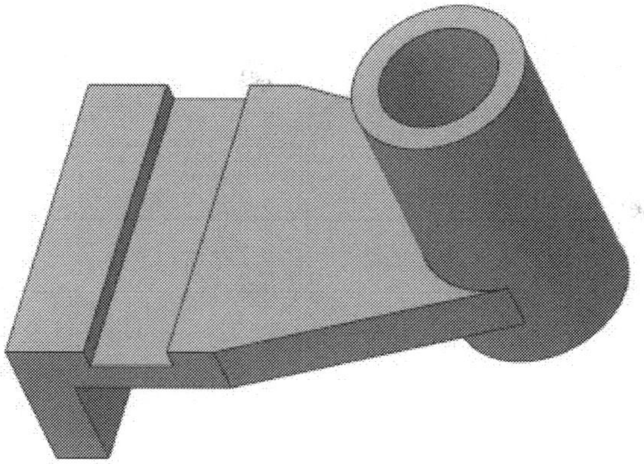

Figure 5-33 Final model of Tutorial 3

Saving the Model

1. Choose the **Save** button from the **File** menu and save the model.

EXERCISES

Exercise 1

Create the model shown in Figure 5-34. The dimensions, front view, and right-side view of the model are shown in Figure 5-35. **(Expected time: 45 min)**

Figure 5-34 Isometric view of the model

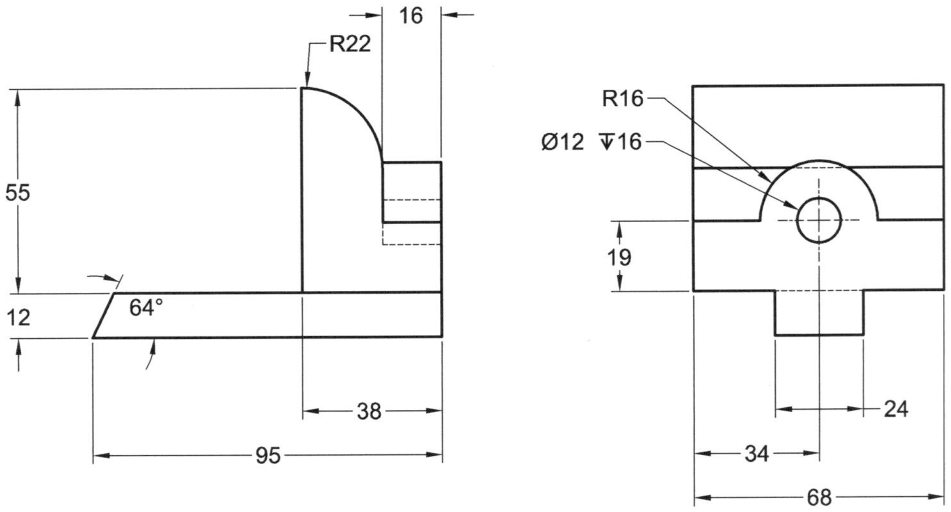

Figure 5-35 *The front and right view of the model*

Exercise 2

Create the model shown in Figure 5-36. The dimensions, the left-side view, auxiliary view, and front view of the model are also shown in Figure 5-37. **(Expected time: 45 min)**

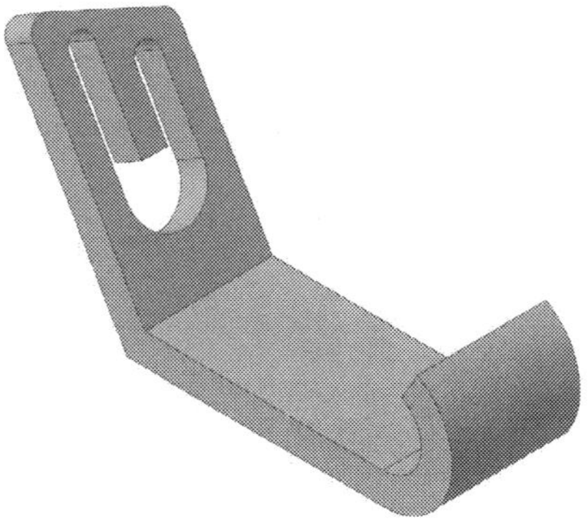

Figure 5-36 *Isometric view of the model*

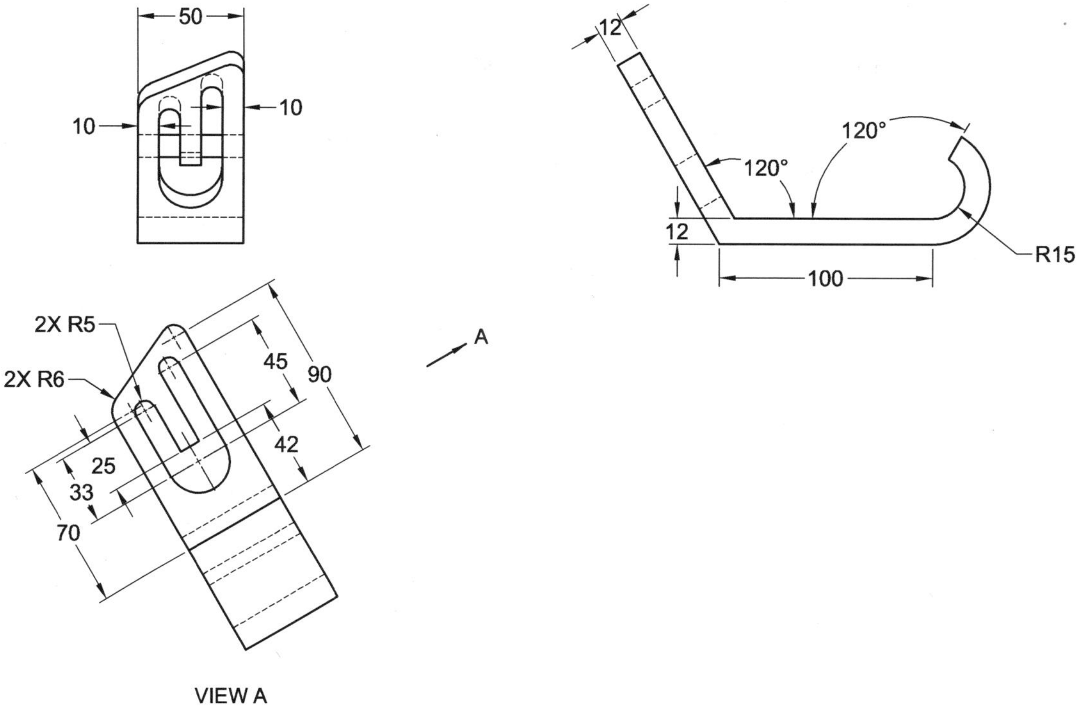

Figure 5-37 *Orthographic views of the model*

This page is intentionally left blank

Chapter 6

Options Aiding Construction of Parts-I

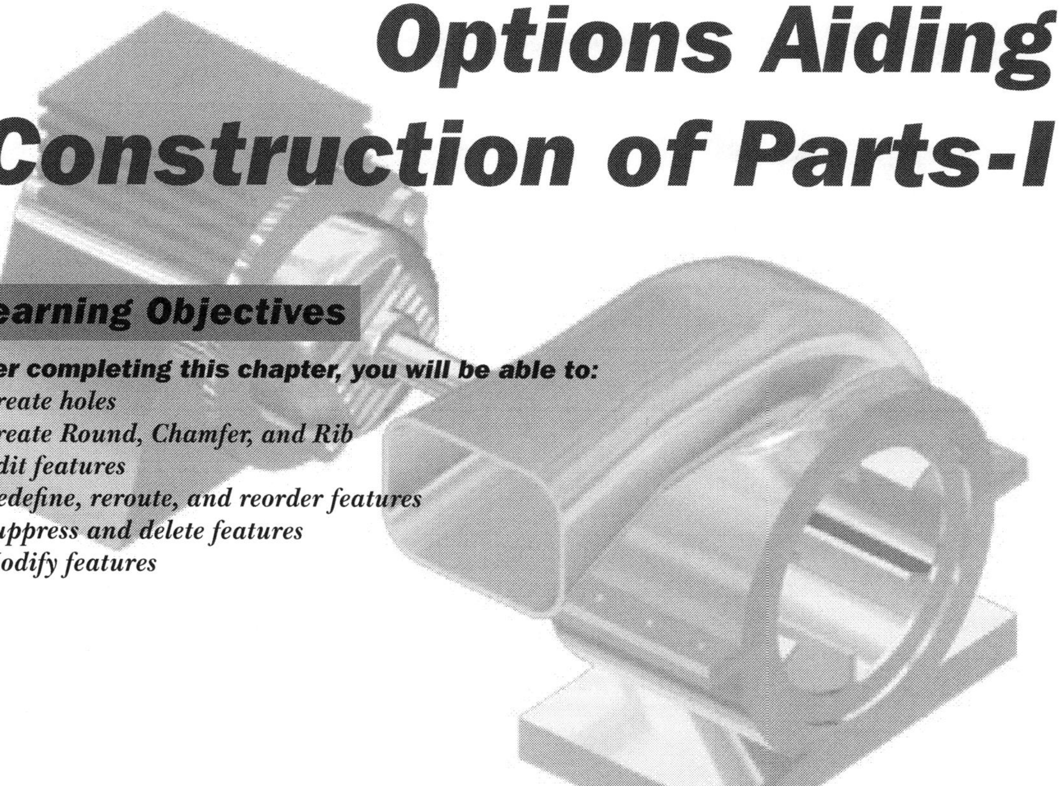

Learning Objectives

After completing this chapter, you will be able to:

- *Create holes*
- *Create Round, Chamfer, and Rib*
- *Edit features*
- *Redefine, reroute, and reorder features*
- *Suppress and delete features*
- *Modify features*

OPTIONS AIDING CONSTRUCTION OF PARTS

This chapter explains the feature creation tools provided in Creo Parametric that help in creating and editing a model. In this chapter, you will learn to create various types of holes that are required in most of the engineering designs. The **Hole** tool enables you to add simple, custom, and industry-standard holes to your models. You can create and modify holes in an easier manner using the options in the **Hole** tool. In this chapter, you will also learn to create rounds, chamfers, and ribs.

CREATING HOLES

Hole feature is similar to the cut features with only difference is that a hole feature does not require a sketch. A hole feature is preferred over other cut features because it requires less dimensioning than that is required in a cut feature. In engineering components, holes can be counterbore, countersink, tapered, or drilled. Creo Parametric allows you to create all such types. Creo Parametric also provides industry standard holes that have standard dimensions.

CREATING ROUNDS

In mechanical engineering, round is a feature which creates a radius or chord on one or more edges, an edge chain, or at the space between surfaces. The radius added to the inside corner is called fillet and to the outside corner is called round. In Creo Parametric, the **Round** tool is used to create a fillet or a smooth rounded transition between two adjacent faces, with a circular or a conic profile. Using the **Round** tool, you can add or remove material, depending on the edge selected.

Figures 6-1 and 6-2 show some examples of rounds. From the figures, it is evident that the geometry of the rounds is tangent to the references selected.

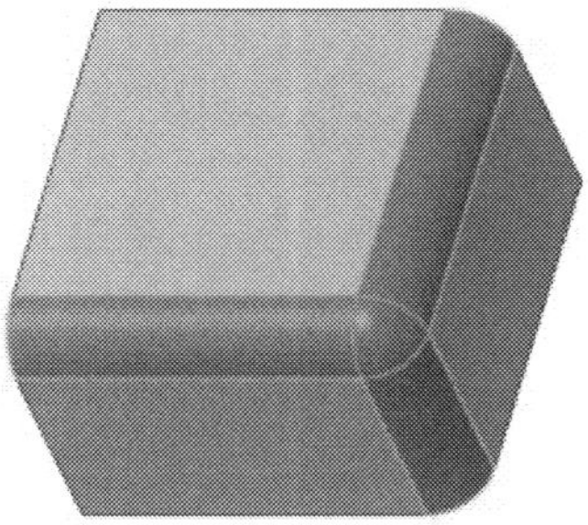

Figure 6-1 *Rounds created on the edges*

Figure 6-2 *Round created on surfaces*

Note

The round created after you select references on the model has a circular cross-section and rolling shape. In Creo Parametric, there is more than one shape that a round can have.

CREATING CHAMFERS

The **Chamfer** tool is used to bevel the edges and corners as per some specified parameters. In Creo Parametric you can create two types of chamfers. The first is the Corner chamfer and the second is the Edge chamfer. Figure 6-3 shows two types of chamfers. To create a chamfer on a corner, choose the **Corner Chamfer** tool from the **Chamfer** drop-down in the **Engineering** group. To create a chamfer on an edge, choose **Chamfer > Edge Chamfer** from the **Engineering** group.

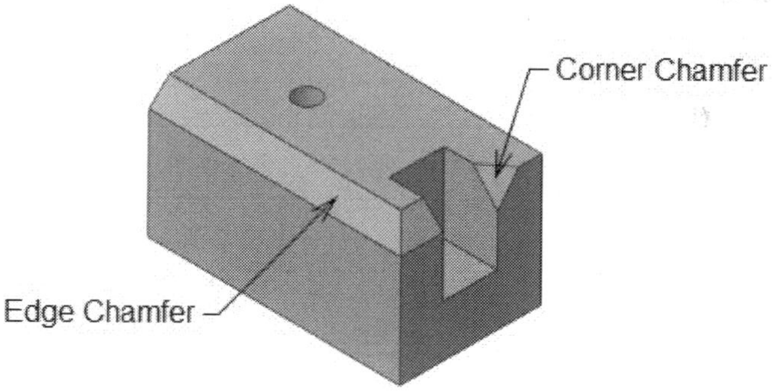

Figure 6-3 *Two types of chamfers*

Note

*When you are creating the corner chamfer, notice that only the **Vertex** filter is available in the Filter drop-down list in the Status Bar. As mentioned earlier, the filters in Creo Parametric narrow the entities available for selection. Therefore, you can easily select the reference entities on the model.*

UNDERSTANDING RIBS

Ribs are defined as thin wall-like structures used to bind the joints together so that they do not fail under an increased load. In Creo Parametric, the section for the rib is sketched as an open section and can be extruded equally in both directions of the sketch plane or on either side. In Creo Parametric, you can create two types of ribs: Trajectory Ribs and Profile Ribs.

EDITING FEATURES OF A MODEL

Most of the designs require editing during or after their creation. As mentioned earlier, Creo Parametric is a parametric and feature-based solid modeling software. Hence, the features constituting a model can be individually edited. Also, you can edit the datums and the features referenced to these datums. As parent-child relationship exists between the two features, therefore the child feature is also modified when the parent feature is modified. For example, if you have created a feature using a datum plane that is at some offset distance, the feature will be automatically repositioned when the offset value of the datum plane is changed.

Suppressing Features

When a model has many features, then suppressing some features decreases its regeneration time. Once the feature is suppressed, it will neither be displayed in the drawing area nor in the drawing views. Note that on suppressing, the feature is not deleted, only its visibility and regeneration is turned off temporarily. You can resume the feature anytime by unsuppressing it using the **Model Tree** or using the **Resume** option from the **Operations** group. As soon as you unsuppress the feature, it will be displayed in the drawing area and also in the drawing views.

TUTORIALS

Tutorial 1

Create the model shown in Figure 6-4. The dimensions, and the front, top, and left-side views of the model are shown in the Figure 6-5. **(Expected time: 45 min)**

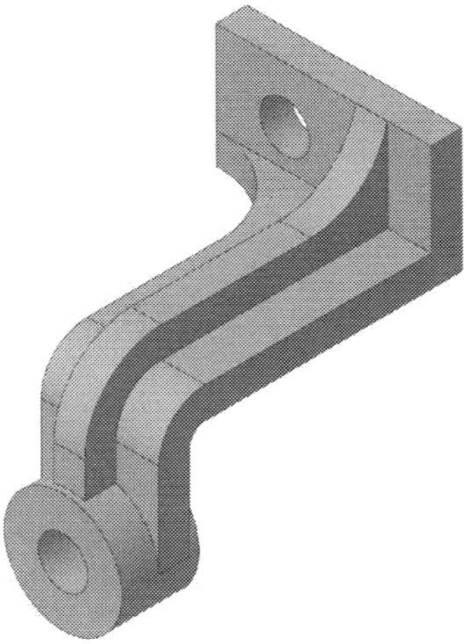

Figure 6-4 *Isometric view of the model*

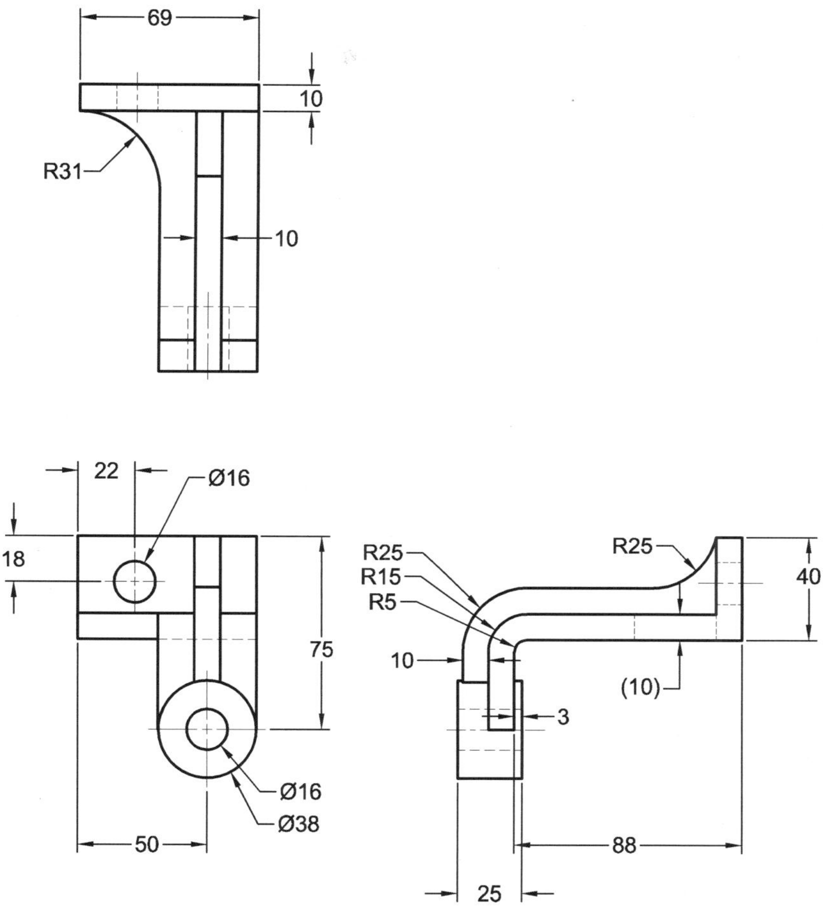

Figure 6-5 *Orthographic views of the model*

The following steps are required to complete this model:

Examine the model and determine the number of features in it, refer to Figure 6-4.

a. Create the base feature on the **FRONT** datum plane, refer to Figures 6-6 and 6-7.
b. Create the second extrude feature on the **TOP** datum plane, refer to Figures 6-8 and 6-9.
c. Create the third extrude feature on the front planar surface of the second feature, refer to Figures 6-10 and 6-11.
d. Create the non symmetrically extruded cylindrical feature on the front face of the third feature, refer to Figures 6-12 and 6-13.
e. Create a hole feature that is coaxial to the cylindrical feature, refer to Figure 6-15.
f. Create the round features, refer to Figures 6-16 and 6-17.
g. Create the last feature, which is the rib, refer to Figures 6-18 and 6-19.

When a Creo Parametric session starts, the first task is to set the working directory. As this is the first tutorial of this chapter, you need to select the Working Directory first. Set *C:\ Creo-5.0\c06* as the Working Directory; the message **Successfully changed to C:\ Creo-5.0\c06 directory** is displayed in the message area.

Starting a New Object File

1. Start a new part file and name it as *c06tut01*.

   The part mode is invoked and three default datum planes are displayed in the drawing area. Also, the **Model Tree** is displayed on the left of the drawing area.

Creating the Base Feature

To create the sketch of the base feature, you first need to select the sketching plane for it. In this model, you need to draw the base feature on the **FRONT** datum plane because the direction of extrusion of this feature is normal to this plane.

1. Select the **FRONT** datum plane as the sketching plane and choose the **Extrude** tool from the **Shapes** group or from the mini popup toolbar; the sketcher environment is invoked.

2. After you enter into the sketcher environment, create the sketch of the base feature and then apply constraints and dimensions to it, as shown in Figure 6-6. Note that in the sketch, the bottom line of the rectangular section coincides with the **TOP** datum plane.

   As is evident from the sketch of the base feature shown in Figure 6-6 that the **RIGHT** datum plane is located at a distance of 50 from the left edge because later in the tutorial, the rib feature will be created on this plane. Refer to Figure 6-5 for the distance required for rib feature. As mentioned earlier, that, by default, the sketch for the rib feature is extruded on both sides of the sketching plane. Therefore, when you create the sketch for the rib feature on the **RIGHT** datum plane, it will extrude on both the sides.

3. After the sketch is completed, choose the **OK** button to exit the sketcher environment; the **Extrude** dashboard is enabled and appears above the drawing area. All model attributes that are selected by default are accepted to create the model.

4. Enter **10** as the depth in the dimension box on the **Extrude** dashboard.

5. Choose the **Saved Orientations** button from the **Graphics** toolbar; a flyout is displayed. Choose the **Default Orientation** option from the flyout; the model orients in its default orientation, that is, the trimetric view.

6. Next, choose the **OK** button from the **Extrude** dashboard to create the base feature, as shown in Figure 6-7.

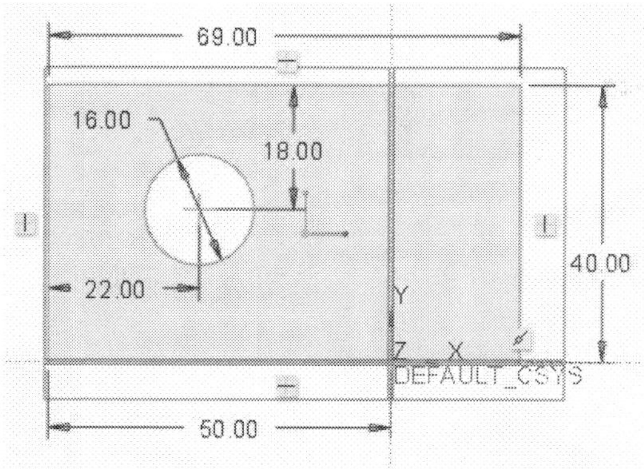

Figure 6-6 *Sketch of the base feature with dimensions and constraints*

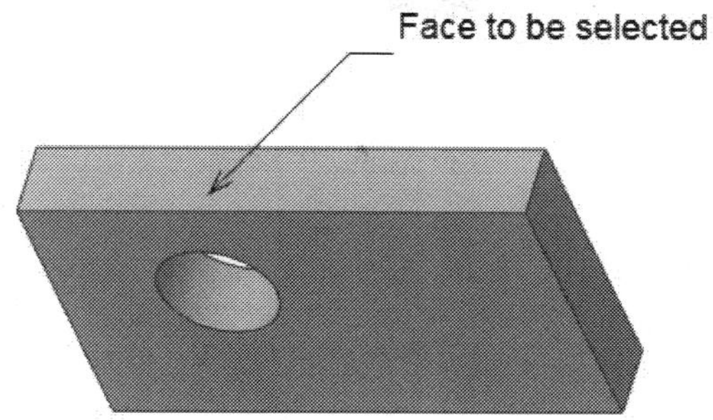

Figure 6-7 *Face of the base feature selected*

Creating the Second Feature

The second feature is also an extruded feature and it will be created on the **TOP** datum plane. Therefore, you need to define the **TOP** datum plane as the sketching plane.

1. Select the **TOP** datum plane as the sketching plane and then choose the **Extrude** tool from the **Shapes** group or from the mini popup toolbar; the sketcher environment is invoked.

2. Create the sketch for the second feature and apply constraints and dimensions to it, as shown in Figure 6-8.

   In Creo Parametric, you can draw sketches by snapping the edge references. To close the sketch, draw a line and aligned it with the bottom edge by snapping the edges of the base feature, refer to Figure 6-8.

Note

If you do not close the section loop by drawing a line or using the edge of the base feature, an arrow is displayed on exiting the sketcher environment. Using this arrow, you can specify the direction in which the material will be added.

3. After completing the sketch, turn the model display to **Shading With Edges** and choose the **OK** button; the **Extrude** dashboard is enabled above the drawing area.

4. Enter **10** in the dimension box in the **Extrude** dashboard and press ENTER.

5. Choose the **OK** button from the **Extrude** dashboard; the second feature is completed and the shaded default trimetric view is displayed, as shown in Figure 6-9.

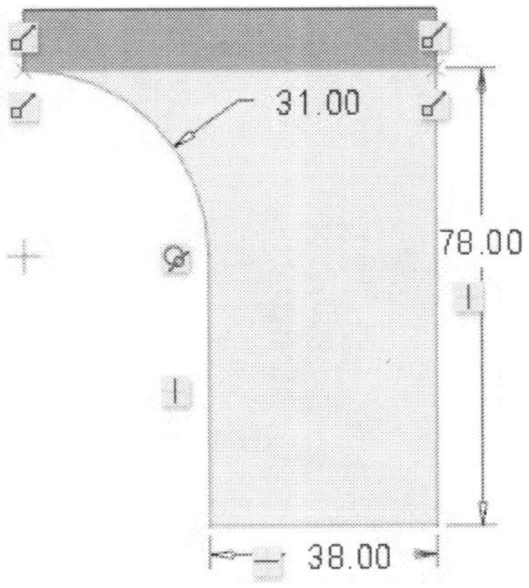

Figure 6-8 Sketch of the second feature with dimensions and constraints

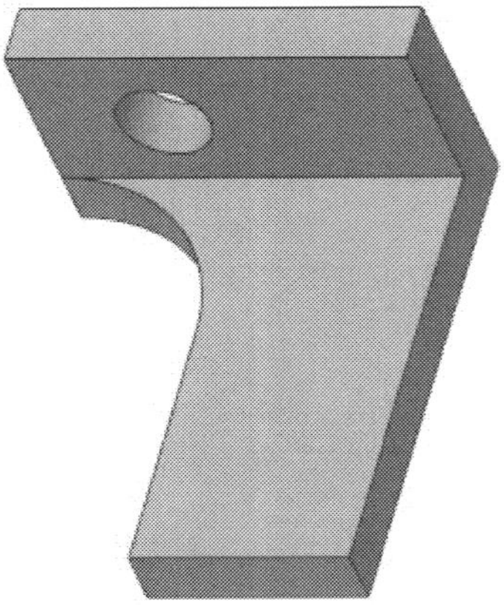

Figure 6-9 The default trimetric view of the completed second feature and the base feature

Creating the Third Feature

The sketch of the third feature will be drawn on the front planar surface of the second feature and it will be extruded to the given depth.

1. Select the face of the second feature as the sketching plane and choose the **Extrude** tool to invoke the sketcher environment, refer to Figure 6-10.

2. Once you enter in the sketcher environment, turn the model display to **No hidden**. Create the sketch for the third feature and apply constraints and dimensions to it, refer to Figure 6-11.

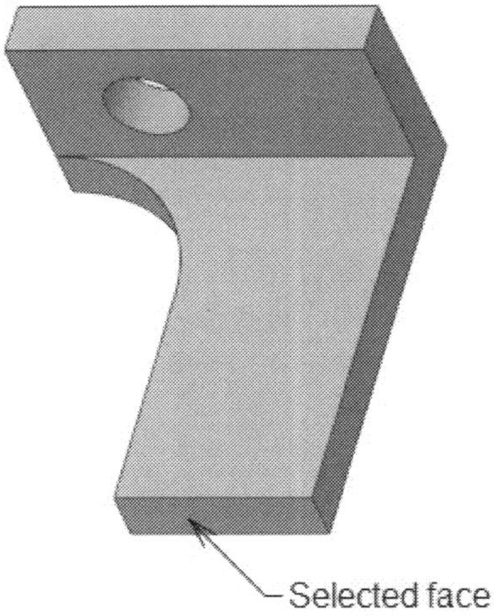

Figure 6-10 Planar surface selected as the sketching plane for the third feature

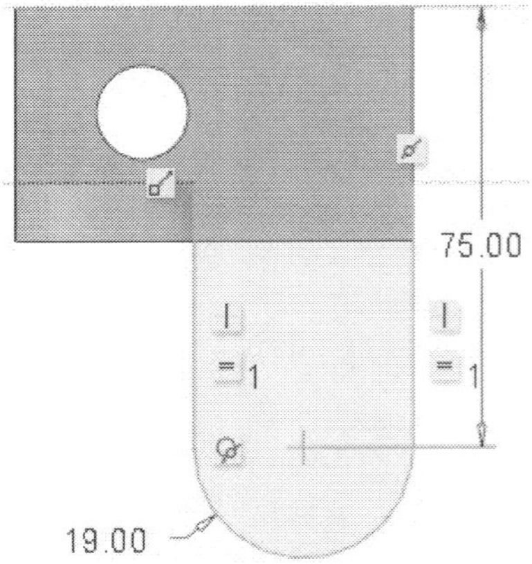

Figure 6-11 Sketch of the third feature with dimensions and constraints

Note

While drawing the sketch of any feature, it is recommended that you first apply the required constraints and then dimension the sketch.

3. Choose the **OK** button from the **Close** group; the **Extrude** dashboard is enabled above the drawing area.

4. Enter **10** in the dimension box in the **Extrude** dashboard.

5. Choose the **OK** button from the **Extrude** dashboard. Turn the model display to **Shading With Edges**. The third feature is completed. You can use the middle mouse button to spin the model to view its different orientations.

Creating the Fourth Feature

The fourth feature of the model is an extruded feature and its sketch is drawn on the front planar surface of the third feature. The extrusion of the cylindrical feature will be created on both sides of the front face. The attributes of the feature, like extrusion on both sides and the depth of extrusion, will be specified after the sketch of the feature is drawn.

1. Select the front planar face of the third feature, shown in Figure 6-12, as the sketching plane.

2. Choose the **Extrude** tool from the **Shapes** tab or from the mini popup toolbar; the sketcher environment is invoked.

3. Choose the **Concentric** button from the **Circle** drop-down in the **Sketching** group to draw circular section for the fourth feature, refer to Figure 6-13. Select the arc using the left mouse button. As you move the mouse, the rubber-band circle changes its size. Move the cursor close to the arc; the cursor snaps to the arc. Use the left mouse button and select a point on the arc. You will notice that the equal radius constraint is applied to the sketch. Now, exit this tool.

4. After the sketch is completed, choose the **OK** button from the **Close** group; the **Extrude** dashboard is enabled above the drawing area.

5. Choose the **Options** option from the **Extrude** dashboard. In the slide-down panel, choose the **Blind** option from both the **Side 1** and **Side 2** drop-down lists.

6. Enter **12** in the dimension box on the right of the **Side 1** drop-down list.

7. Enter **13** in the dimension box that is on the right of the **Side 2** drop-down list.

8. Choose the **OK** button from the **Extrude** dashboard to create fourth feature.

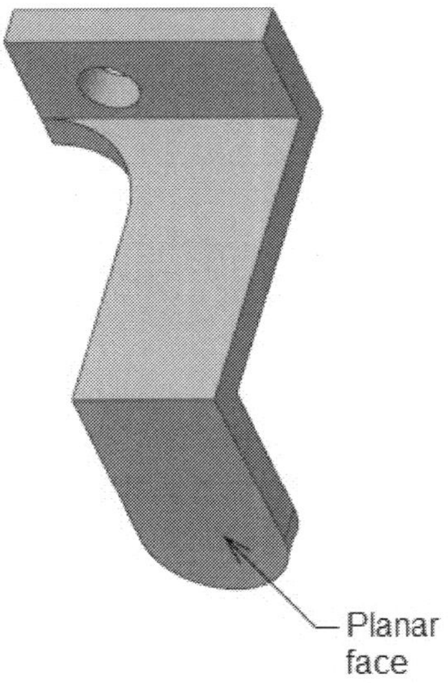

Figure 6-12 *Front planar face selected*

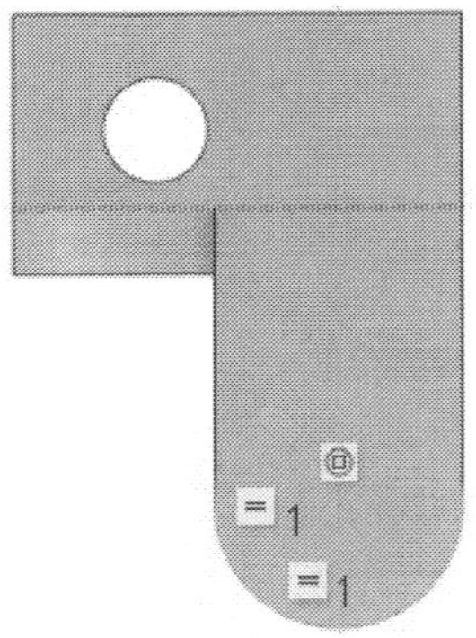

Figure 6-13 *Sketch for the fourth feature*

Creating the Hole Feature

The hole feature is created using the **Hole** dashboard. The hole will be placed coaxial to the fourth cylindrical feature.

Note

For creating a coaxial hole, you need to select the axis of the circular feature. To do so, you need to turn on the display of the axis by selecting the ***Axis Display*** *check box from the* ***Datum Display Filters*** *drop-down list in the* ***Graphics*** *toolbar, if it is not displayed.*

1. Choose the **Hole** tool from the **Engineering** group; the **Hole** dashboard is displayed and the **Create simple hole** tool is chosen by default.

2. Choose the **Placement** tab from the **Hole** dashboard; a slide-down panel is displayed.

3. Select the front face of the cylindrical feature, as shown in Figure 6-14, to place the hole. As you select the front face of the cylindrical feature, a preview of the hole is displayed in the drawing area. Now, you need to specify reference for placing the hole.

4. Press the CTRL key and select the axis of the cylindrical feature from the drawing area.

   Note that the **Coaxial** option is automatically selected in the drop-down list of the slide-down panel of the **Placement** tab.

5. In the diameter dimension box on the **Hole** dashboard, type **16** and press ENTER.

6. In the depth flyout of the **Hole** dashboard, choose the **Drill to intersect with all surfaces** button.

7. Choose the **OK** button from the **Hole** dashboard; the hole is created and trimetric view of the shaded model with the hole is displayed, as shown in Figure 6-15.

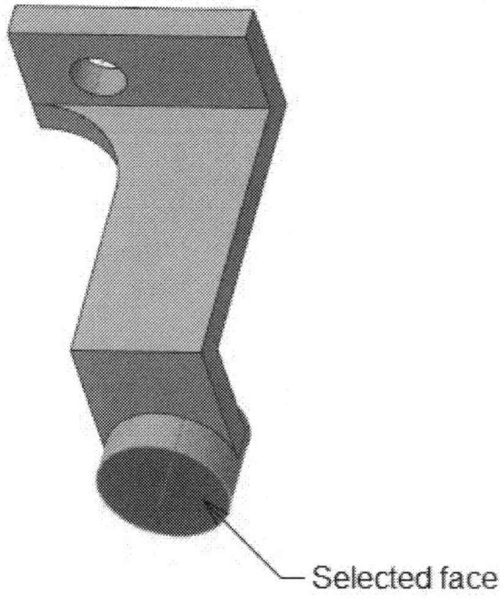

Figure 6-14 *Planar face selected for creating hole*

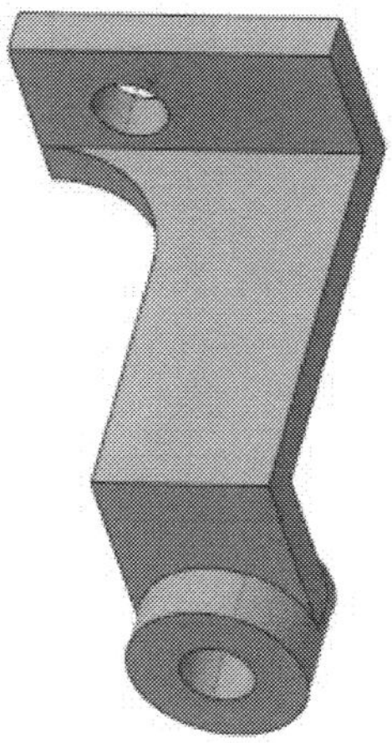

Figure 6-15 *The default trimetric view*

Creating Two Round Features

Now, the two round features need to be created. Both the rounds have different radii and will be created by defining two sets.

1. Choose the **Round** button from the **Engineering** group; the **Round** dashboard is displayed.

2. Choose the **Sets** tab to display the slide-down panel. Let this slide-down panel remains open so that you can view the selections made on the model.

3. Select the **Edge** option from the Filter drop-down list in the Status Bar. Spin the model and select the edge shown in Figure 6-16.

   A preview of the round is created on the edge and the default value of the radius is displayed.

4. Double-click on the default radius value that is displayed on the preview of the round. In the edit box that appears, enter the value **5** and press ENTER; the first round is created.

5. After spinning the model, select the edge shown in Figure 6-17. You will notice that in the slide-down panel, **Set 1** and **Set 2** appear. This indicates that the second set has been defined.

   A preview of the round is created on the edge and the default value of the radius is displayed on the round geometry.

6. Double-click on the default radius value that is displayed in the preview of the second round. In the edit box that appears, type the value **15** and press ENTER.

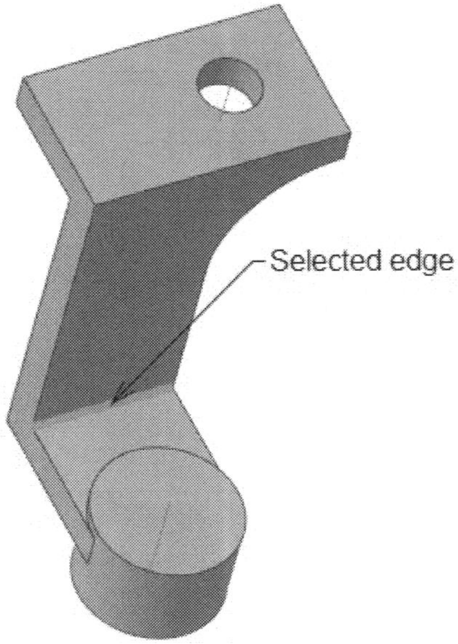

Figure 6-16 *Faces for creating a round of radius **5***

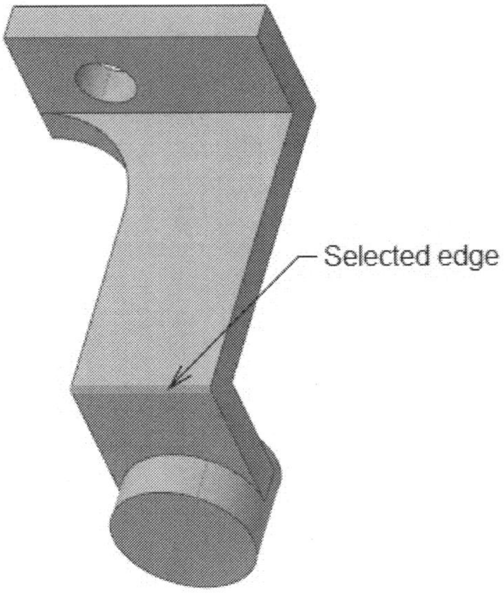

Figure 6-17 *Edges selected for creating a round of radius **15***

7. Choose the **OK** button from the **Round** dashboard.

   The second round is created. In the **Model Tree**, the two rounds created appear as a single feature.

Creating the Rib Feature

1. A rib feature is always sketched from the side view. Turn on the display of the datum planes from the **Datum Display** drop-down list in the **Graphics** toolbar if it is turned off.

   The location for the rib from the **RIGHT** datum plane was calculated while sketching the base feature. The section for the rib feature will be drawn on the **RIGHT** datum plane.

2. Select the **RIGHT** datum plane and then choose the **Profile Rib** tool from the **Rib** drop-down in the **Engineering** group; the sketcher environment is invoked.

3. Draw the open sketch for the rib feature and apply required constraints and dimensions, as shown in Figure 6-18.

4. After the sketch is completed, choose the **OK** button. On doing so, an arrow pointing in the direction of material addition is displayed on the sketch. As the section for the rib feature is open, Creo Parametric allows you to specify the direction where the material should be added. Therefore, you need to change the direction of the arrow toward the model, if it is not already changed.

5. Enter **10** in the dimension box in the **Profile Rib** dashboard.

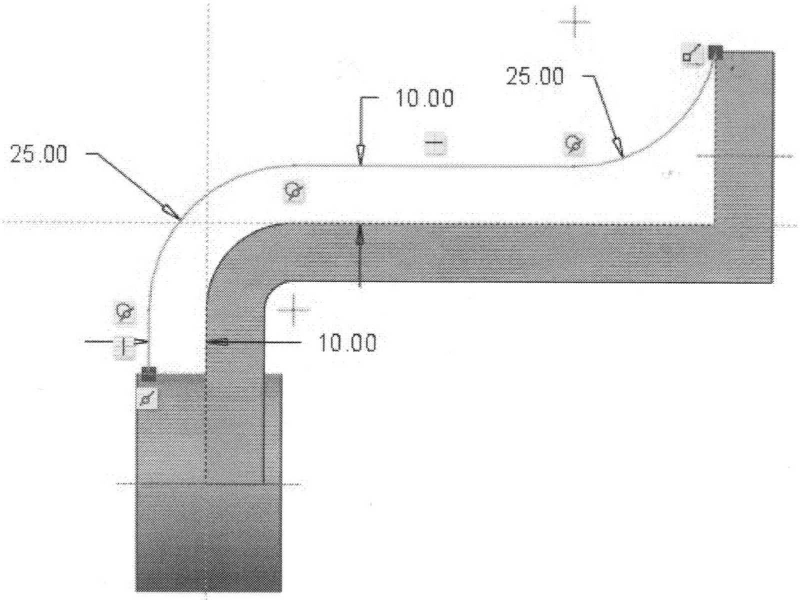

***Figure 6-18** Open sketch of the rib feature with dimensions and constraints*

6. Choose the **OK** button from the **Profile Rib** dashboard to complete creating the model and exit this feature creation tool.

   All features in the model have been created and the model is now complete. The trimetric shaded view of the completed model is shown in Figure 6-19.

Saving the Model

1. Choose the **Save** button from the **File** menu and save the model. The order of feature creation can be seen from the **Model Tree** shown in Figure 6-20.

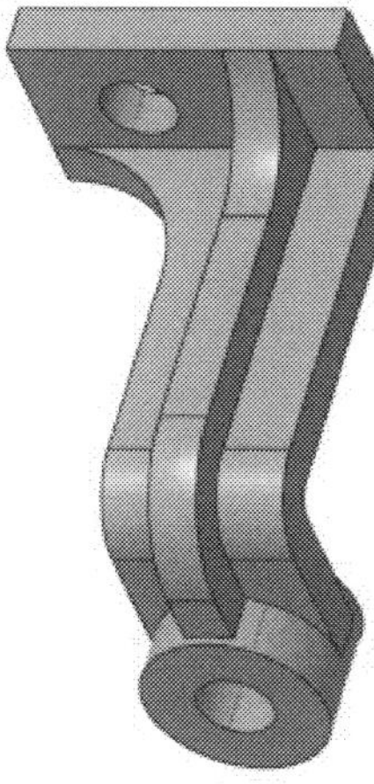

***Figure 6-19** The default trimetric view of the model*

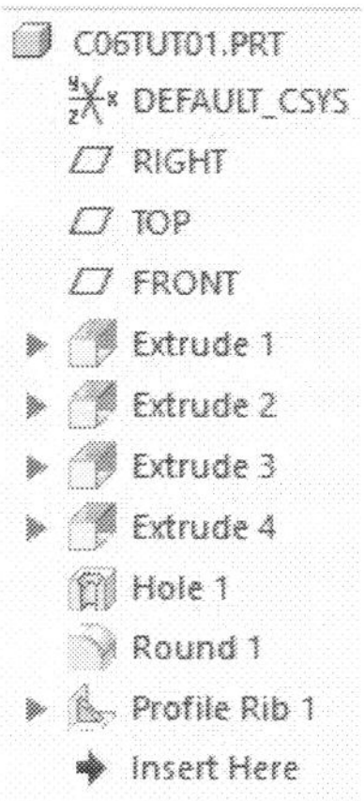

***Figure 6-20** The **Model Tree** for Tutorial 1*

Tutorial 2

In this tutorial, you will create the model shown in Figure 6-21. The dimensions of the model are shown in Figure 6-22. **(Expected time: 30 min)**

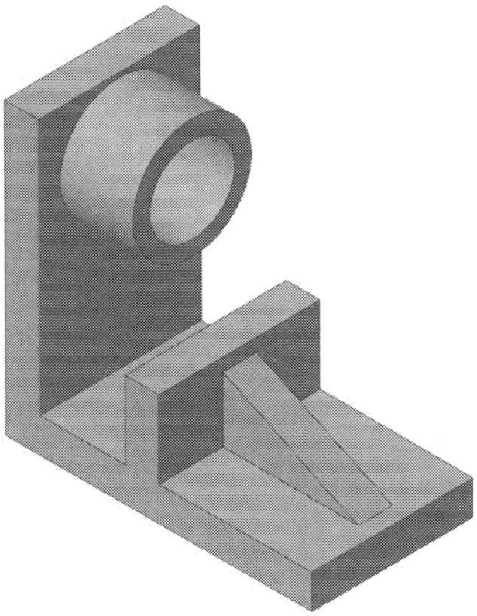

Figure 6-21 *Isometric view of the solid model*

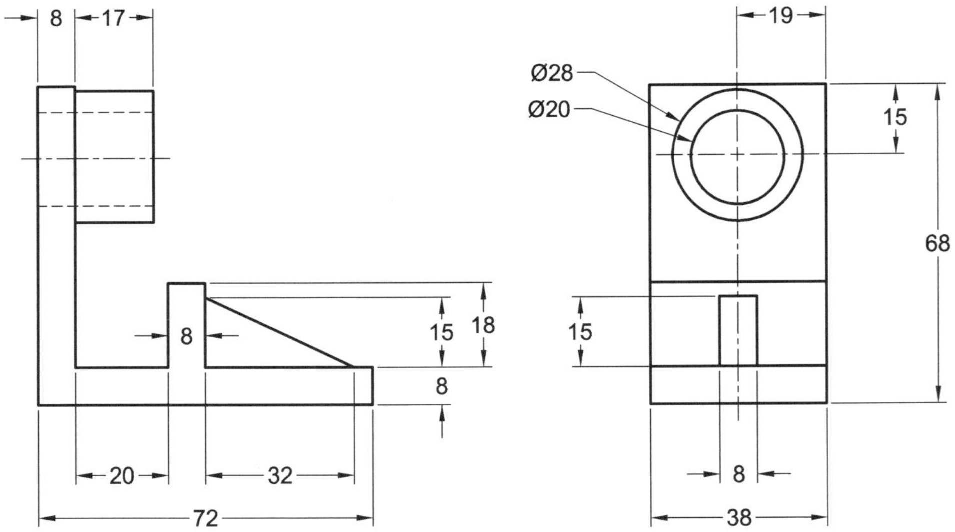

Figure 6-22 *Front and right view of the model*

The following steps are required to complete this model:

Examine the model and determine the number of features in it, refer to Figure 6-21.

a. Create the base feature, refer to Figures 6-23 and 6-24.
b. Create the cylindrical feature, refer to Figures 6-25 through 6-27.

c. Create the hole feature coaxially on the cylindrical feature, refer to Figure 6-28.
d. Create the rib feature, refer to Figures 6-29 and 6-30.

The working directory has already been selected in Tutorial 1 and therefore, you do not need to select it again. However, if you need to change the working directory, choose **File > Select Working Directory** and then select *c06* in the **Select Working Directory** dialog box.

Starting a New Object File

1. Start a new part file and name it as *c06tut02*.

The three default datum planes and the **Model Tree** appear in the drawing area.

Creating the Base Feature

1. Select the **FRONT** datum plane from the drawing area and choose the **Extrude** tool from the mini popup toolbar; the sketcher environment is invoked automatically.

2. In the sketcher environment, create the sketch of the base feature and apply constraints and dimensions to it, as shown in Figure 6-23.

3. After the sketch is completed, choose the **OK** button and exit the sketcher environment; the **Extrude** dashboard is enabled and appears above the drawing area.

4. Enter **38** as the depth in the dimension box that is present on the **Extrude** dashboard; the default trimetric view of the base feature is displayed, as shown in Figure 6-24.

5. Choose the **OK** button from the **Extrude** dashboard to exit the feature creation tool.

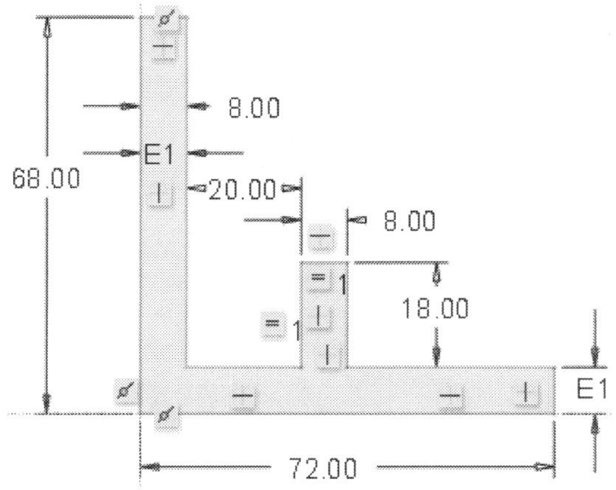

Figure 6-23 *Sketch of the base feature with dimensions and constraints*

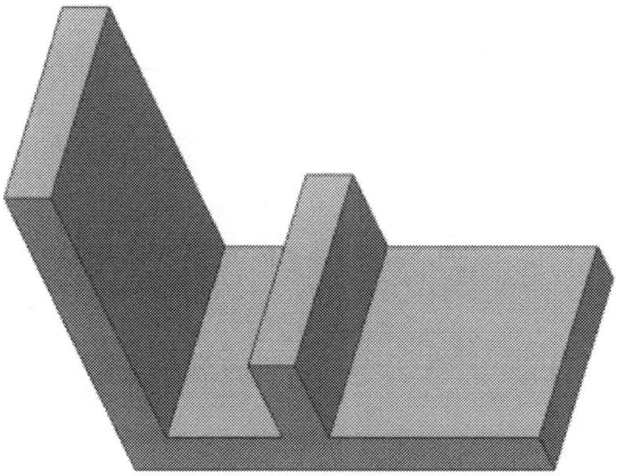

Figure 6-24 *The default trimetric view of the base feature*

Note

*In this tutorial, the base feature can also be created by extruding it on both sides of the sketching plane that is the **FRONT** datum plane. This would reduce the step required to create the datum plane, which in turn is used to create the rib feature. However, to familiarize you with creating embedded datum planes, the base feature is extruded on one side of the sketching plane.*

Creating the Second Feature

The second feature is a cylindrical feature that is sketched on the planar surface of the base feature which is shown in Figure 6-25.

1. Choose the **Extrude** tool from the **Shapes** group.

2. Choose the **Placement** tab and then from the slide-down panel, choose the **Define** button; the **Sketch** dialog box is displayed.

3. Select the face of the base feature as the sketching plane, refer to Figure 6-25.

4. Select the **TOP** datum plane from the drawing area and then select the **Top** option from the **Orientation** drop-down list.

5. Choose the **Sketch** button from the **Sketch** dialog box to enter into the sketcher environment.

6. Draw the sketch for the second feature. Apply and modify the dimensions, as shown in Figure 6-26.

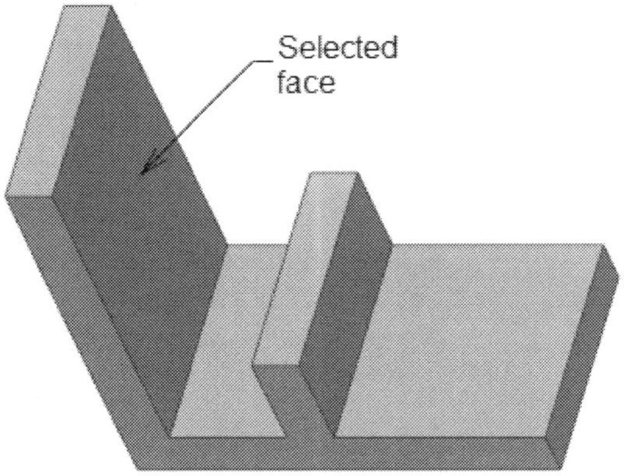

Figure 6-25 *The face of the base feature selected as the sketching plane*

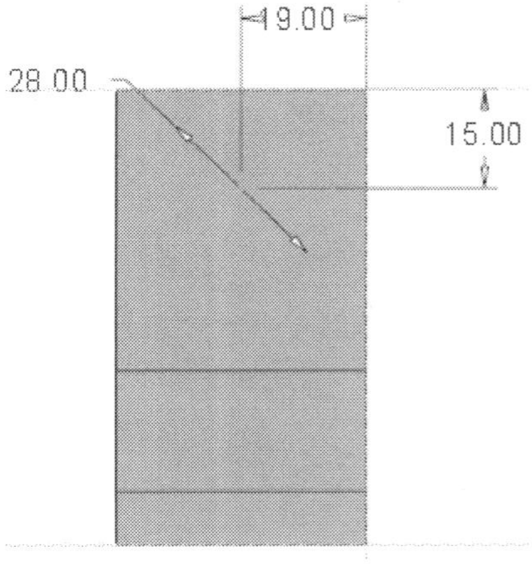

Figure 6-26 *Sketch and dimensions of the second feature*

7. Turn the display of the model to **Shading With Edges**. Exit the sketcher environment by choosing the **OK** button; the **Extrude** dashboard is enabled and appears above the drawing area.

8. Enter **17** in the dimension box in the **Extrude** dashboard.

9. Choose the **OK** button from the **Extrude** dashboard to exit the feature creation tool. The trimetric view of the model with the second feature is shown in Figure 6-27.

Creating the Hole Feature

The hole feature will be created using the **Hole** dashboard. The coaxial hole will be created on the cylindrical feature. The axis of the cylindrical feature will be used as the axial reference to create the coaxial hole.

1. Choose the **Hole** tool from the **Engineering** group in the **Ribbon**; the **Hole** dashboard is displayed. The **Create simple hole** tool in the **Hole** dashboard is chosen by default.

2. Choose the **Placement** tab from the **Hole** dashboard; a slide-down panel is displayed.

3. Select the front face of the cylindrical feature to place the hole.

   As you select the front face of the cylindrical feature, the preview of the hole is displayed in the drawing area. Now, you need to specify the reference for the placement of hole.

4. Hold-down the CTRL key and select the axis of the cylindrical feature from the drawing area.

   The **Coaxial** option is selected automatically in the drop-down list of the slide-down panel.

5. In the diameter edit box of the **Hole** dashboard, enter **20** and press ENTER.

6. From the depth flyout on the **Hole** dashboard, choose the **Drill to intersect with all surfaces** button.

7. Choose the **OK** button from the **Hole** dashboard; the hole is created and the trimetric view of the shaded model with the hole is displayed, as shown in Figure 6-28.

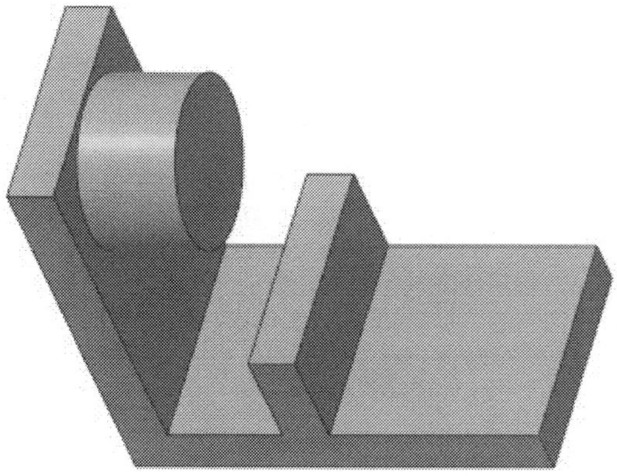

Figure 6-27 *Model with the second feature*

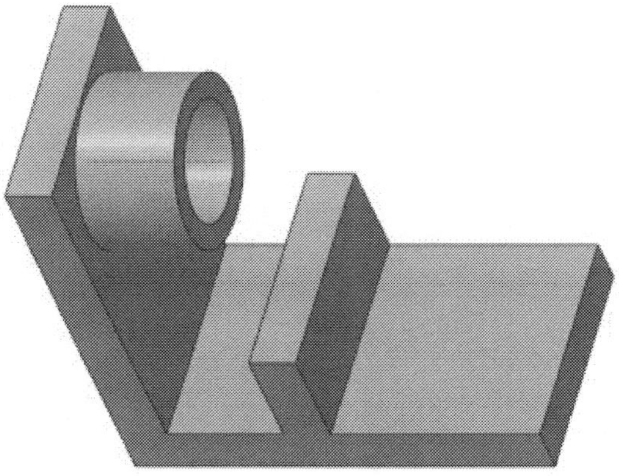

Figure 6-28 *Model with the hole feature*

Creating the Rib Feature

To create the rib feature, an internal datum plane is created. As mentioned earlier, rib features are always drawn from the side view.

1. Choose the **Profile Rib** tool from the **Rib** drop-down in the **Engineering** group.

2. Choose the **References** tab; a slide-down panel is displayed. Choose the **Define** button; the **Sketch** dialog box is displayed.

3. Choose the **Plane** tool from the **Datum** group of the **Model** tab in the **Profile Rib** dashboard; the **Datum Plane** dialog box is displayed. You may need to move the **Sketch** dialog box to bring the **Plane** button into view.

4. Select the axis of the hole, press the CTRL key, and select the **FRONT** datum plane. You can view your selections in the **References** collector.

5. Choose the **Offset** button in the **References** collector; a drop-down list appears on the right of the reference. Select the **Parallel** option from this drop-down list.

6. Choose the **OK** button from the **Datum Plane** dialog box to exit it.

   A datum plane that passes through the selected axis and is parallel to the **FRONT** datum plane will be created. This datum plane is selected automatically as the sketching plane. Now, you need to select the reference plane.

7. Select the **TOP** datum plane and then select the **Top** option from the **Orientation** drop-down list.

8. Choose the **Sketch** button; the system takes you to the sketcher environment. Choose the **No Hidden** option from the **Display Style** drop-down in the **Graphics** toolbar.

9. Draw the open sketch of the rib feature and apply dimensions, as shown in Figure 6-29. Exit the sketcher environment by choosing the **OK** button; the **Profile Rib** dashboard is enabled and appears above the drawing area.

10. Choose the **Flip** button on the **References** slide-down panel to flip the direction of the arrow; the preview of the rib is displayed on the model. Choose this button only when the preview of the rib is not displayed. Alternatively, click on the arrow to flip the direction.

11. Enter **8** as the thickness of the rib in the edit box in the **Profile Rib** dashboard. Next, choose the **OK** button.

    The trimetric view of the complete model with the rib feature is shown in Figure 6-30.

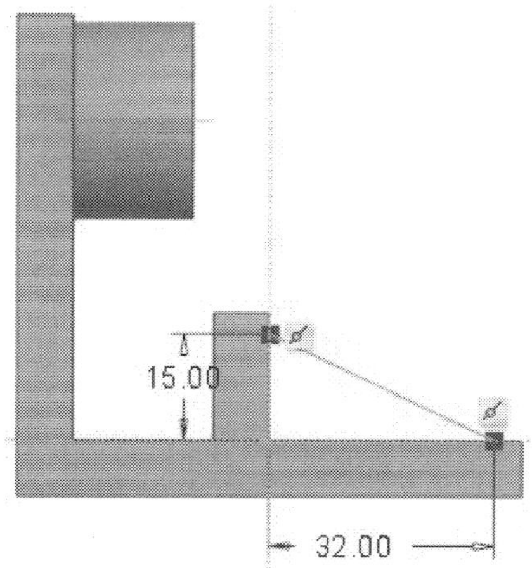

Figure 6-29 Sketch for the rib feature with the model display set to ***No hidden***

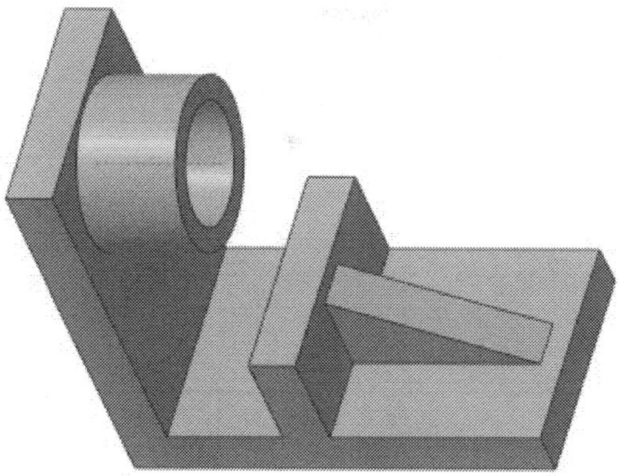

Figure 6-30 The default trimetric view of the final model

Saving the Model

1. Choose the **Save** button from the **File** menu to save the model and then close the active window.

EXERCISES

Exercise 1

Create the model shown in Figure 6-31. The dimensions and front and top views of the model are shown in Figure 6-32. **(Expected time: 45 min)**

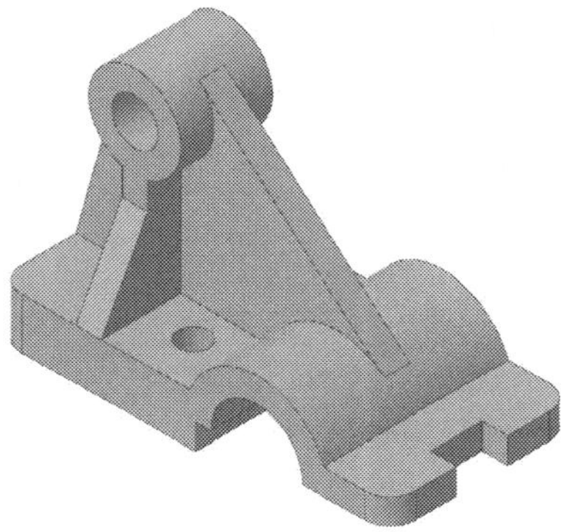

Figure 6-31 Isometric view of the model

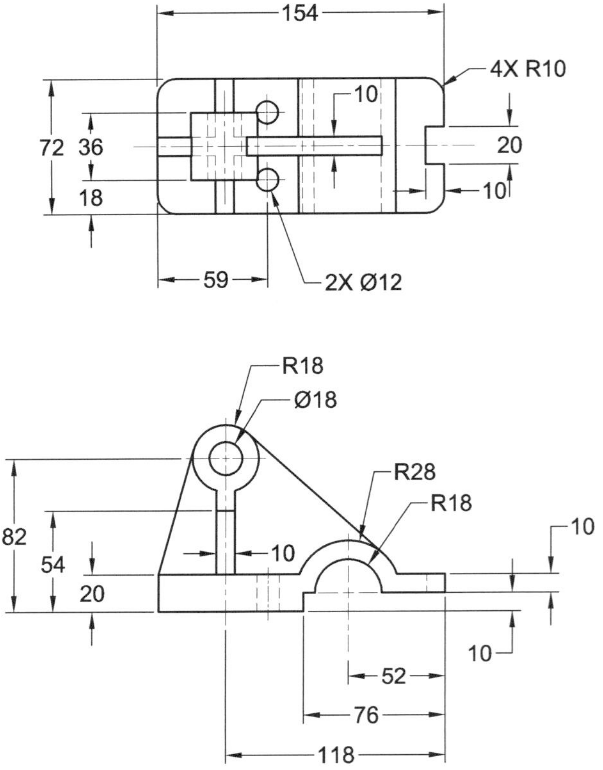

Figure 6-32 *Orthographic views of the model*

Exercise 2

Create the model shown in Figure 6-33. The dimensions and front and right-side views of the model are shown in Figure 6-34. **(Expected time: 30 min)**

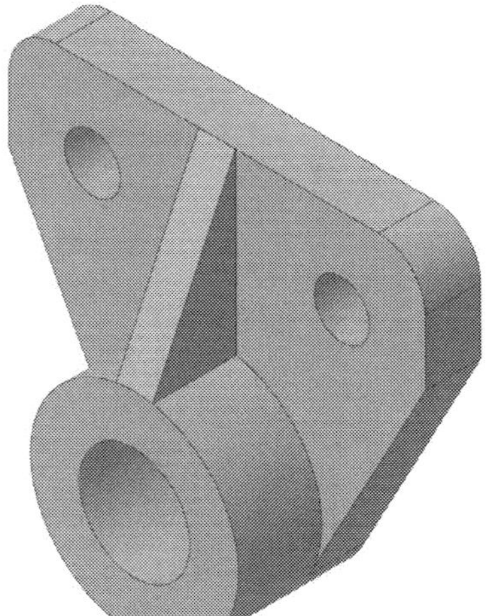

Figure 6-33 *Isometric view of the model*

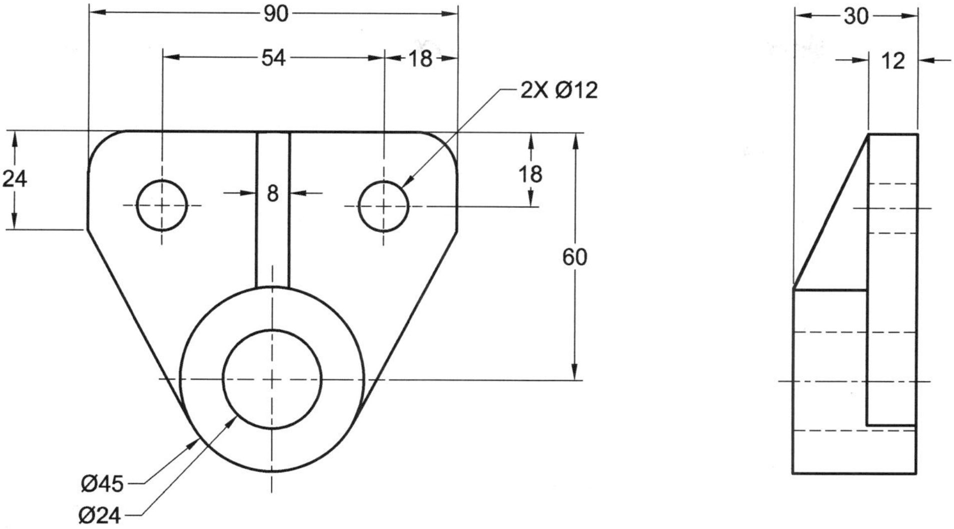

Figure 6-34 *Front and right view of the model*

This page is intentionally left blank

Chapter 7

Options Aiding Construction of Parts-II

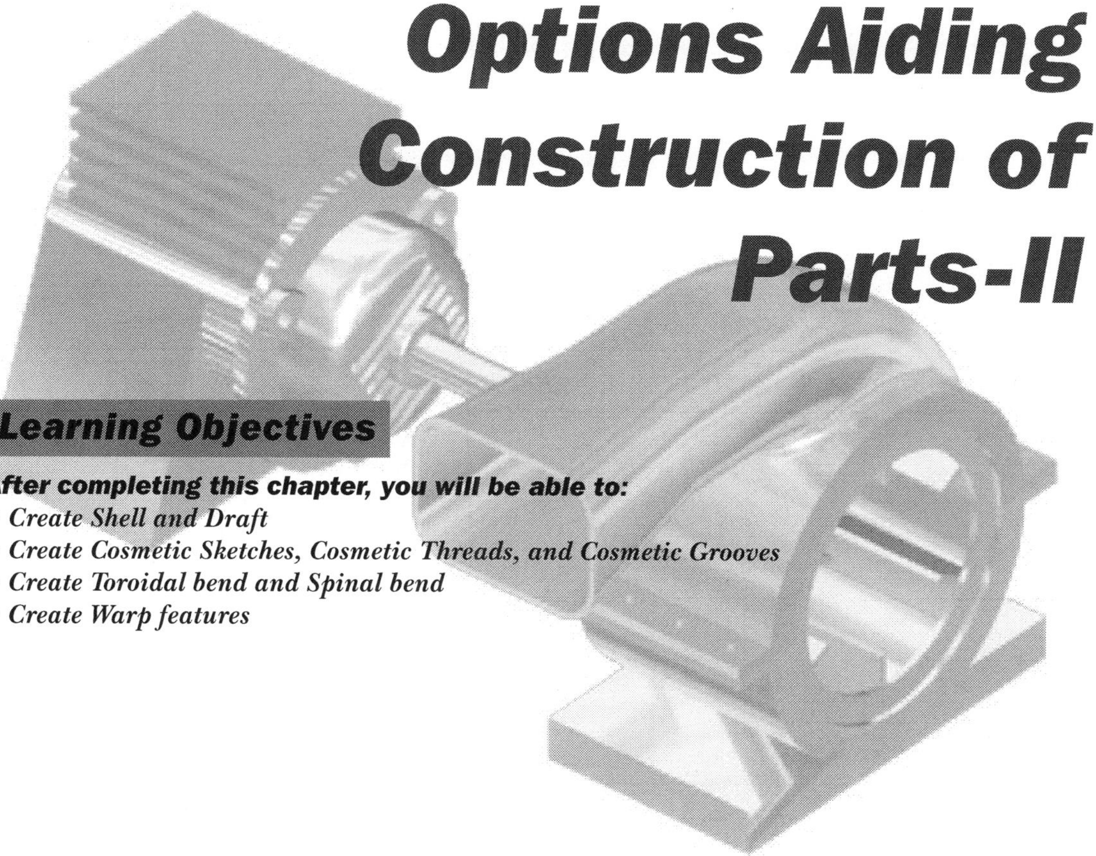

Learning Objectives

After completing this chapter, you will be able to:

- *Create Shell and Draft*
- *Create Cosmetic Sketches, Cosmetic Threads, and Cosmetic Grooves*
- *Create Toroidal bend and Spinal bend*
- *Create Warp features*

OPTIONS AIDING CONSTRUCTION OF PARTS

In this chapter, you will learn about the feature creation tools such as **Shell** and **Draft**, provided in Creo Parametric. Also, you will learn to create cosmetic features such as cosmetic sketch, cosmetic thread, and cosmetic groove.

SHELL FEATURE

Ribbon: Model > Engineering > Shell

The **Shell** tool scoops out material from the model and at the same time removes the selected faces, leaving behind a thin model with some specified wall thickness. If you do not select a surface to remove, a closed shell is created as a part completely hollow from inside.

CREATING DRAFT FEATURES

Ribbon: Model > Engineering > Draft drop-down> Draft

The **Draft** tool adds an angle to individual surfaces or to series of surfaces. The value of the angle is always between -89.9° to +89.9°. One of the applications of draft features is found in molds and castings where a taper is required to separate the casting from the mould or vice versa.

TOROIDAL BEND

Ribbon: Model > Engineering drop-down > Toroidal Bend

The **Toroidal Bend** tool is used to provide a toroidal (revolved) shape to solids, quilts, or datum curves. The example of a plate after creating the toroidal bend is shown in Figure 7-1. To create the toroidal bend, first you will create a rectangular plate and cut the profile on it, as shown in Figure 7-2. Then, using the **Toroidal Bend** tool, you will bend the rectangular plate through an angle of 360 degrees, refer to Figure 7-1.

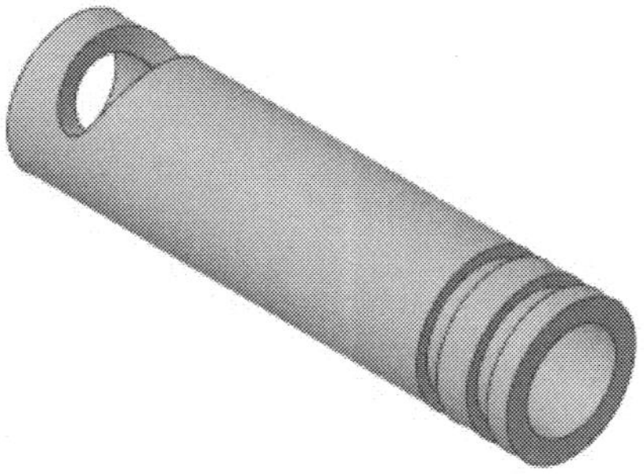

***Figure 7-1** The plate after creating the toroidal bend*

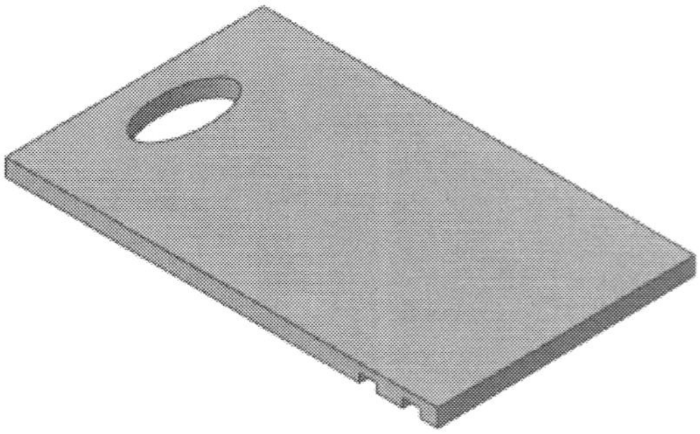

***Figure 7-2** Rectangular plate with a cut profile*

SPINAL BEND

Ribbon: Model > Engineering > Spinal Bend

The **Spinal Bend** tool is used to bend solids or quilts about a curved spine by continuously repositioning cross-sections along the curve. Cross-sections, which are perpendicular to an axis, are repositioned perpendicular to the sketched spine without any distortion. Any kind of compression or distortion in the geometry is created along the trajectory of the spine.

COSMETIC SKETCH

Cosmetic sketches are used to sketch entities on the surface of a model for visualization of stamped features such as company logos and serial numbers. The method to create cosmetic sketches is similar to the method of creating typical sketches. Cosmetic sketches cannot be used for removing or adding material to the model. You can also use imported entities as cosmetic sketch.

COSMETIC THREAD

Cosmetic threads are used to create a lightweight representation of threads on a model. In general, a cosmetic thread is used to represent the thread symbol. Cosmetic threads can be applied to cylinders, cones, and planes. The surface selected for the placement determines whether the thread will be external or internal.

COSMETIC GROOVE

Cosmetic groove is a projected cosmetic feature. The cosmetic groove is created by making a sketch and projecting it onto a surface.

WARP

Ribbon: Model > Editing > Warp

The **Warp** tool helps you study the design variations during the conceptual design stage. Using the **Warp** tool, you can manipulate the form and shape of solids, quilts, facets, and curves. The **Warp** tool can be used only in the **Part** mode.

TUTORIALS

Tutorial 1

In this tutorial, you will create the model shown in Figure 7-3. To perform this tutorial, you need to download the zipped file named as *c07_creo_5.0_input* from the **Input Files** section of the CADSOFT website *http://www.cadsofttech.com*. The complete path for downloading the file is:

Textbooks > CAD/CAM > Creo Parametric > Creo Parametric 5.0 for Novices > Input Files

(Expected time: 30 min)

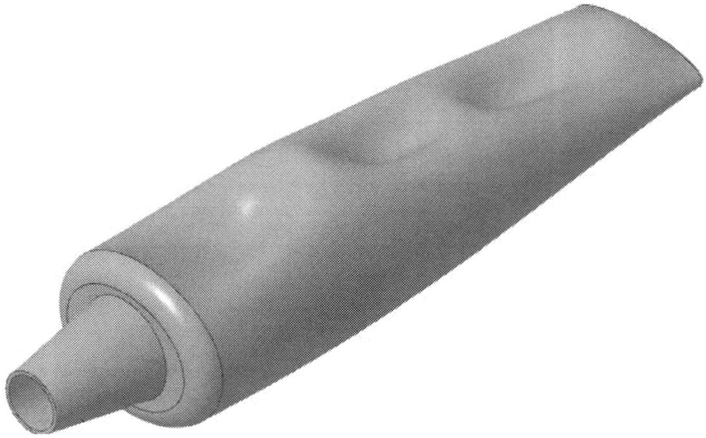

Figure 7-3 *The resulting model for Tutorial 1*

The following steps are required to complete this tutorial:

a. Create the sketch for warp feature 7-3.
b. Create the warp features using the **Spine** tool, refer to Figures 7-4 through 7-7.

When you start the Creo Parametric session, the first task is to set the working directory. Make sure the required working directory is selected.

Opening the File

1. Start a new session of Creo Parametric 5.0.

2. Select the **Open** button in the **Home** tab; the **File Open** dialog box will be displayed. Browse to the location where you have stored the downloaded file and select the **Open** button in the dialog box to open the model in the Part environment.

Creating the Sketch for the Warp Feature

You need to use the **Spline** tool to create deformation in the tube. Therefore, you first need to create the curve which will be used to create the warp feature.

1. Choose the **Sketch** tool from the **Datum** group; the **Sketch** dialog box is displayed.

2. Choose the **RIGHT** plane as the sketching plane and select the **Top** option from the drop-down list in the **Orientation** tab. Next, choose the **Sketch** button to enter the sketching environment.

3. Choose the **Spline** tool from the **Sketching** group and draw the spline curve shown in Figure 7-4. Note that the dimensions of the spline are not important as this sketch will be used as reference for deciding the shape after deformation.

4. Choose the **OK** button to exit the sketcher environment.

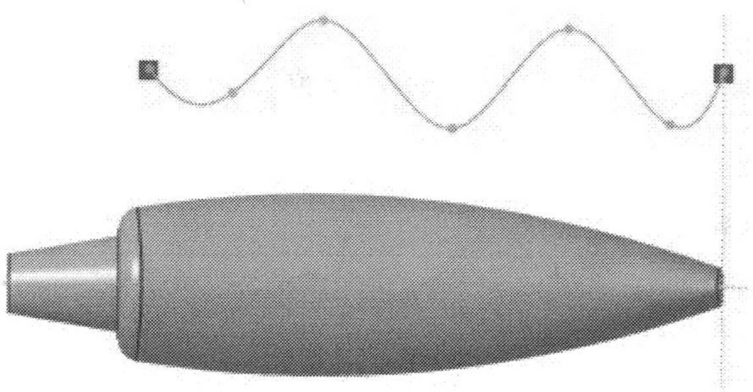

Figure 7-4 Sketched spline curve

Applying the Warp Transformation by Using the Spine Tool

1. Choose the **Warp** tool from the expanded **Editing** group; the **Warp** dashboard is displayed.

2. Choose the **References** tab; the **Geometry** collector is enabled by default. Select the model from the drawing area.

3. Click in the **Direction** selection box; you are prompted to select a plane or a coordinate system to define the direction of the warp.

4. Select the **TOP** plane from the drawing area; you are prompted to select the deformation tool.

5. Choose the **Spine** tool from the **Warp** dashboard; you are prompted to select a curve to define the deformation.

6. Select the sketched spine as the reference curve for the warp transformation. The model similar to the one shown in Figure 7-5 is displayed in the drawing area.

7. Select the control points on the reference curve and drag them until the required spinal deformation is obtained. The position of the control points after carrying out deformation is shown in Figure 7-6.

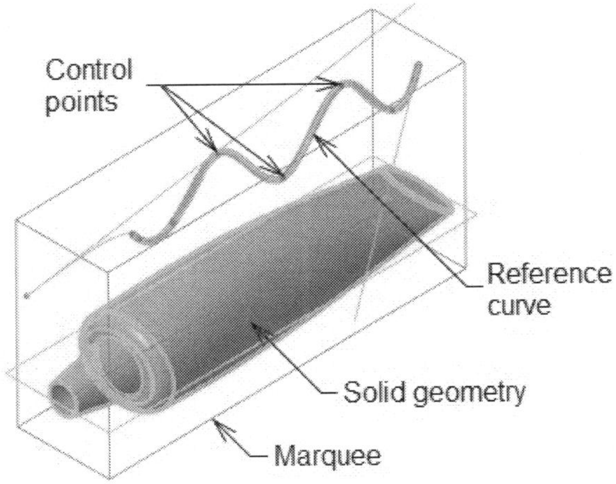

Figure 7-5 Figure showing the parameters for the warp transformation

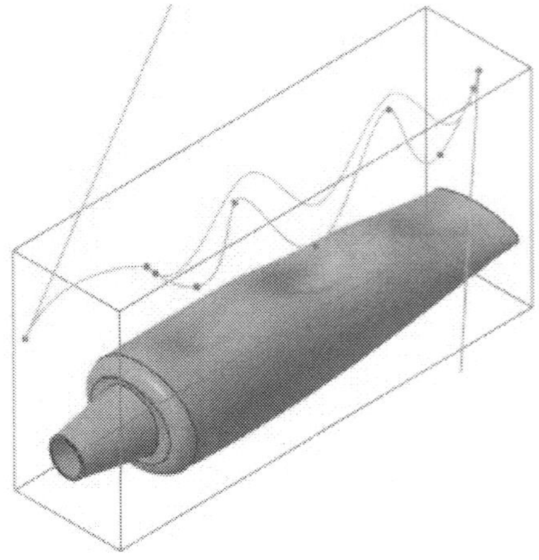

Figure 7-6 Figure showing the deformed model

Note

The control points for creating the deformation may be randomly selected based on the profile required.

8. Choose the **OK** button to create the feature. Figure 7-7 shows the final model after applying the warp transformation and hiding the spline created.

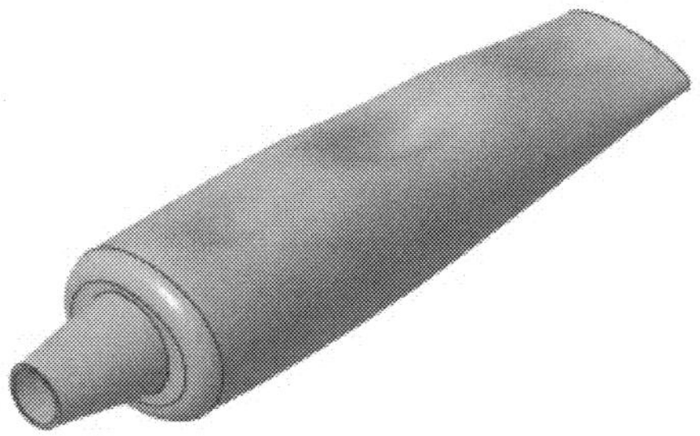

Figure 7-7 Final model after applying the warp transformation

Saving the Model

1. Choose the **Save** button from the **File** menu to save the model.

Tutorial 2

In this tutorial, you will create the model shown in Figure 7-8. **(Expected time: 30 min)**

Figure 7-8 Model for Tutorial 3

The following steps are required to complete this tutorial:

a. Create the extrude feature, refer to Figures 7-9 and 7-10.
b. Create the second extrude feature, refer to Figures 7-11 and 7-12.
c. Create the pattern of the tread, refer to Figure 7-13.
d. Create the toroidal bend feature, refer to Figures 7-14 and 7-15.
e. Create the mirror of the model, refer to Figure 7-16.

Starting a New Object File

Start a new part file and name it as *C07_TUT02*.

Creating the First Extrude Feature

1. Draw the sketch for the first extrude feature in the **FRONT** plane, as shown in Figure 7-9.

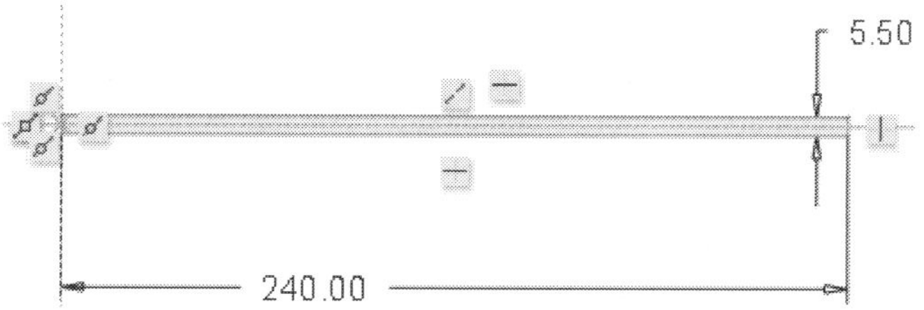

Figure 7-9 Sketch of the extrude feature

2. Extrude it to a length of **1500**, refer to Figure 7-10.

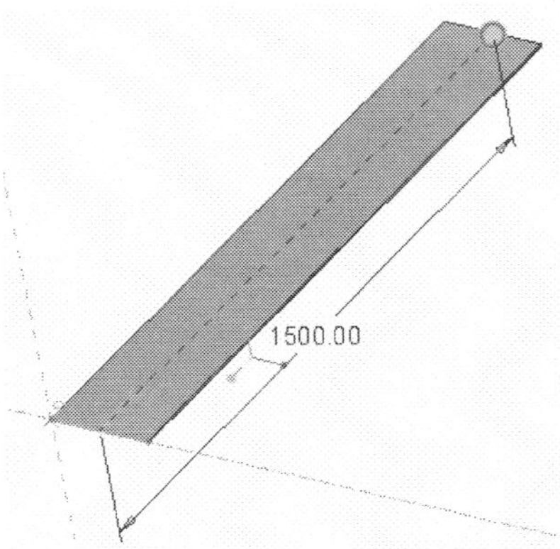

Figure 7-10 Preview of the first extrude feature

Creating the Second Extrude Feature

1. Select the top face of the model from the drawing area.
2. Choose the **Extrude** tool from the **Shapes** group; the **Extrude** dashboard is displayed.
3. Draw the sketch for the second extrude feature, as shown in Figure 7-11, and select the **OK** button.
4. Enter **8** in the **Enter the depth value for side 1** edit box and choose the **OK** button from the **Extrude** dashboard.
5. Specify the correct direction of cut and choose the **OK** button to complete the creation of the feature.

   The model after creating the second extrude feature is shown in Figure 7-12. Constraints visibility have been turned off for clarity.

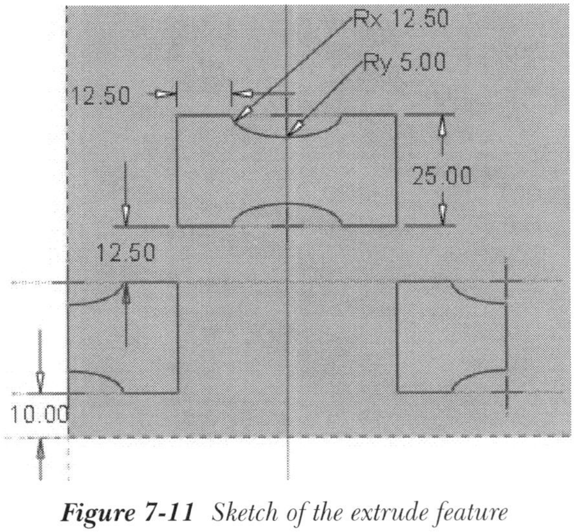

Figure 7-11 Sketch of the extrude feature

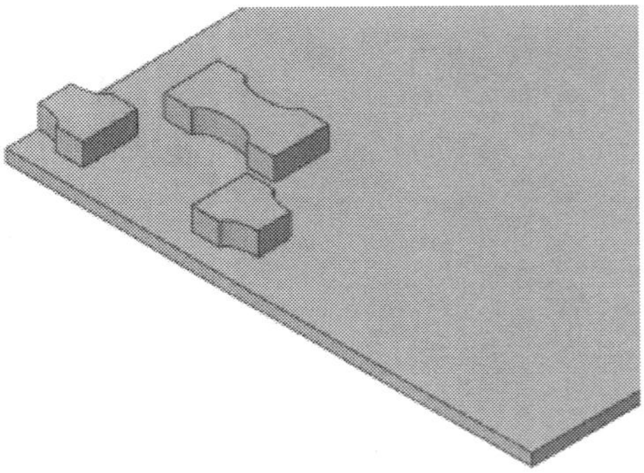

Figure 7-12 Model showing the second extrude feature

Creating the Pattern of the Tread

1. Select the extruded feature created in the previous section from the **Model Tree** or the drawing area.

2. Choose the **Pattern** tool from the **Pattern** drop-down in the **Editing** group; the **Pattern** dashboard is displayed and you are prompted to select a plane, flat face, linear curve, coordinate system axis, or axis to define the first direction.

3. Click in the Direction reference collector in the **Pattern** dashboard and select the **Front** plane from the drawing area or from the **Model Tree**.

4. Enter **75** in the **Enter the spacing between pattern members in the first direction** edit box.

5. Choose the **OK** button from the dashboard to exit from the **Pattern** tool.

   The model after creating the pattern is shown in Figure 7-13.

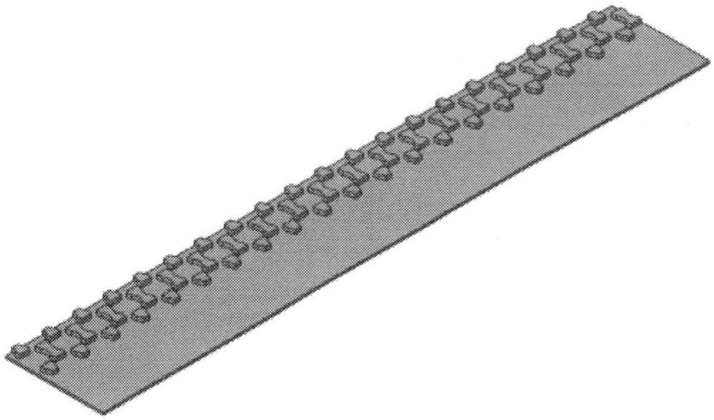

Figure 7-13 *Model after creating the pattern of the tread*

Note

You will learn about ***Pattern*** *tool in detail in next chapter.*

Creating the Toroidal Bend Feature

1. Choose the **Toroidal Bend** tool from the **Engineering** group; the **Toroidal Bend** dashboard is displayed.

2. Choose the **References** tab; the **References** slide-down panel is displayed. Select the **Solid Geometry** check box and choose the **Define** button from the slide-down panel; the **Sketch** dialog box is displayed.

3. Select the front face of the base feature as the sketching plane. Next, choose the **Sketch** button from the **Sketch** dialog box to accept the default orientation and enter the sketcher environment.

4. Draw the section for the toroidal bend feature, as shown in Figure 7-14.

5. Choose the **Geometry Coordinate System** button from the **Datum** group and place the reference coordinate system at the origin, as shown in Figure 7-14.

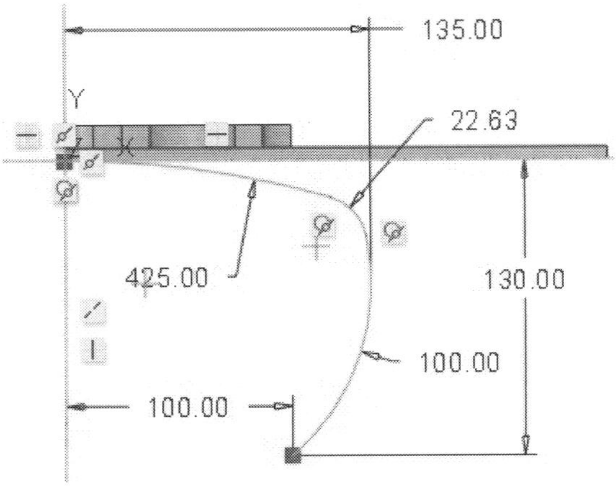

Figure 7-14 *Section created for the toroidal bend feature*

6. Choose the **OK** button from the **Sketch** dashboard to exit the sketcher mode; the **Toroidal Bend** dashboard is displayed again. Choose the **Options** tab and select the **Standard** radio button, if it is not already selected from the **Curve Bend** area of the slide-down panel.

7. Click on the **Bend Radius** option; a drop-down list is displayed. Choose the **360 degrees Bend** option from the drop-down list; you are prompted to select two parallel planes to define the length of the bend.

8. Select the front face and then the back face of the base feature; preview of the toroidal bend feature is displayed.

9. Choose the **OK** button; the model similar to the model shown in Figure 7-14 is created in the drawing area.

Creating the Mirror of the Model

1. Choose the **Mirror** tool from the **Editing** group; the **Mirror** dashboard is displayed. Next, click in the Mirror plane collector and select the **Right** datum plane from the **Model Tree**.

2. Choose the **References** tab; the **References** slide-down panel is displayed. Click in the **Mirror items** collector and then click on the **C07_TUT2** from the **Model Tree** and choose the **OK** button from the dashboard.

   Figure 7-15 shows the final model after mirroring the part.

Figure 7-15 Model after creating the toroidal bend feature

Figure 7-16 Model after mirroring the part

Saving the Model

1. Choose the **Save** button from the **File** menu to save the model.

EXERCISES

Exercise 1

In this exercise, you will create the model shown in Figure 7-17. The dimensions for the model are shown in Figure 7-18. The hints provided are suggestive. **(Expected time: 20 min)**

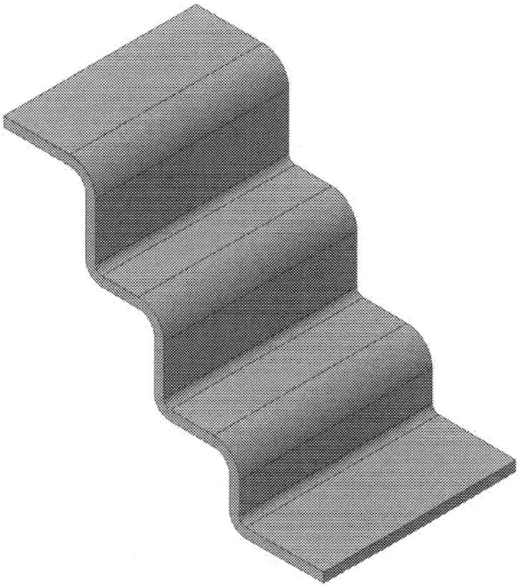

Figure 7-17 The resulting model for Exercise 1

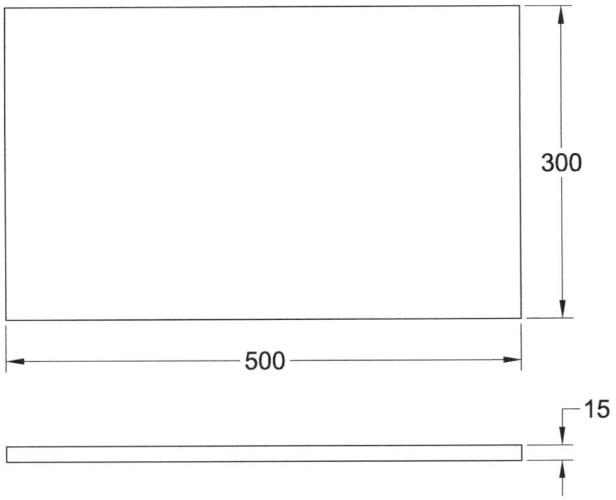

Figure 7-18 Drawing of the base feature for Exercise 1

Hint

1. Create the sketch of the base feature which is a rectangle of dimensions **500 X 300** and extrude it to a depth of **15**.
2. Create a new datum plane at a distance of **250** from the **RIGHT** datum plane.
3. Choose the **Spinal Bend** tool from the **Engineering** group and select the **FRONT** datum plane as the sketching plane.
4. Create the trajectory for the **Spinal Bend** which should be tangent to the bottom face of the base feature. The shape of the trajectory needs to be in the form of steps with the corners rounded. The dimensions of the trajectory can be assumed.
5. Exit from the sketcher environment; you will be prompted to select a plane to define the volume of the bend.
6. Select the **DTM1** datum plane. The model similar to the one shown in Figure 7-17 is displayed in the drawing area.

Exercise 2

In this exercise, you will create the model shown in Figure 7-19. The dimensions for the base feature are shown in Figure 7-20. **(Expected time: 20 min)**

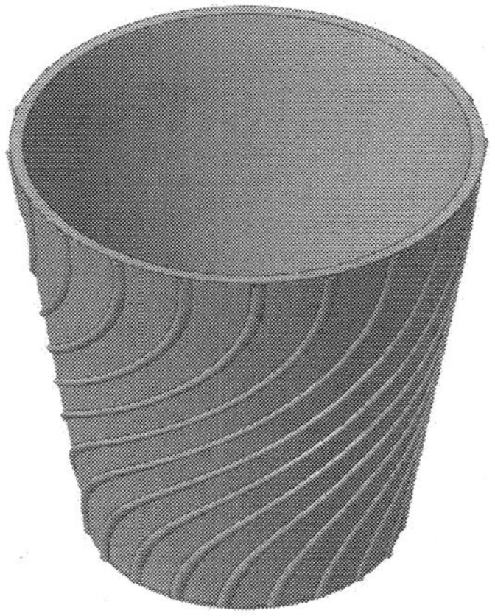

Figure 7-19 *The resulting model for Exercise 2*

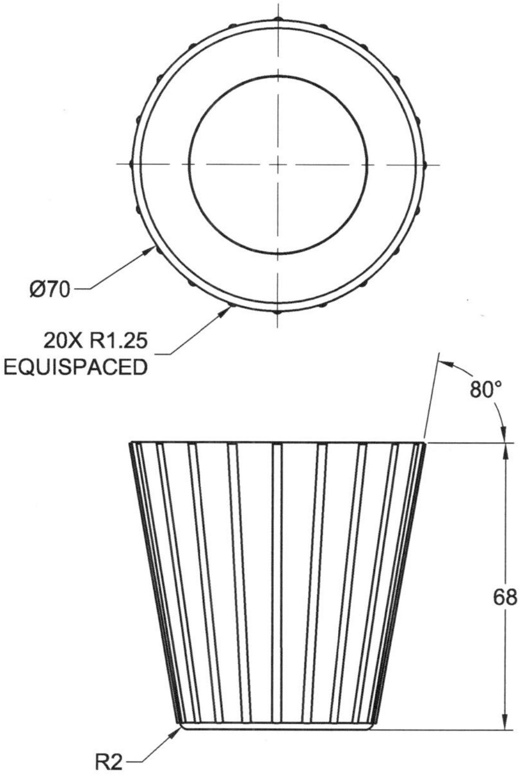

Figure 7-20 *Drawing of the base feature for Exercise 2*

This page is intentionally left blank

Chapter 8

Options Aiding Construction of Parts-III

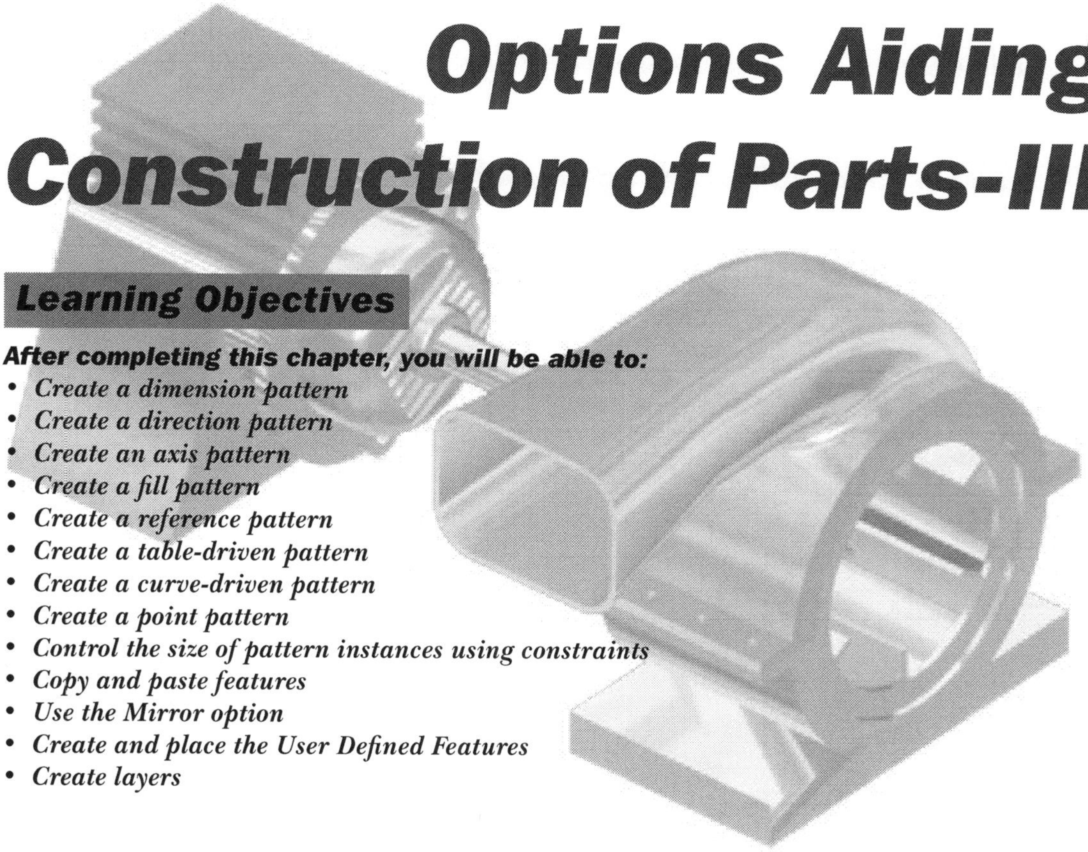

Learning Objectives

After completing this chapter, you will be able to:

- *Create a dimension pattern*
- *Create a direction pattern*
- *Create an axis pattern*
- *Create a fill pattern*
- *Create a reference pattern*
- *Create a table-driven pattern*
- *Create a curve-driven pattern*
- *Create a point pattern*
- *Control the size of pattern instances using constraints*
- *Copy and paste features*
- *Use the Mirror option*
- *Create and place the User Defined Features*
- *Create layers*

INTRODUCTION

In this chapter, you will learn about various methods for duplicating existing features. In Creo Parametric, you can duplicate a feature by using the **Pattern**, **Copy**, and **Mirror** tools.

You will also learn to create user defined features and simplified representation in solid models.

CREATING FEATURE PATTERN

A pattern consists of multiple instances of a feature. Patterns are used to create incremental array of features in one or two directions from a single feature called the parent feature or the leader. When a pattern is created, the leader also becomes a part of the pattern. When you pattern a feature, you need to specify the total number of features to be created including the one being patterned and also the increment in the dimensions, if required. You can pattern several features at a time by creating a local group of the selected features. You can also apply move or rotational transformations to a pattern, group pattern, or to a pattern of a pattern.

COPYING AND PASTING FEATURES

The **Copy**, **Paste**, and **Paste Special** tools allow you to duplicate and place features, geometries, curves, and edge chains within the same model or in other models. The instances created using the **Copy** and **Paste** tools can be independent, partially dependent, or fully dependent on the original features and geometries. You can also apply move or rotate transformation to the pasted instances.

MIRRORING A GEOMETRY

Ribbon: Model > Editing > Mirror

The **Mirror** tool is used to copy features and geometries about a planar face. The copy created by using the **Mirror** tool can be independent or dependent on the parent geometry.

USER-DEFINED FEATURES (UDF)

Creo Parametric allows you to reuse existing geometry while creating new design models. This is done with the help of user defined features (UDFs). A user defined feature (UDF) is a group of selected features, all their dimensions, relations between the selected features, and a list of references for placing the UDF on a model. User defined features save considerable amount of time in designing of parts by establishing a library of commonly used geometry.

LAYERS

Layers provide you an effective way to manage and organise model or assembly items. With the help of layers, you can group items, such as features, datum planes, parts in an assembly, and even other layers so that you can perform operations on those items collectively. Layers enable you to simplify geometry selection by temporarily hiding or displaying specific model features or assembly components in the drawing area. Layers can also be used to perform actions, such as suppressing all the items in a layer at once. You can also create filters in a layer so that the items that fulfill the conditions of that filter are automatically added to the layer.

TUTORIALS

Tutorial 1

In this tutorial, you will create the model shown in Figure 8-1. The orthographic views of the model are shown in Figure 8-2. **(Expected time: 30 min)**

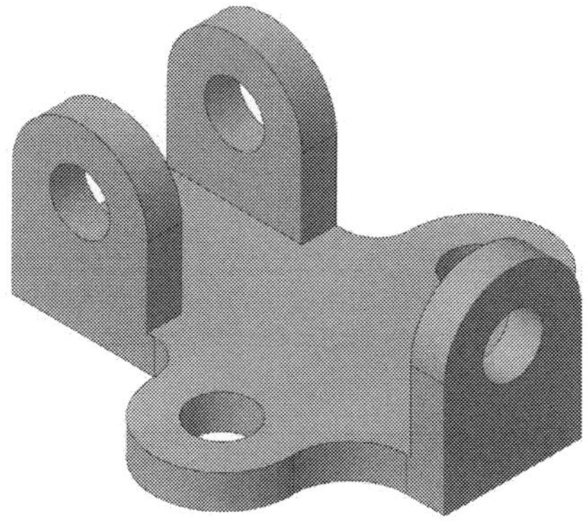

Figure 8-1 *Solid model for Tutorial 1*

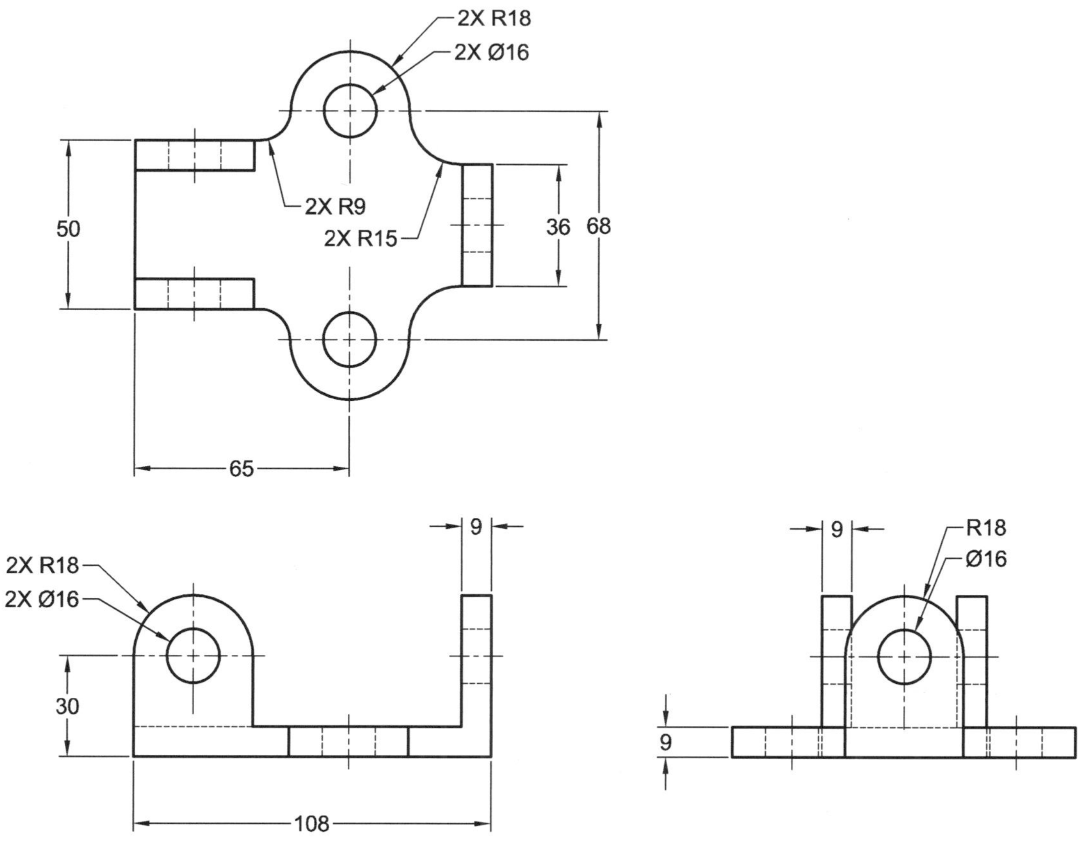

Figure 8-2 *Orthographic views of the solid model*

Examine the model to determine the number of features in it. The model consists of four features, refer to Figure 8-1.

The following steps are required to complete this tutorial:

a. Create the base feature on the **TOP** datum plane, refer to Figures 8-3 and 8-4.
b. Create the second feature on the right face of the base feature, refer to Figures 8-5 through 8-7.
c. Create the third feature, refer to Figures 8-8 and 8-9.
d. Create the fourth feature by mirroring the third feature, refer to Figure 8-10.

After starting Creo Parametric session, the first task is to set the working directory. Since it is the first tutorial of this chapter, you need to create a folder with the name *c08*, if it does not exist, and then set it as working directory.

Starting a New Object File

1. Start a new part file and then name it as *c08tut1*.

The three default datum planes are displayed in the drawing area. Also, the **Model Tree** is displayed in the drawing area.

Creating the Base Feature

To create a sketch for the base feature, you need to select the **TOP** datum plane as the sketching plane.

1. Choose the **Extrude** tool from the **Shapes** group of the **Model** tab; the **Extrude** dashboard is displayed above the drawing area.

2. Choose the **Placement** tab; the slide-down panel is displayed. From the slide-down panel, choose the **Define** button; the **Sketch** dialog box is displayed.

3. Select the **TOP** datum plane as the sketching plane. Next, select the **RIGHT** datum plane as the reference plane and set its orientation to **Right**, if not set by default.

4. Choose the **Sketch** button to enter the sketcher environment.

5. Next, create the sketch of the base feature and apply constraints and dimensions to it, as shown in Figure 8-3.

   Note that in the sketch, the bottom half of the sketch is mirrored to create the top half of the sketch. This is evident from the constraints of symmetry applied to the sketch in Figure 8-3.

6. After the sketch is completed, choose the **OK** button to exit the sketcher environment; the **Extrude** dashboard is enabled and displayed above the drawing area.

7. Enter **9** as the value of depth in the dimension box in the **Extrude** dashboard.

8. Choose the **OK** button from the **Extrude** dashboard.

   Now, the base feature is completed and you need to create the second feature. The default trimetric view of the base feature is shown in Figure 8-4.

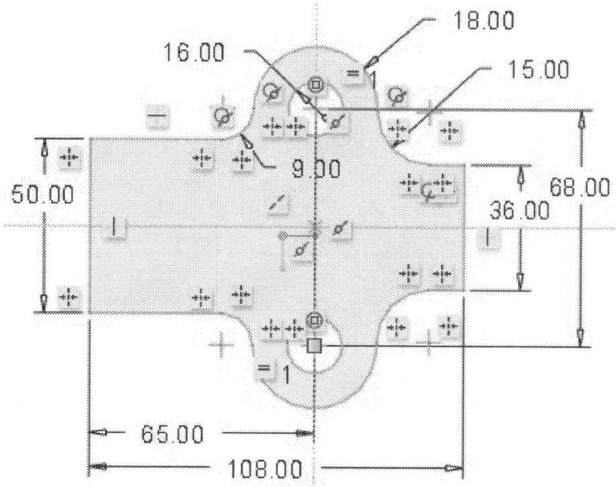

Figure 8-3 Sketch of the base feature

Figure 8-4 The default trimetric view of the base feature

Note

The two holes in the base feature are sketched while drawing the sketch for the base feature. These holes are integrated with the base feature. Therefore, the base feature is created as a single feature that includes two holes. The other method is to create the two holes separately on the base feature by using the ***Hole*** *dashboard. When you create the features separately, the total number of features created will be three.*

Creating the Second Feature

The second feature is also an extruded feature. You will create the second feature on the right face of the base feature. Therefore, you need to select the right face as the sketching plane.

1. Choose the **Extrude** tool from the **Shapes** group; the **Extrude** dashboard is displayed above the drawing area.

2. Choose the **Placement** tab to display the slide-down panel. Then, choose the **Define** button from the slide-down panel; the **Sketch** dialog box is displayed.

3. Select the right face of the base feature as the sketching plane.

4. If required, choose the **Flip** button to reverse the direction of the pink arrow.

5. Select the **TOP** datum plane as reference and then select the **Top** option from the **Orientation** drop-down list.

6. Choose the **Sketch** button from the **Sketch** dialog box to enter the sketcher environment.

7. Choose the **Line Chain** tool from the **Sketching** group and draw a vertical line starting from the edge of the base feature, as shown in Figure 8-5. Notice that the cursor automatically snaps the edge of the base feature while creating the vertical line.

8. Complete the sketch of the second feature and then add dimensions and constraints, as shown in Figure 8-5. After completing the sketch, choose the **OK** button from the **Close** group; the **Extrude** dashboard is enabled.

9. Enter **9** as the value of depth in the dimension box in the **Extrude** dashboard.

10. Choose the **OK** button from the **Extrude** dashboard; the default shaded trimetric view of the model after creating the second feature is shown in Figure 8-6.

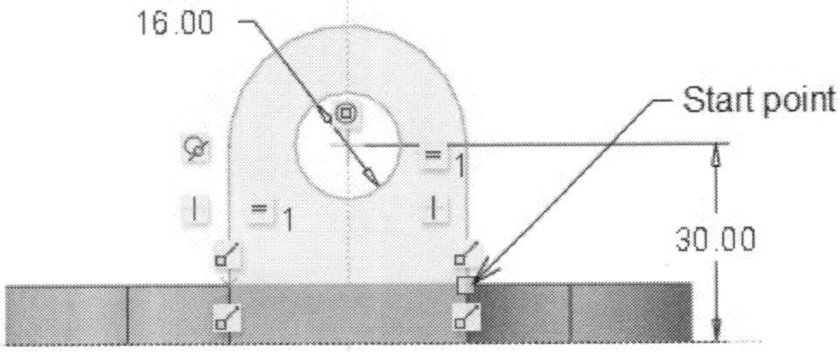

Figure 8-5 *Sketch of the second feature with dimensions and constraints*

Figure 8-6 *Model after creating the second feature*

Creating the Third Feature

Now, you need to create the third feature.

1. Select the second feature from the **Model Tree** and then choose the **Copy** tool from the **Operations** group.

2. Select the face shown in Figure 8-7 and then choose the **Paste** tool from the **Operations** group; the **Sketch** dialog box is displayed and you are prompted to select a reference, surface, or plane to define the view orientation.

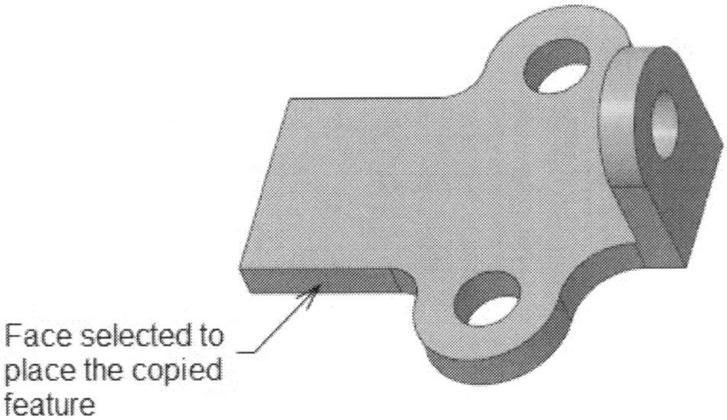

Figure 8-7 *Reference plane selected to paste the copied feature*

3. Choose the **Sketch** button; the sketch of the second feature is attached to the cursor, as shown in Figure 8-8.

4. Click anywhere in the drawing area to place the sketch. Notice that some weak dimensions and constraints are applied to the sketch.

5. Add required dimensions and constrains to the sketch and then choose the **OK** button from the **Close** group; the **Extrude** dashboard is enabled and displayed above the drawing area.

6. Click on the pink arrow in the drawing area to flip the direction of extrusion and then enter **9** as the value of depth in the dimension box in the **Extrude** dashboard. Next, choose the **OK** button from the **Extrude** dashboard. The model after creating the third feature is shown in Figure 8-9.

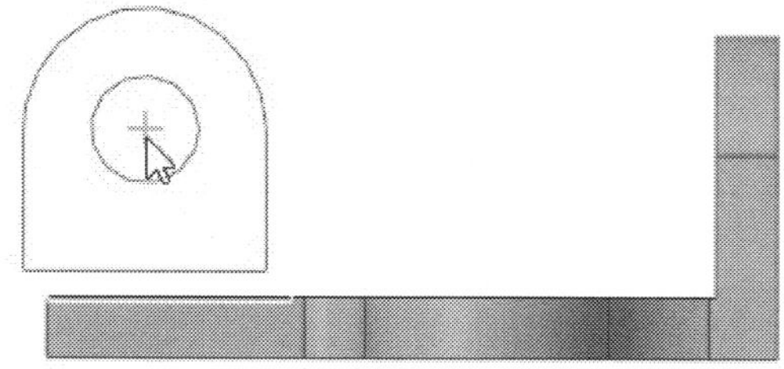

Figure 8-8 *Sketch of the second feature attached to the cursor*

Figure 8-9 *The model after creating the third feature*

Creating the Fourth Feature

The fourth feature can be created by sketching and extruding it to a given depth. You can also create this feature by placing a mirrored copy of the third feature at the required location. In this tutorial, you will use the second method because it consumes less time.

1. Select the third feature from the **Model Tree** and choose the **Mirror** tool from the **Editing** group of the **Model** tab; the **Mirror** dashboard is displayed and you are prompted to select the mirror plane.

2. Select the **FRONT** datum plane as the mirror plane.

3. Choose the **OK** button from the **Mirror** dashboard. The third feature is mirrored about the **FRONT** datum plane. The trimetric view of the final model is shown in Figure 8-10.

4. Choose the **Save** button from the **File** menu and save the model. The order of feature creation can be seen from the **Model Tree** shown in Figure 8-11. Note that the feature id numbers in your model may be different from the ones shown in this figure.

Figure 8-10 Default trimetric view of the model

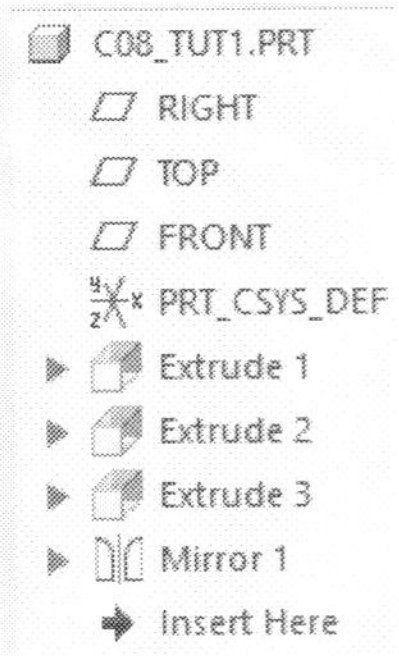

Figure 8-11 *The* ***Model Tree*** *for Tutorial 1*

Tutorial 2

In this tutorial, you will create the model shown in Figure 8-12. The orthographic views of the model are shown in Figure 8-13. **(Expected time: 30 min)**

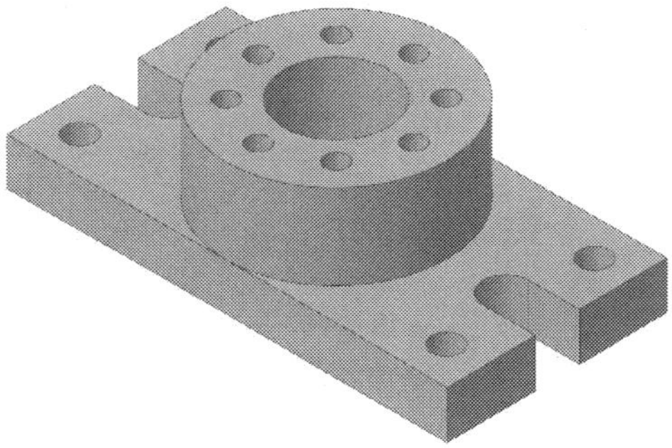

Figure 8-12 Isometric view of the solid model

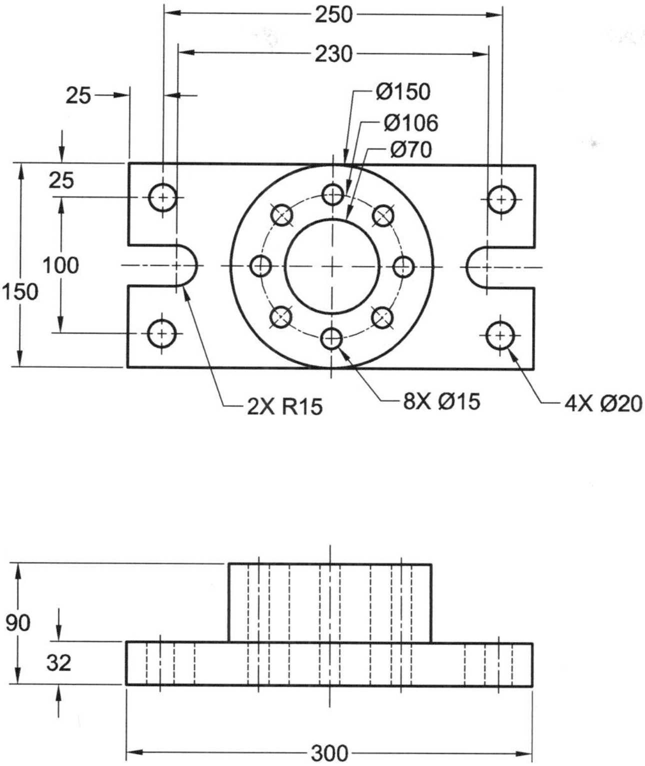

Figure 8-13 *Top and front views of the solid model*

Examine the model and determine the number of features in it. The model consists of five features, refer to Figure 8-12.

The following steps are required to complete this tutorial:

a. Start a new file and create the base feature on the **TOP** datum plane, refer to Figures 8-14 and 8-15.
b. Create the cylindrical extrude feature, refer to Figures 8-16 and 8-17.
c. Create the hole feature coaxial with the cylindrical feature, refer to Figure 8-18.
d. Create the hole feature on the top planar surface of the base feature and then pattern it, refer to Figures 8-19 and 8-20.
e. Create the hole feature on the top planar surface of the cylindrical feature, refer to Figure 8-21.
f. Create a rotational pattern of this hole, refer to Figure 8-22.

If required, set the working directory to the *c08* folder.

Starting a New Object File

1. Start a new part file and name it as *c08tut2*.

The three default datum planes are displayed in the drawing area.

Creating the Base Feature

To create sketch for the base feature, you first need to select the sketching plane for the base feature. In this model, you need to draw the base feature on the **TOP** datum plane. This is because the direction of extrusion of this feature is perpendicular to the **TOP** datum plane.

1. Choose the **Extrude** tool from the **Shapes** group; the **Extrude** dashboard is displayed.
2. Invoke the **Sketch** dialog box by using the **Extrude** dashboard.
3. Select the **TOP** datum plane as the sketching plane.
4. Select the **RIGHT** datum plane and then select the **Right** option from the **Orientation** drop-down list, if it is not selected by default.
5. Choose the **Sketch** button to enter the sketcher environment.
6. Create the sketch of the base feature and apply required constraints and dimensions to it, as shown in Figure 8-14. For better visibility of the sketch, you can turn off the display of constraints.

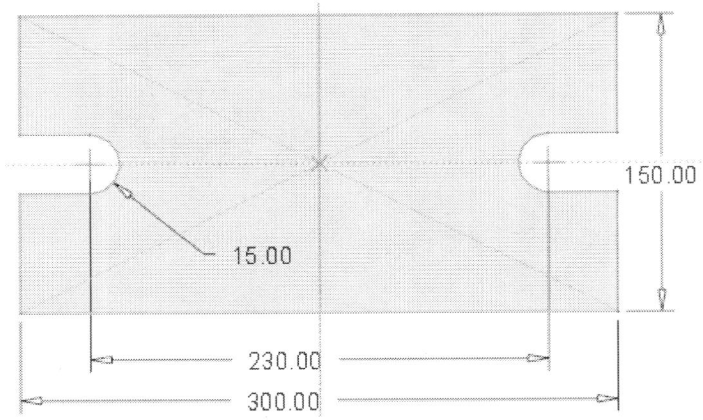

Figure 8-14 *Sketch of the base feature with dimensions and constraints*

7. After the sketch is complete, choose the **OK** button to exit the sketcher environment; the **Extrude** dashboard is enabled and displayed above the drawing area.
8. Enter **32** as the depth value in the dimension box of the **Extrude** dashboard and then choose the **OK** button from the dashboard.

   The base feature is completed and now you need to create the second feature. The default trimetric view of the base feature is shown in Figure 8-15.

Figure 8-15 Default trimetric view of the base feature

Creating the Second Feature

The second feature is also an extruded feature. You need to create this feature on the top face of the base feature. Therefore, you need to define the top face as the sketching plane for the second feature.

1. Choose the **Extrude** tool from the **Shapes** group; the **Extrude** dashboard is displayed.
2. Invoke the **Sketch** dialog box by using the **Extrude** dashboard.
3. Select the top face of the base feature as the sketching plane.
4. Select the **RIGHT** datum plane and then select the **Right** option from the **Orientation** drop-down list, if it is not selected by default.
5. Choose the **Sketch** button to enter the sketcher environment.
6. Create the sketch of the second feature and then dimension it, as shown in Figure 8-16.
7. After creating the sketch, choose the **OK** button; the **Extrude** dashboard is enabled.
8. Enter **58** in the dimension box of the **Extrude** dashboard and choose the **OK** button. The model similar to the one shown in Figure 8-17 is displayed in the drawing area.

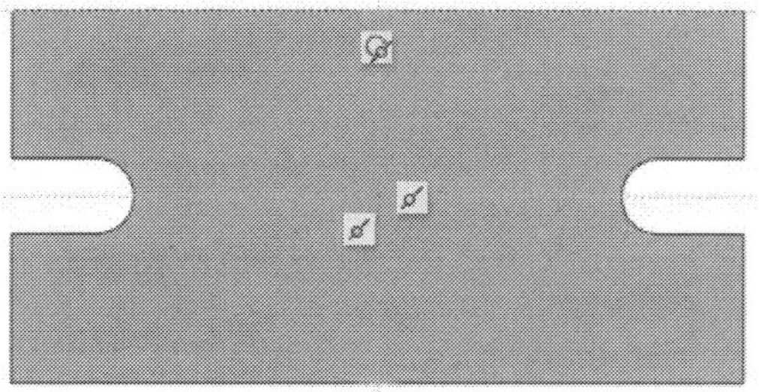

Figure 8-16 Sketch of the cylindrical feature with diameter of the cylinder

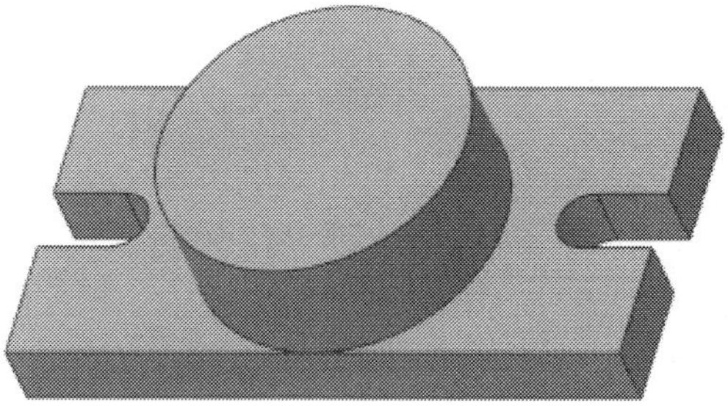

Figure 8-17 Default trimetric view of the cylindrical feature

Creating the Third Feature

The third feature is a through hole that is coaxial to the cylindrical feature. You need to create the hole feature by using the **Hole** dashboard.

1. Choose the **Hole** tool from the **Engineering** group; the **Hole** dashboard is displayed. By default, the **Create simple hole** button is chosen in the **Hole** dashboard.

2. Create a hole of diameter **70**, as shown in Figure 8-18. Refer to Figure 8-13 for specifying the placement parameters.

Creating the Fourth Feature

The fourth feature is a through hole. You need to create this feature on the top planar surface of the base feature by using the **Hole** dashboard.

1. Choose the **Hole** tool from the **Engineering** group; the **Hole** dashboard is displayed. By default, the **Create simple hole** button is chosen in the **Hole** dashboard.

2. Create the hole, as shown in Figure 8-19, by specifying the placement parameters (refer to Figure 8-16).

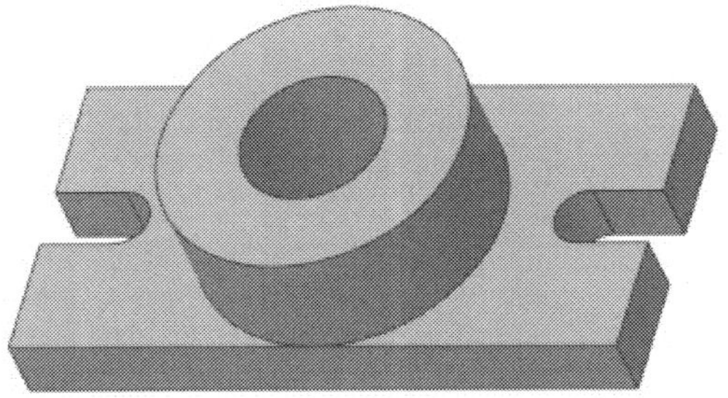

Figure 8-18 *Coaxial hole on the cylindrical feature*

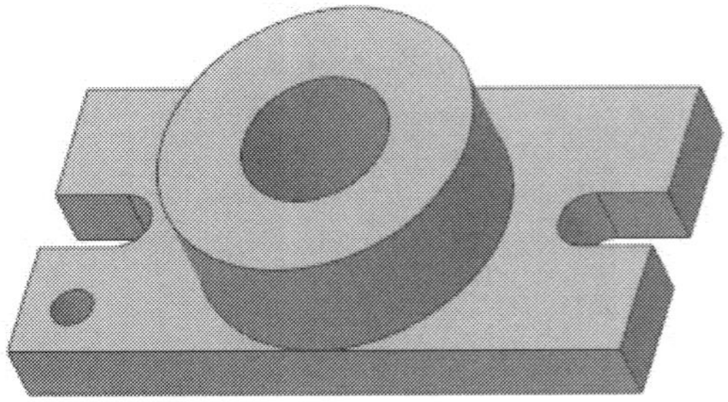

Figure 8-19 *Hole on the base feature*

Patterning the Hole Feature

Next, you need to create a rectangular pattern of the hole feature that is created on the base feature. You can also create individual holes but it is time-consuming and increases the number of features. Therefore, it is recommended that you create a rectangular pattern of the hole feature.

1. Select the hole feature and then choose the **Pattern** tool from the **Editing** group; the **Pattern** dashboard is displayed and you are prompted to select the dimensions to be changed in the first direction.

2. Select the **Identical** option from the **Options** slide-down panel.

   Here you need to use the **Identical** option because the feature on which the pattern is created does not intersect the pattern.

3. Select the dimension value **25** from the drawing area. Since both the dimensions displayed in the drawing area have the value **25**, select the dimension **25** that is along the shorter side of the base feature. After you have selected the dimension in the first direction, an edit box is displayed. Enter **100** in the edit box.

4. Hold down the right mouse button to display the shortcut menu. Choose the **Direction 2 Dimensions** option from the shortcut menu.

5. Select the dimension value **25** that is along the longer side of the base feature; an edit box is displayed.

6. Enter **250** in the edit box. Note that the number of instances, **2**, is specified by default in the instances edit boxes of the dashboard.

7. Choose the **OK** button from the **Pattern** dashboard; a rectangular pattern of the hole feature is displayed, as shown in Figure 8-20.

Creating a Hole on the Cylindrical Feature

You need to create a hole on the cylindrical feature diametrically by using the **Hole** dashboard.

1. Choose the **Hole** tool from the **Engineering** group; the **Hole** dashboard is displayed. By default, the **Create simple Hole** tool is chosen in the **Hole** dashboard.

2. Choose the **Placement** tab from the **Hole** dashboard to display the slide-down panel.

3. Select the top face of the cylindrical feature as the placement plane.

4. From the drop-down list in the slide-down panel, select the **Diameter** option.

5. Click in the **Offset References** collector and select the axis of the cylindrical feature.

6. Enter **106** in the second dimension box on the right of the **Diameter** option in the **Offset References** collector.

7. Use the CTRL key+left mouse button and select the **FRONT** datum plane from the drawing area. Enter the value **90** in the dimension box of the **Offset References** collector.

8. Enter **15** as the diameter of the hole in the diameter dimension box in the **Hole** dashboard.

9. Choose the **Drill to intersect with all surfaces** option from the depth flyout in the **Hole** dashboard.

10. Choose the **OK** button from the **Hole** dashboard; the hole is created, as shown in Figure 8-21.

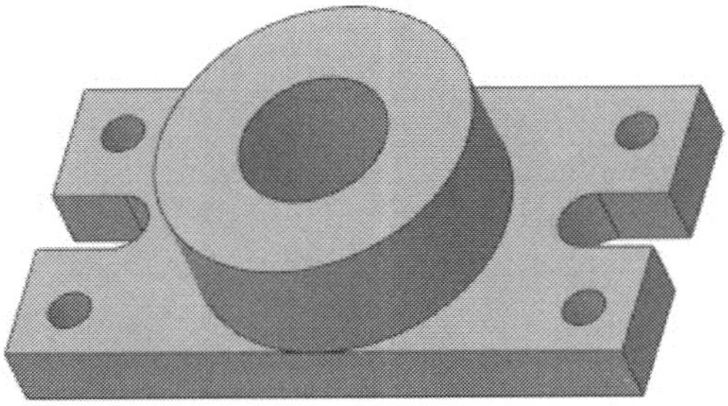

Figure 8-20 *Rectangular pattern of the hole feature*

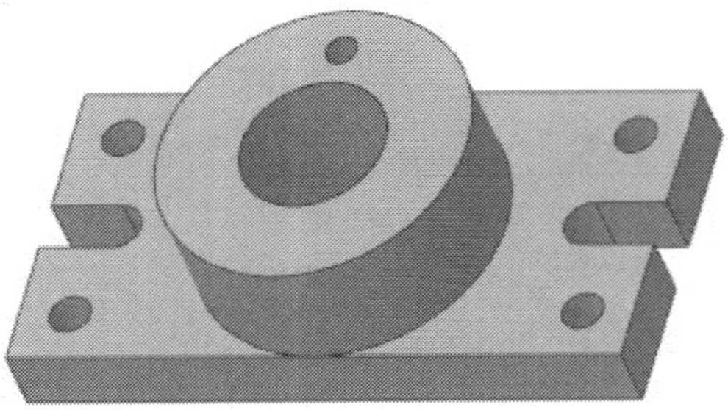

Figure 8-21 *Diametrical hole on the cylindrical feature*

Creating the Rotational Pattern of the Hole Feature

As the creation of the remaining holes individually on the cylindrical feature is time-consuming, you need to create a rotational pattern of the hole feature.

1. Select the hole feature and then choose the **Pattern** tool from the **Editing** group; the **Pattern** dashboard is displayed. Also, the dimensions of the hole feature are displayed in the drawing area and you are prompted to select dimensions to vary in the first direction.

2. Select the **Identical** option from the **Options** slide-down panel.

3. Select the angular dimension **90** from the model; an edit box is displayed.

4. Enter **45** in the edit box and press ENTER. Now, you need to specify the number of instances of the hole feature in the pattern.

5. Enter **8** in the **1** edit box and press ENTER. Choose the **OK** button from the dashboard. The rotational pattern is created and the model is completed, as shown in Figure 8-22. You can see the order of feature creation in the **Model Tree**, as shown in Figure 8-23.

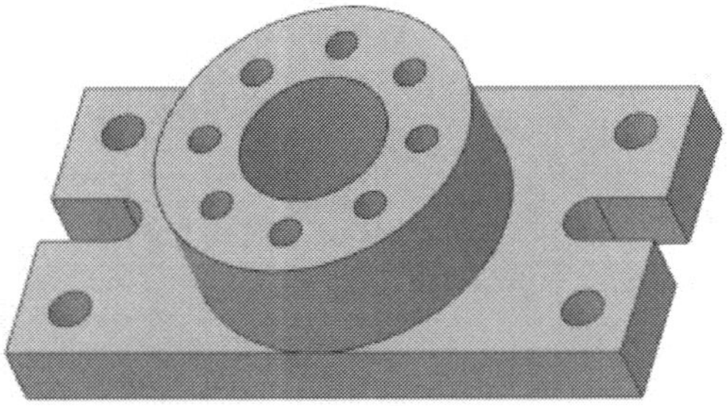

Figure 8-22 *The complete model*

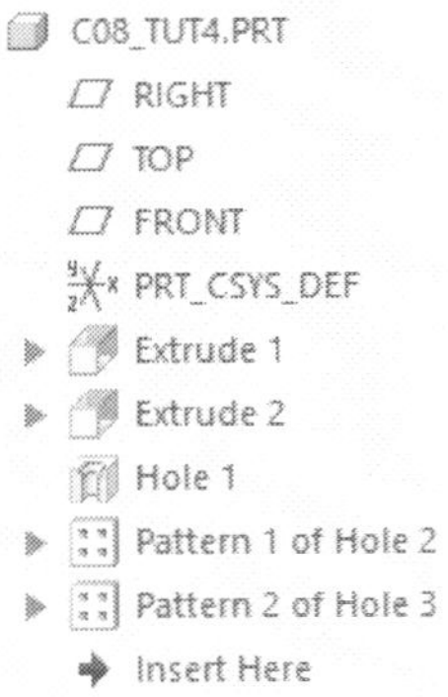

Figure 8-23 *The* ***Model Tree*** *for Tutorial 2*

Saving the Model

You need to save the model because you may need it later.

1. Choose the **Save** button from the **File** menu and save the model.

EXERCISES

Exercise 1

Create the model shown in Figure 8-24. The orthographic views of the model are shown in Figure 8-25. **(Expected time: 30 min)**

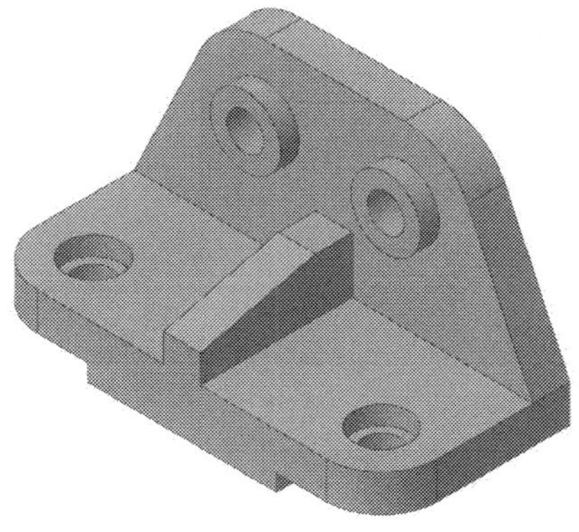

Figure 8-24 Isometric view of the model

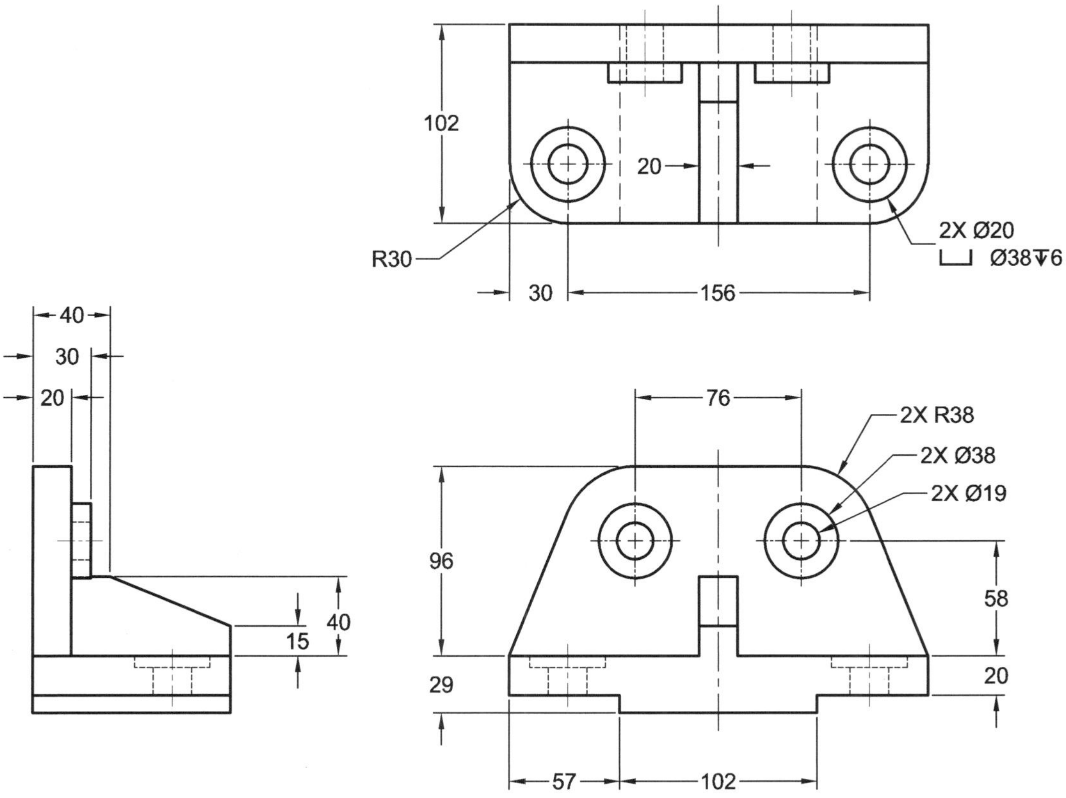

Figure 8-25 Orthographic views of the solid model

Exercise 2

Create the model shown in Figure 8-26. The top, front, and isometric views are shown in Figure 8-27. **(Expected time: 30 min)**

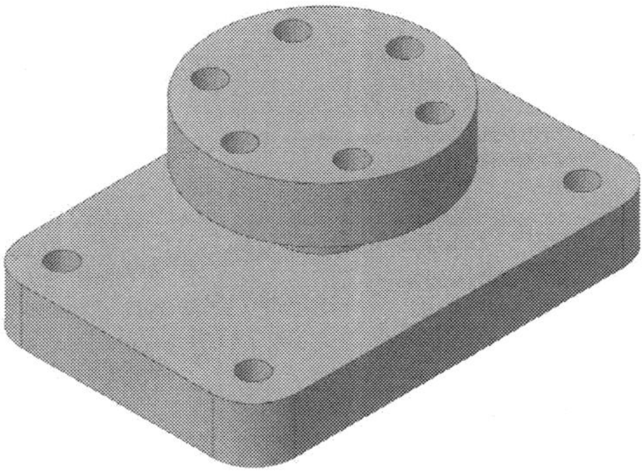

Figure 8-26 *Isometric view of the solid model*

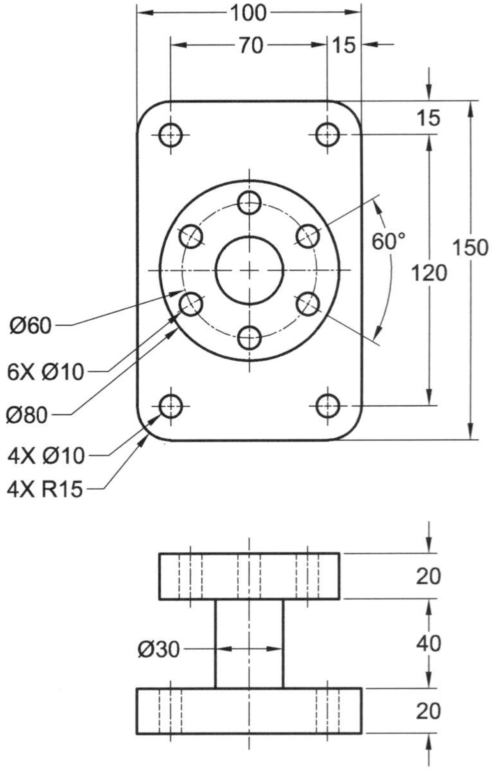

Figure 8-27 *Orthographic views of the solid model*

Chapter 9

Advanced Modeling Tools

Learning Objectives

After completing this chapter, you will be able to:

- *Create sweep features*
- *Create features by using sweep cut*
- *Create variable section sweep*
- *Create helical sweep*
- *Create volume helical sweep*
- *Create blend*
- *Create rotational blend*
- *Use blend vertex in blend features*

ADVANCED MODELING TOOLS

In this chapter, you will learn about the tools that are used to create complex features. These tools are called advanced modeling tools. The advanced modeling tools available in Creo Parametric are **Sweep**, **Helical Sweep**, **Volume Helical Sweep**, **Blend**, **Swept Blend**, and **Rotational Blend** which are discussed next.

SWEEP FEATURES

Sweeping is a process of creating solid and surface geometries by moving an open or a closed section along an open or closed trajectory. A sweep feature consists of a single constant or variable section which is swept along one or more trajectories. Trajectory is the path along which a section is swept. The trajectory for a sweep feature can be either sketched or selected. The order of operation is to first create a trajectory and then a section.

Sweep

Ribbon: Model > Shapes > Sweep drop-down > Sweep

The **Sweep** tool is similar to the **Extrude** tool. The only difference is that in case of the **Extrude** tool, the feature is extruded in a direction normal to the sketching plane, but in case of the **Sweep** tool, the section is swept along the sketched or selected trajectory. The sketching tools available in the sketcher environment are used for sketching the trajectory.

Helical Sweep

Ribbon: Model > Shapes > Sweep drop-down > Helical Sweep

The **Helical Sweep** tool is used to create helical sweep features by sweeping a section along a helical trajectory. The main use of this tool is to create the helical springs and threads.

Volume Helical Sweep

Ribbon: Model > Shapes > Sweep drop-down > Volume Helical Sweep

The **Volume Helical Sweep** tool is used to create a volume helical cut. The process of this tool resembles machining processes, particularly turning and milling. Using this tool, a rotating 3D object formed by a revolved sketch, moves along a helix axis and removes material from the part.

BLEND FEATURES

A blend feature is created using a minimum two planar sections joined at their vertices with transitional surfaces to form a continuous feature. Depending upon the direction of blending, three types of blend features can be created in Creo: Parallel blend, Swept blend, and Rotational blend.

Blend

Ribbon: Model > Shapes drop-down > Blend

The **Blend** tool is used to create a feature that has varying cross-sections. To create a blend feature using the **Blend** tool, all sections should lie on parallel planes.

Swept Blend

Ribbon: Model > Shapes > Swept Blend

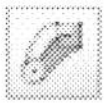

The **Swept Blend** tool is used to create a blend that is a combination of sweep and blend features. In this type of blend, multiple sections are placed along the main trajectory and the geometry is created by blending the sections while sweeping along the trajectory. The sections can vary in shape and size along the swept blend. A swept blend can have only two trajectories: origin trajectory and secondary trajectory. To define a trajectory of the swept blend, you can select a sketched curve, a chain of datum curves, or edges.

Rotational Blend

Ribbon: Model > Shapes drop-down > Rotational Blend

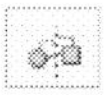

The rotational blends are created by blending the sections that are rotated about an axis of revolution. If the first section contains an axis of revolution or a centerline, it will be automatically selected as the axis of revolution. If the first sketch does not contain an axis of revolution or centerline, you can select a datum axis, part edge, or sketched entity as the axis of revolution. To create a rotational blend feature, all the sections must lie in planes that intersect the axis of rotation.

TUTORIALS

Tutorial 1

In this tutorial, you will create the model shown in Figure 9-1. The orthographic views of the model are shown in Figure 9-2. **(Expected time: 25 min)**

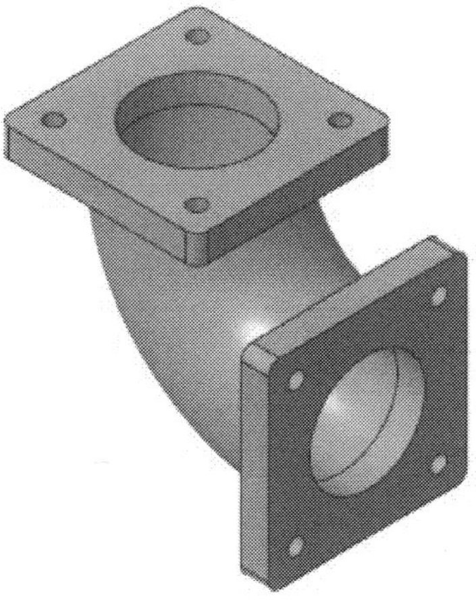

Figure 9-1 *Isometric view of the model*

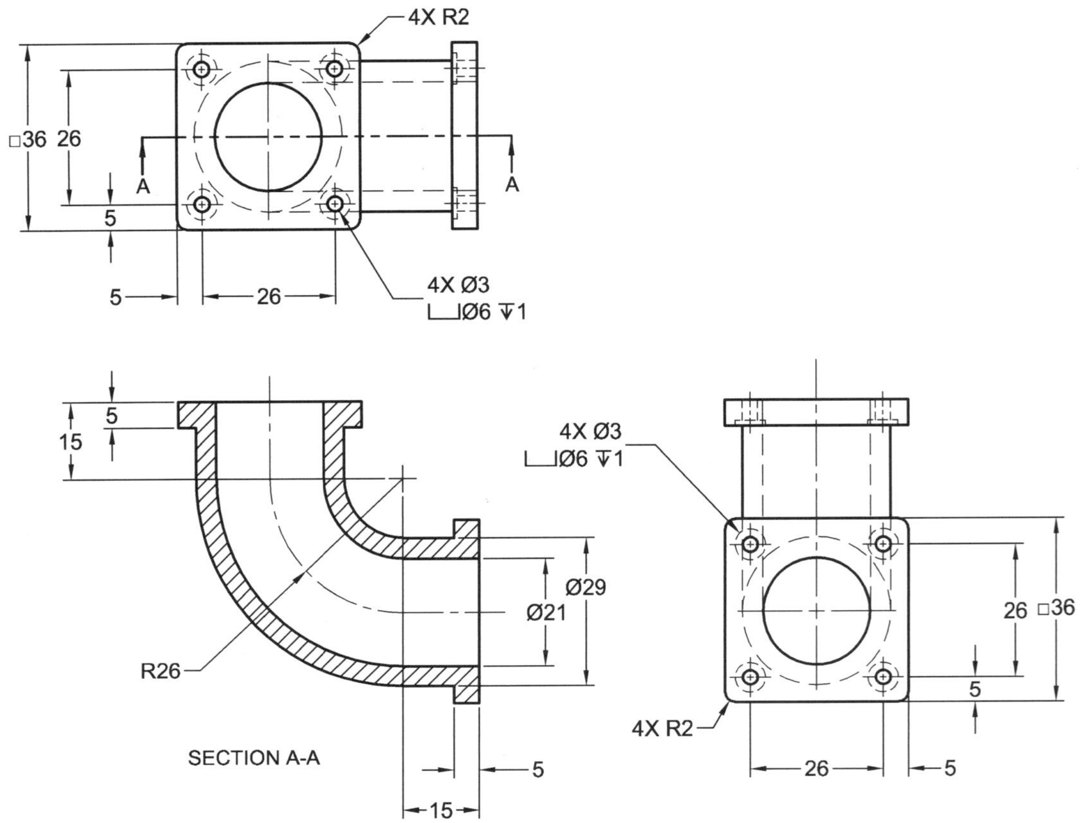

Figure 9-2 Orthographic views of the model

Examine the model and determine the number of features in it. The model consists of six features, refer to Figure 9-1.

The following steps are required to complete this tutorial:

a. Start a new file and create the base feature, refer to Figures 9-6 through 9-8.
b. Create the shell feature of given thickness, refer to Figure 9-9.
c. Create the third and fourth extrude features on the two ends of the sweep feature respectively, refer to Figures 9-10 and 9-11.
d. Create the counterbore hole on the third and fourth features, refer to Figures 9-12 and 9-13.
e. Pattern the counterbore holes, refer to Figure 9-14.

When the Creo Parametric session starts, the first task is to set the working directory. As this is the first tutorial of the chapter, you need to create a folder *c09* if it has not been created earlier and set it as the Working Directory.

Starting a New Object File

1. Start a new part file and name it as *c09tut1*.

The three default datum planes as well as the **Model Tree** are displayed in the drawing area.

Sketching the Trajectory

You need to sketch the trajectory of the sweep feature on the **FRONT** datum plane.

1. Invoke the **Sketch** tool from the **Datum** group. Next,select the **FRONT** datum plane as the sketching plane.

2. Select the **Top** datum plane as the reference plane from the **Reference** collector.

3. Select the **Top** option from the **Orientation** drop-down list and then choose the **Sketch** button from the **Sketch** dialog box.

   After you have selected the planes for orientation, the system takes you to the sketcher environment.

4. Draw the sketch of the trajectory and apply dimension, as shown in Figure 9-3.

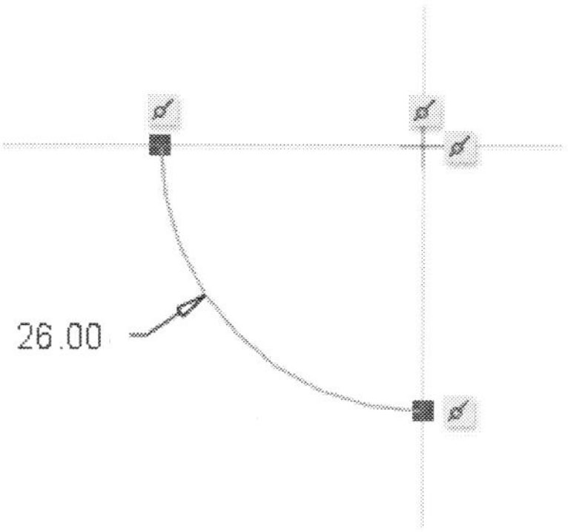

***Figure 9-3** The trajectory drawn for sweep feature*

Invoking the Sweep Tool

You need to invoke the **Sweep** tool in the **Shapes** group from the **Ribbon**.

1. Choose the **Sweep** tool from the **Sweep** drop-down in the **Shapes** group of the **Ribbon**; the **Sweep** dashboard is displayed.

2. Click on the trajectory from the drawing area; the trajectory gets highlighted and an arrow is displayed on it. This arrow displays the direction of sweep.

Tip

*To change the start point on the trajectory, click on the point where you want the start point. When the point is highlighted in red color, press and hold the right mouse button to invoke a shortcut menu. From the shortcut menu, choose the **Start Point** option.*

Drawing the Section for the Base Feature

1. After selecting the trajectory from the drawing area, choose the **Create or edit sweep section** button from the dashboard; your sketching plane is automatically oriented. Also you are prompted to draw the cross-section for the sweep feature.

2. Choose the **Center and Point** button from the **Sketching** group of the sketcher environment and create a circle such that the center of the circle lies at the intersection of the two infinite perpendicular lines.

   Note that, when you draw the circle, the cursor snaps to the intersection point of the cross.

Modifying the Dimensions of the Section

1. Double-click on the dimension and modify the diameter dimension to **29**. As you enter the dimension; the sketch gets modified and refits on the screen.

2. Choose the **OK** button to close the sketcher environment.

Creating the Base Feature

After closing the sketcher environment, preview of the base feature will be displayed in the drawing area.

1. Press CTRL+ D keys to orient the model in default orientation, refer to Figure 9-4. The default orientation in Creo Parametric is **Trimetric**.

2. Now, choose the **OK** button from the **Sweep** dashboard to exit it.

Creating the Shell Feature

After creating the sweep feature, you need to create the next feature called the shell feature.

Note

*Instead of using the **Shell** tool, two concentric circles can be drawn while drawing the section for the sweep feature in order to obtain the desired hollow feature. Alternatively, the **Sweep > Create a thin feature** option can be used to obtain the same hollow feature. In this tutorial, you will use the **Shell** option.*

1. Choose the **Shell** tool from the **Engineering** group; the **Shell** dashboard is displayed. Also, you are prompted to select the surfaces to be removed.

2. Select one end surface of the sweep feature and then by using CTRL+left mouse button, select the other end surface. The two selected surfaces are highlighted in green.

3. Enter **4** as the thickness value of the shell in the dimension edit box present on the **Shell** dashboard and then press ENTER.

4. Choose the **OK** button to create the shell feature and exit the **Shell** dashboard.

The default trimetric view of the shell feature is shown in Figure 9-5. You can use the middle mouse button to view the model from various angles.

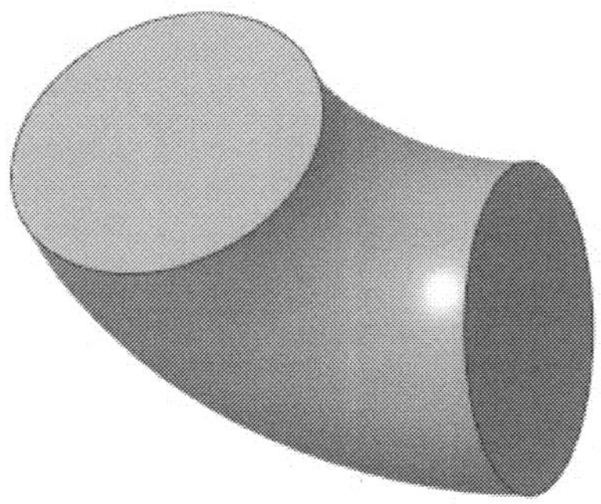

Figure 9-4 *The default view of the sweep feature*

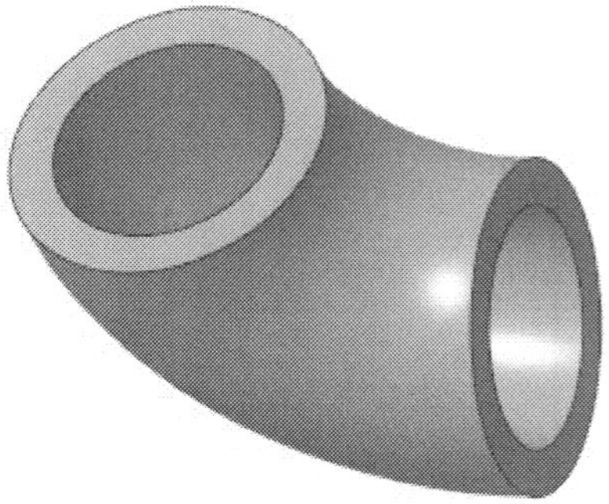

Figure 9-5 *Model after applying the shell feature*

Creating Extrude Features

The next feature is a protrusion feature with a depth of 5 and need to be created at both ends of the sweep feature. While drawing the circle for the sketch of the extrude feature, remember to use the edge of the shell in order to create a hole in the extruded feature as well.

1. Choose the **Extrude** tool from the **Shapes** group; the **Extrude** dashboard is displayed.
2. Choose the **Placement** tab from the dashboard to display the slide-down panel and invoke the **Sketch** dialog box. Select the top face of the sweep feature as the sketching plane.
3. Select the **RIGHT** datum plane and then select the **Right** option from the **Orientation** drop-down list. Then, enter the sketcher environment.
4. Draw the sketch of the extrude feature and then apply constraints and dimensions to it, as shown in Figure 9-6.
5. Create the extruded feature having an extrusion depth of 5. Similarly, create the next extruded feature at the other end of the sweep feature.

   Alternatively, you can use the **Copy** and **Paste** tools to save the time for creating the second protrusion.

   The protrusion features created at both ends of the sweep feature are shown in Figure 9-7.

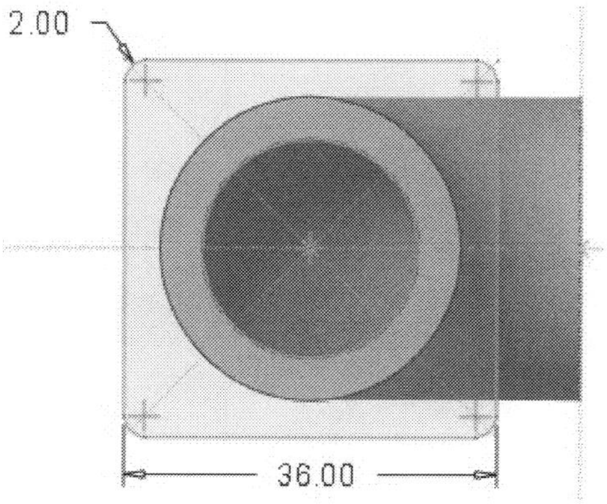

Figure 9-6 *Sketch with dimensions*

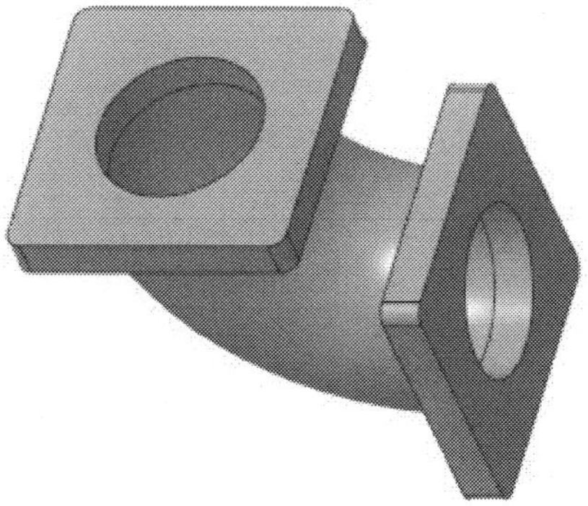

Figure 9-7 *Two extruded features created at both the ends of the sweep feature*

Creating the Hole Feature

After creating the extruded features at both ends of the sweep feature, you need to create the counterbore holes. One hole is to be created on each extruded surface and then they will be patterned on individual planes separately to create the remaining three instances.

1. Choose the **Hole** tool from the **Engineering** group; the **Hole** dashboard is displayed.

2. Choose the **Use standard hole profile as drill hole profile** button from the **Hole** dashboard; two buttons are added to the dashboard.

3. Next, choose the **Adds counterbore** button from the dashboard to make the hole counterbore. Also, choose **Drill up to next surface** from the depth flyout.

4. Choose the **Shape** tab from the dashboard to enter the counterbore parameters of the hole, as shown in Figure 9-8.

5. Select the back face of the second extruded feature for the placement of hole; a preview of the hole appears in the drawing area.

   Now, you need to specify the placement parameters for the hole. For this purpose, refer to Figure 9-2 and look for dimensions that can help in placing the hole. The two edges are used to dimension the hole.

6. Choose the **Placement** tab and click in the **Offset References** collector to turn it green.

7. Select the front edge of the back face and then use CTRL+left mouse button to select the left edge of the third feature for specifying linear references. The hole is at a distance of **5** from both the edges. The default dimensions appear on the hole.

   You can also drag the green handles and place them on the two edges.

8. Double-click on the dimensions and modify the linear distance to **5**.

9. Choose the **OK** button from the **Hole** dashboard; a hole is created on the selected face.

10. Create another hole on the fourth feature by following the same procedure as discussed earlier.

> **Tip**
> *You can also create a counterbore hole by sketching its profile. To do so, choose the* ***Use sketch to define drill hole profile*** *button from the dashboard. Next, choose the* ***Activates Sketcher to create section*** *button to activate the sketcher environment.*

The trimetric view of the model completed so far is shown in Figure 9-9.

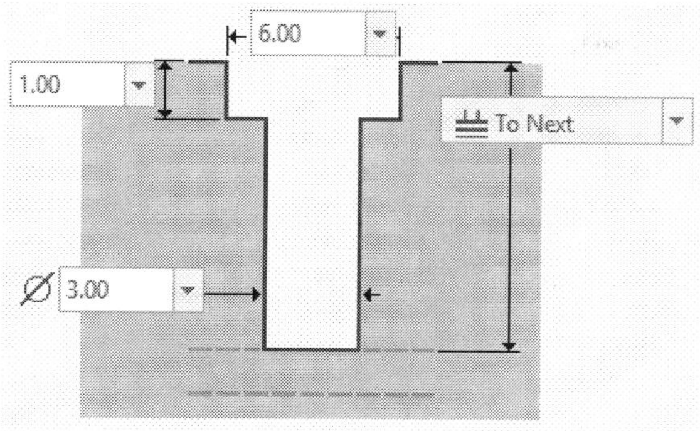

Figure 9-8 Parameters of the counterbore hole

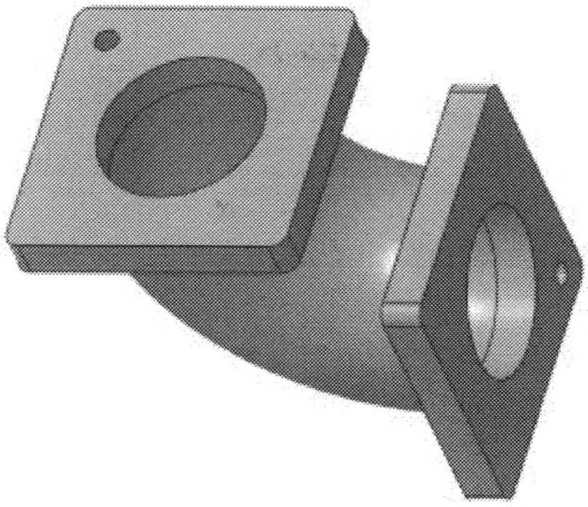

Figure 9-9 One hole each on the two extruded features

Creating the Pattern of Holes

After one hole is placed on each of the two faces, you need to pattern the holes.

1. Select the hole feature from the **Model Tree** or from the drawing area and hold the right mouse button to invoke the shortcut menu. Choose the **Pattern** option from the shortcut menu; the **Pattern** dashboard is displayed.

2. Specify the required parameters in the **Pattern** dashboard and create the pattern of the hole.

   Similarly, create the pattern of the hole on the other extruded feature as well. The default trimetric view of the complete model is shown in Figure 9-10.

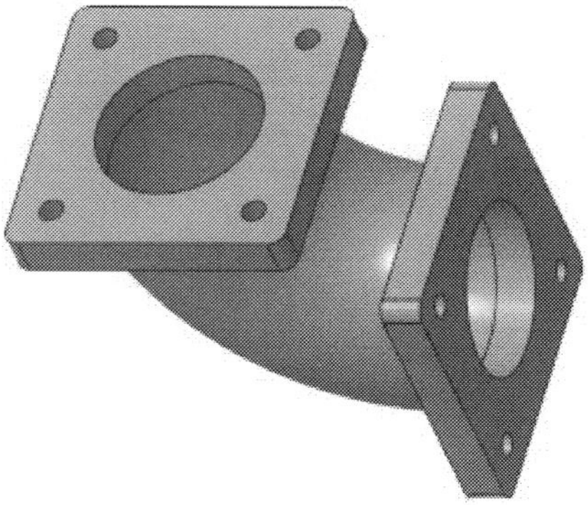

Figure 9-10 The trimetric view of the complete model

Saving the Model

1. Choose the **Save** option from the **File** menu and then save the model.

Note

As discussed in the earlier chapters, the ***Model Tree*** *is used to get an idea about the order of feature creation. Therefore, while creating a model, the order of feature creation will remain same but the id numbers of the features may be different.*

Tutorial 2

In this tutorial, you will create the spring shown in Figure 9-11. Figure 9-12 shows the front view of the spring with its dimensions. **(Expected time: 10 min)**

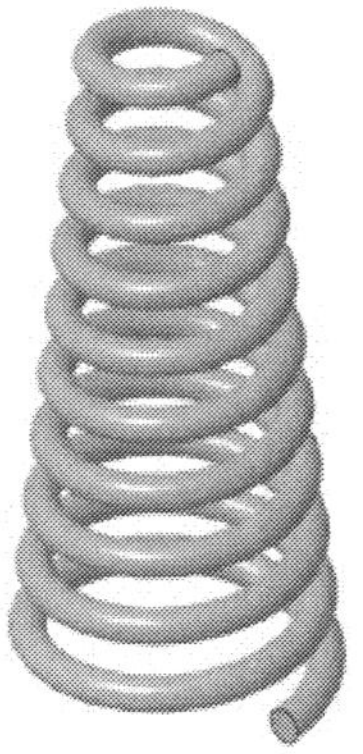

Figure 9-11 Isometric view of the spring

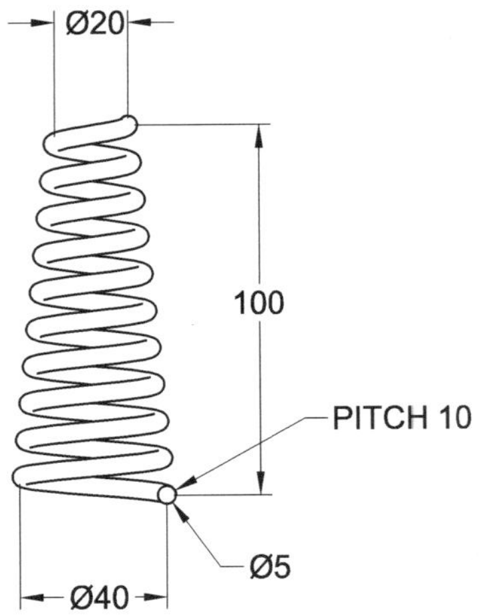

Figure 9-12 Front view with dimensions of the spring

Examine the spring and determine the specifications of the spring, refer to Figure 9-13. The following steps are required to complete this tutorial:

a. Draw the trajectory using the sketcher environment tools and apply dimensions to it, refer to Figures 9-11 and 9-12.
b. Specify the pitch of the spring.
c. Draw the section of the spring, refer to Figure 9-14.

Make sure that the required working directory is selected.

Starting a New Object File

1. Start a new part file and name it as *c09tut2*.

The three default datum planes and the **Model Tree** are displayed in the drawing area.

Creating the Helical Sweep Feature

The spring that you have to create is a right-handed spring of constant pitch.

1. Choose **Helical Sweep** from the **Sweep** drop-down of the **Shapes** group; the **Helical Sweep** dashboard is displayed. Select the **Through axis of revolution** radio button from the **References** tab if not selected by default.

2. Make sure that the **Use right handed rule** button is selected in the dashboard.

3. Select the **Keep constant section** radio button from the **Options** tab.

4. Select the **Define** radio button from the **References** tab; the **Sketch** dialog box is displayed and you are prompted to select a sketching plane.

5. Select the **FRONT** datum plane from the drawing area and then choose the **Sketch** button from the **Sketch** dialog box.

6. Once you enter the sketcher environment, draw the sketch of the trajectory and dimension it, as shown in Figure 9-13. As is evident from Figure 9-13, the endpoint of the trajectory is aligned with the **TOP** datum plane.

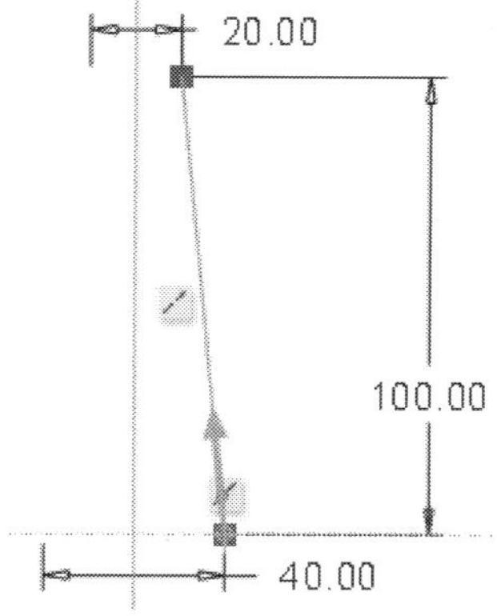

Figure 9-13 *Sketch of the trajectory*

You need to draw a geometry center line in the sketch about which the spring will be rotated. This is the axis of the spring.

7. After you complete the sketch of the trajectory, choose the **OK** button from the dashboard; the **Helical Sweep** dashboard is displayed and a default pitch value of spring is displayed on the trajectory.

8. Enter **10** in the **Pitch** edit box and choose the **Create or edit sweep section** button from the dashboard. You will notice that two pink lines of infinite length crossing each other perpendicularly appear on the screen. The intersection point of these lines is the start point of the trajectory.

9. Draw the section of the spring such that the center of the circle coincides with the intersection of the two perpendicular lines. Then, dimension it, as shown in Figure 9-14.

10. After completing the sketch, choose the **OK** button to exit the sketcher environment.

11. Choose the **OK** button from the **Helical Sweep** dashboard; the spring is created, as shown in Figure 9-15.

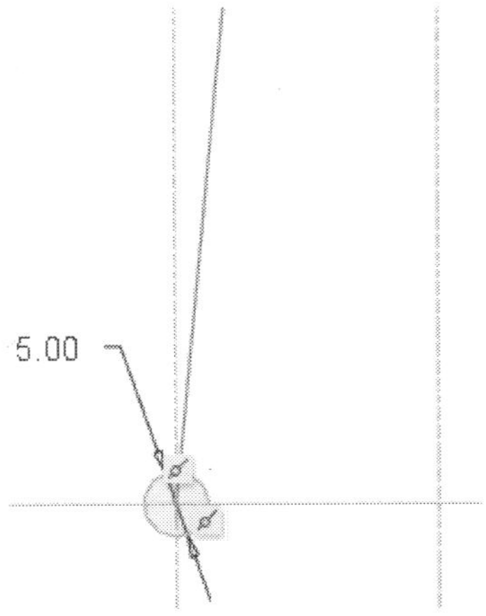

Figure 9-14 *Sketch of the section*

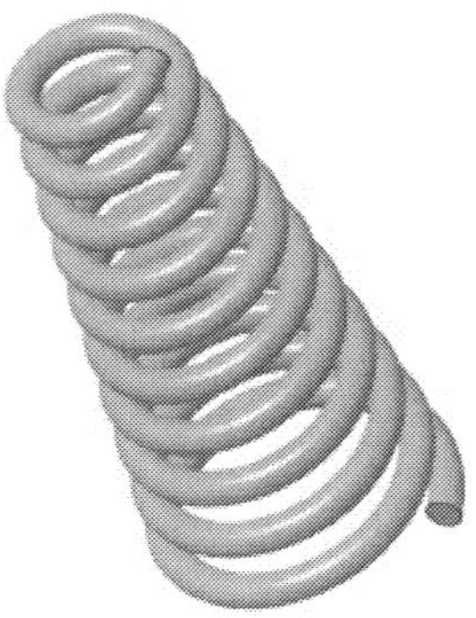

Figure 9-15 *Trimetric view of the helical spring*

Saving the Model

1. Choose the **Save** button from the **File** menu and save the model.

EXERCISES

Exercise 1

Create the model of the Helical Gear shown in Figure 9-16. Its views and dimensions are shown in Figure 9-17. **(Expected time: 45 min)**

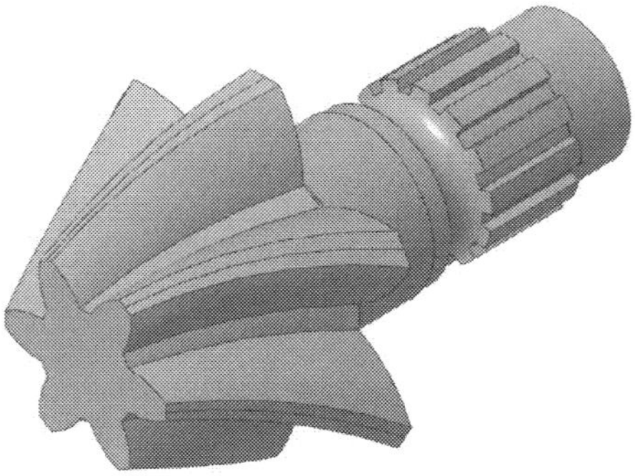

Figure 9-16 *Model of the Helical Gear*

NOTE:

1. CREATE A BLEND FEATURE WITH 3 SKETCHES. SIZE OF THE SECOND SKETCH IS 75% OF THE FIRST SKETCH AND, SIZE OF THE THIRD SKETCH IS 50% OF THE FIRST SKETCH.

2. THE PLANE FOR CREATING THE SECOND SKETCH IS 45 UNITS FROM THE FIRST SKETCH. THE PLANE FOR CREATING THE THIRD SKETCH IS 95 UNITS FROM THE FIRST SKETCH.

3. ROTATION ANGLE FOR SECOND AND THIRD SECTION IS 15° AND 30° FROM THE FIRST SECTION RESPECTIVELY.

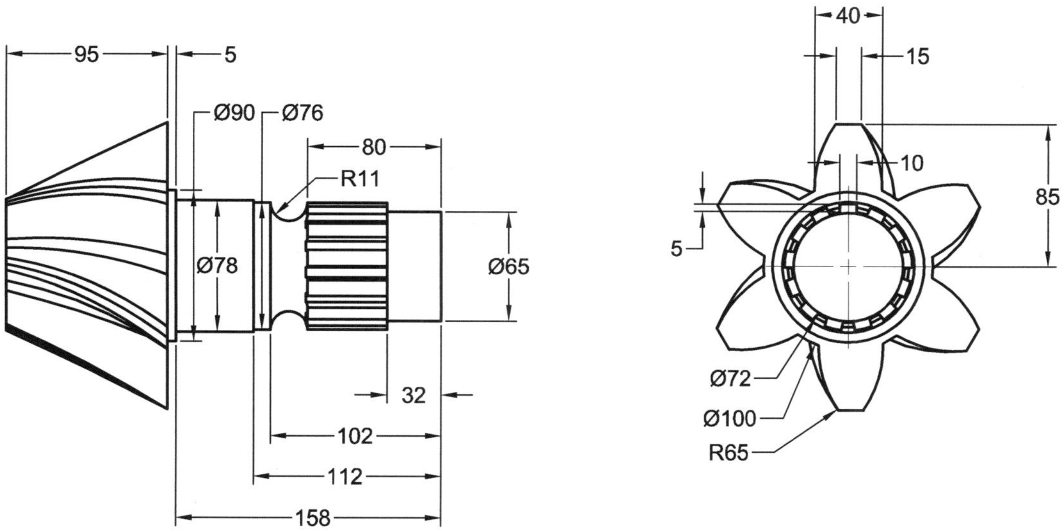

Figure 9-17 *Views and Dimensions for the model*

Hints to create a Helical Gear is discussed next:

1. Choose the **Sketch** tool from the **Datum** group of the **Model** tab in the **Ribbon** and draw the sketch, as shown in Figure 9-18. Next, choose the **OK** button to exit the sketching environment.

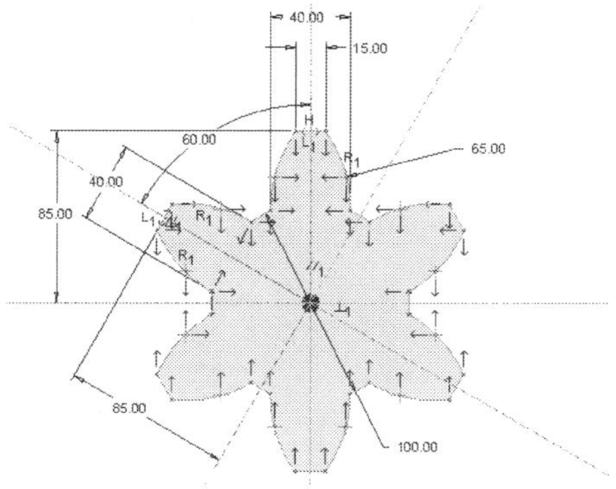

Figure 9-18 *Sketches for the Swept Blend feature (Section 1)*

2. Next, choose the **Sketch** tool from the **Datum** group and choose the **Project** tool from the **Sketching** group and choose the **Loop** option from the **Type** menu. Next, select Section 1 from the drawing area; the section will be projected.
3. Next, select the projected section and choose the **Rotate Resize** tool from the **Editing** group and scale the selected sketch by 75% and rotate it by 15 degrees.
4. Similarly, draw the third sketch at an offset distance of 50 and scale the sketch by 50% and rotate it by 15 degrees from the second sketch.
5. Now, choose the **OK** button; a blend feature is created, as shown in Figure 9-19.

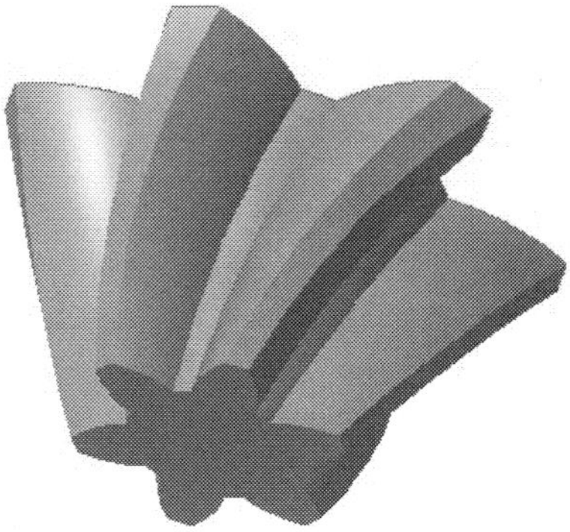

Figure 9-19 *Preview of the Blend feature*

Exercise 2

In this exercise, you will create the model of a carburetor cover shown in Figure 9-20. Figure 9-21 shows the sectioned top view, sectioned front view, sectioned right view, and the sectioned bottom view with dimensions. **(Expected time: 45 min)**

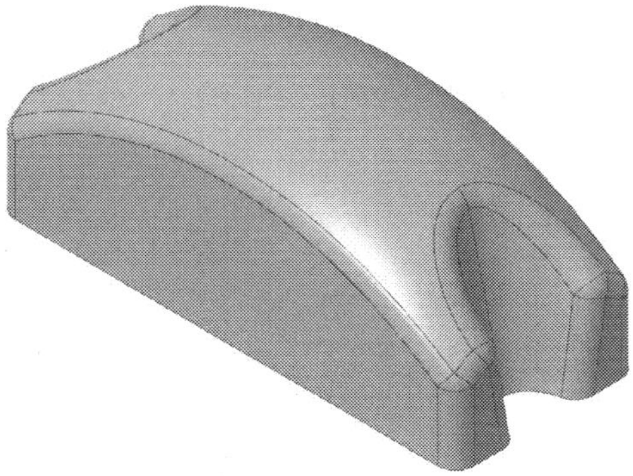

Figure 9-20 *Isometric view of the carburetor cover*

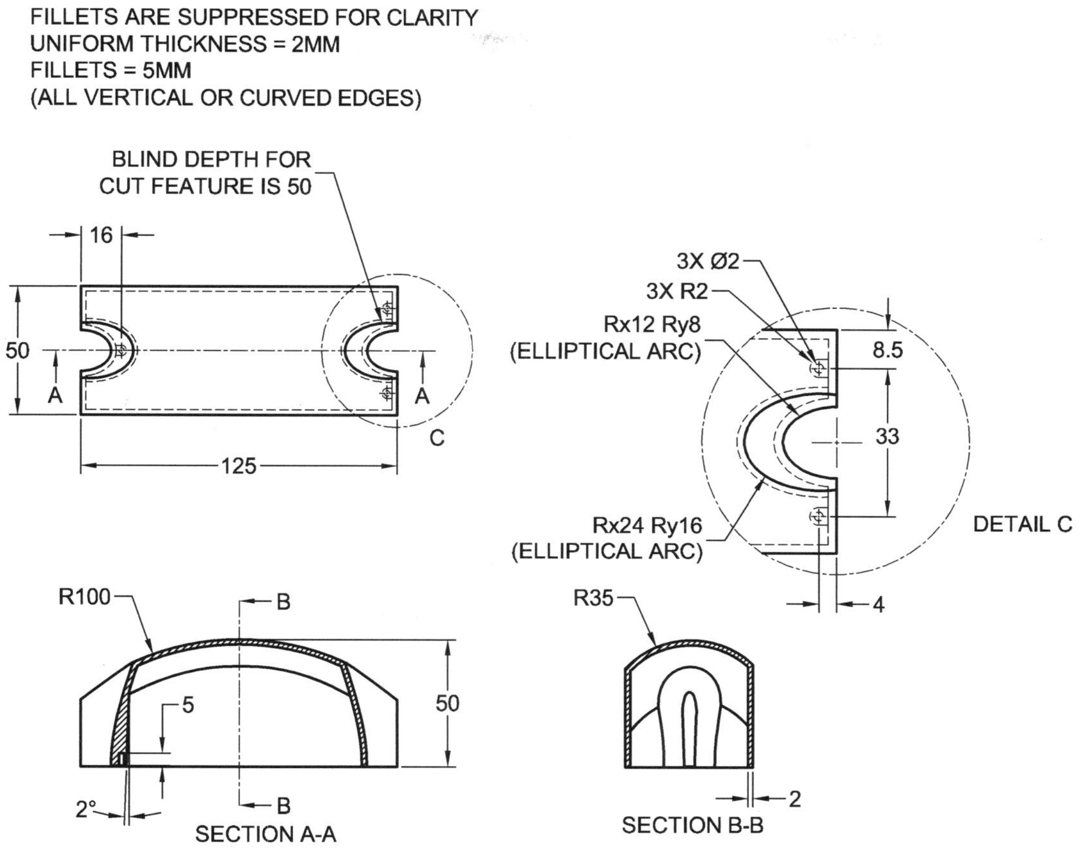

Figure 9-21 *Orthographic views of the carburetor cover*

Hints

1. Create the sketch of the base feature that includes a rectangle of 125x50 and then extrude it.
2. Invoke the sketcher environment, select the **FRONT** datum plane as the sketching plane for sketching a trajectory.
3. Sketch a trajectory by using the **3-Point / Tangent End** button from the **Sketching** group. The start point and the endpoint of the arc should be at a distance of 18 units from the bottom of the base feature, and the radius of the arc should be 100.
4. Exit the sketcher environment by using the **OK** button.
5. Choose the **Sweep** tool from the **Shapes** group of the **Ribbon**, select the trajectory, and choose the **Remove material** button.
6. Select the **Merge Ends** check box from the **Options** tab.
7. Choose the **Create or edit a sweep section** option to create the section for the sweep feature. Choose the **3-Point / Tangent End** button from the **Sketching** group. The arc created should be tangent to the reference lines, and the endpoints of the arc need to be aligned with the edges of the base feature with a radius of 35.
8. Choose the **Done** button from the **Sweep** dashboard.

9. Choose the **Blend** tool in the **Shapes** group of the **Ribbon** and choose the **Remove material** button from dashboard.
10. Choose the **Sections** tab from the dashboard. Choose the **Sketched sections** radio button and click on the **Define** button to define the sketching plane.
11. Select the bottom face of the base feature as the sketching plane.
12. Select the **RIGHT** datum plane as the reference plane and then select the **Right** option from the **Orientation** drop-down list.
13. Create an ellipse of Rx12 and Ry8 by using the **Ellipse** button.
14. Choose the **OK** button to exit the sketcher environment. Enter **50** in the offset value edit box for second section.
15. Create another ellipse of Rx24 and Ry16.
16. Exit the sketcher environment and choose the **OK** button to complete the feature creation.
17. Mirror the cut feature about the **RIGHT** datum plane.
18. Create a round feature on all edges except the edges enclosing the bottom planar surface of the base feature.
19. Choose the **Shell** tool from the **Engineering** group. Remove the bottom face of the base feature.
20. Using the bottom face of the base feature as the sketching plane, create three protrusion features that are the supporting structures for the screws. Extrude these features by using the **Extrude up to next surface** button.
21. Use the **Round** tool to round the edges.

Chapter 10

Assembly Modeling

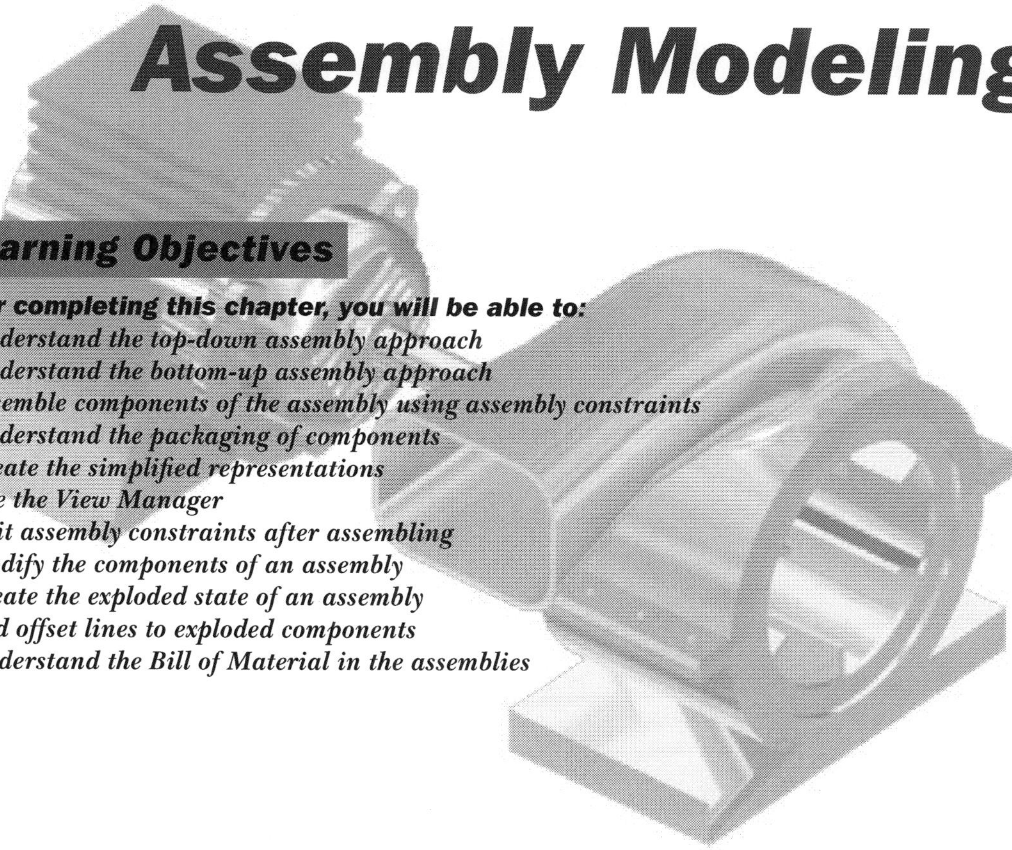

Learning Objectives

After completing this chapter, you will be able to:

- *Understand the top-down assembly approach*
- *Understand the bottom-up assembly approach*
- *Assemble components of the assembly using assembly constraints*
- *Understand the packaging of components*
- *Create the simplified representations*
- *Use the View Manager*
- *Edit assembly constraints after assembling*
- *Modify the components of an assembly*
- *Create the exploded state of an assembly*
- *Add offset lines to exploded components*
- *Understand the Bill of Material in the assemblies*

ASSEMBLY MODELING

An assembly is defined as a design consisting of two or more components bonded together at their respective working positions using the assembly constraints.

IMPORTANT TERMS RELATED TO THE ASSEMBLY MODE

Before proceeding further in this chapter, it is very important for you to understand the following terms.

Top-down Approach

This is the method of assembling the components in which the components of the assembly are created in the same assembly file. In this type of assembly modeling approach, the components are created in the assembly file and then assembled using the assembly constraints.

Bottom-up Approach

This is the method of assembling the components that are created as separate parts in the **Part** mode. Once all parts of an assembly have been created, you will create a new assembly file and then assemble the parts using the assembly constraints available in the **Assembly** mode. Since the assembly file has information related only to the assembling of components, its size is small and therefore requires less hard disk space.

Placement Constraints

The placement constraints are the constraints that are used to rigidly bind the components of the assembly to their respective positions in the assembly. These constraints are also called as assembly constraints. Generally, these constraints are used in combinations and you can constrain upto six degrees of freedom of a component.

Package

Packaged components are unconstrained components in an assembly. It is the state in which the component being assembled is not fully constrained and therefore it is not rigidly placed at its actual location.

3D Dragger

The 3D Dragger is used to move or rotate the component to be assembled. The 3D Dragger can be made visible or hidden by using the **3D Dragger** button available in the dashboard. By using the 3D Dragger, you can move or rotate a component to set its orientation. It has three arrow handles and 3 circular handles, refer to Figure 10-1. The arrow handles are used to move the component and the circular handles are used to rotate the component.

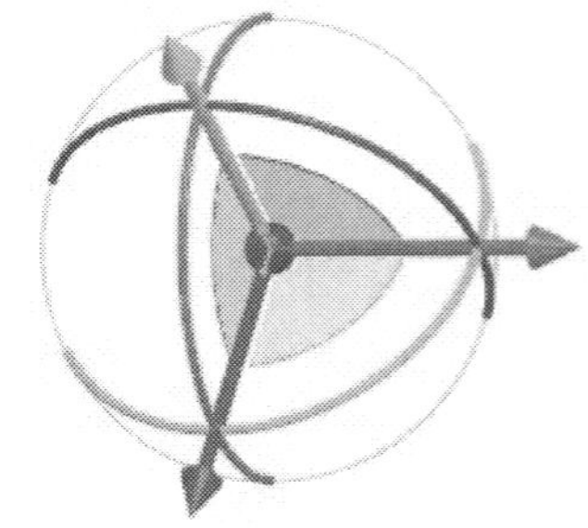

Figure 10-1 *The 3D Dragger*

SIMPLIFIED REPRESENTATIONS

The complexity of individual components and number of parts can reduce the regeneration, retrieval, and display time of the assembly. To speed up the regeneration and display of the assembly, you can temporarily remove a complicated and unrelated component or subassembly from the assembly. This process of removing some components or changing the type of display from the current display is called **Simplified Representation**. A simplified representation consists only of components that have been selected for specific representation types.

CREATING THE EXPLODED STATE

Ribbon: Model > Model Display > Edit Position

In assembly design, the components assembled inside other components may not be visible. This could be misleading and may confuse the viewer as the component cannot be viewed. To avoid this confusion, generally an exploded view is provided along with the assembled view. An exploded view is a state in which all the components move from their original position so that they are visible

THE BILL OF MATERIALS

Ribbon: Model > Investigate > Bill of Materials

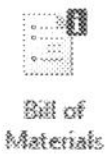

The Bill of Materials or BOM is the tabular representation of all components of the assembly, along with the information associated with them. The information can be the material of the components, the additional note with the components, and so on.

GLOBAL INTERFERENCE

Ribbon: Analysis > Inspect Geometry > Global Interference drop-down > Global Interference

Global Interference is a type of analysis that can be conducted in the **Assembly** and **Drawing** modes. Using this analysis, you can check the total volume overlap of the two assembled parts. This volume can be used to calculate the compressive force working on the two assembled parts.

SHRINKWRAP FEATURES

Shrinkwrap features contain a collection of associatively copied surfaces and datums that represent the exterior shape of a referenced part or assembly. Because the surfaces are copied associatively, the Shrinkwrap feature updates when the source assembly gets modified. Shrinkwrap models load faster than complex models which improves system performance when working with large assemblies.

TUTORIALS

Tutorial 1

In this tutorial, you will create all components of the Shock assembly and then assemble them, as shown in Figure 10-2. Also, you will create an exploded state of the assembly, as shown in Figure 10-3. The BOM is shown in Figure 10-2. The dimensions of the components are shown in Figures 10-4 through 10-9. **(Expected time: 2 hrs)**

Figure 10-2 The Shock assembly

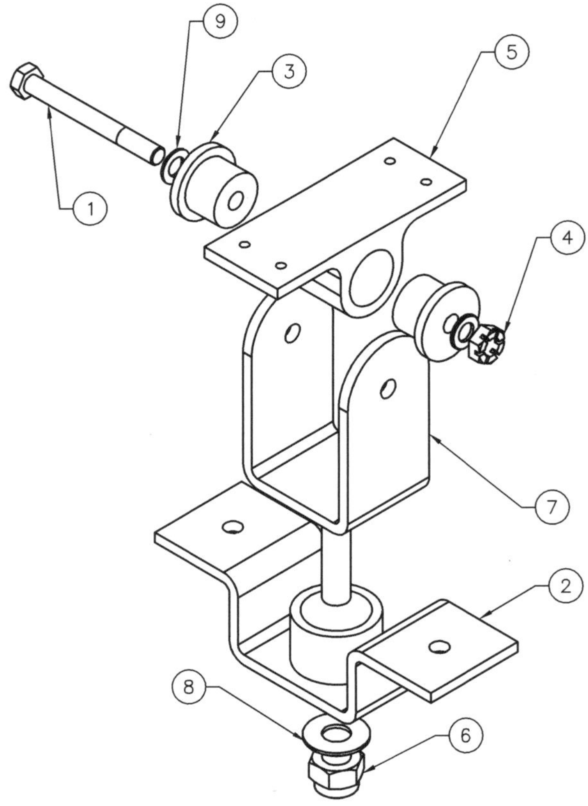

ITEM NO	PART NUMBER	QTY
1	BOLT_M48X5	1
2	BRACKET	1
3	BUSHING	2
4	CASTLE_NUT	1
5	PIVOT	1
6	SELF_LOCKING_NUT	1
7	U-SUPPORT	1
8	WASHER	1
9	WASHER_ISO7090	2

Figure 10-3 The exploded state with Bill of Materials of the Shock Assembly

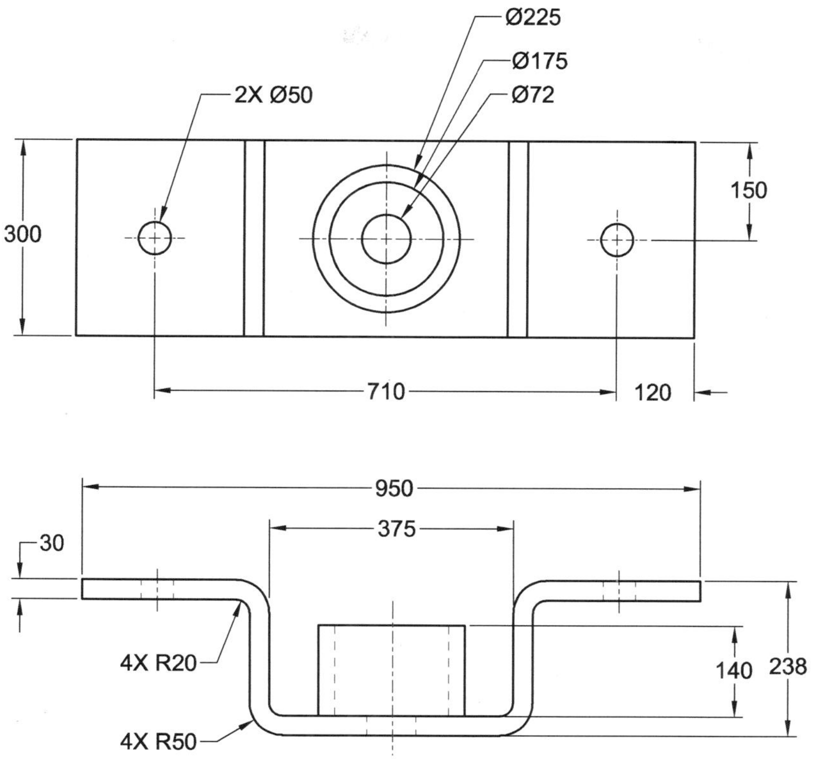

Figure 10-4 Dimensions of the Bracket

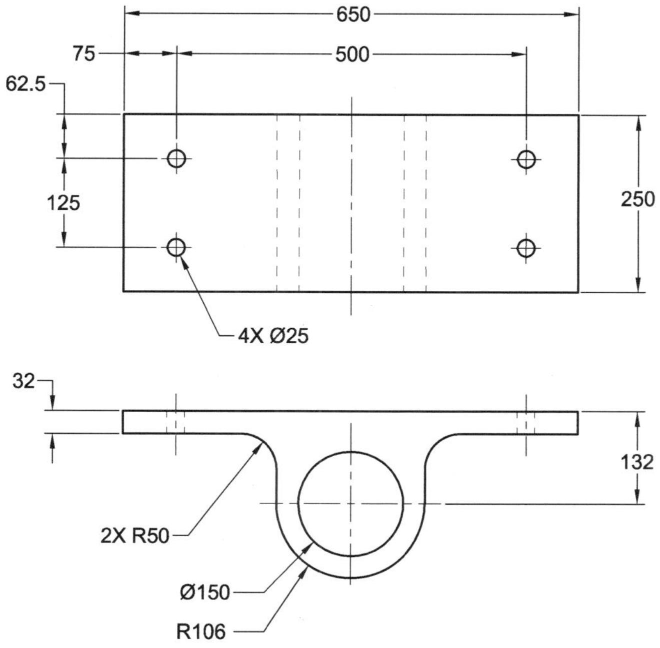

Figure 10-5 Dimensions of the Pivot

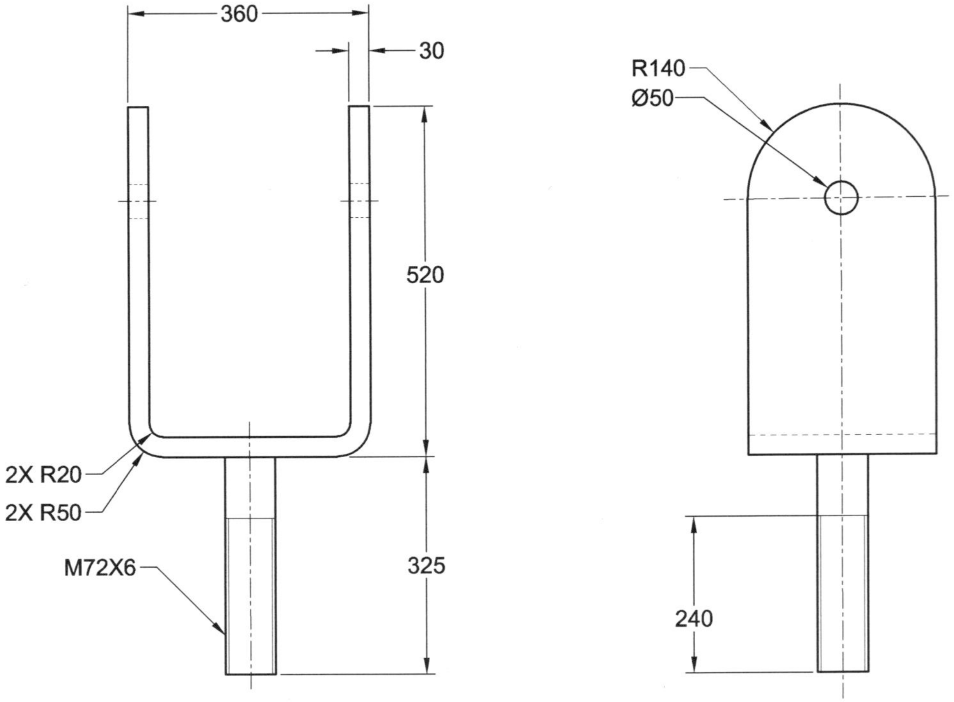

Figure 10-6 Dimensions of the U-Support

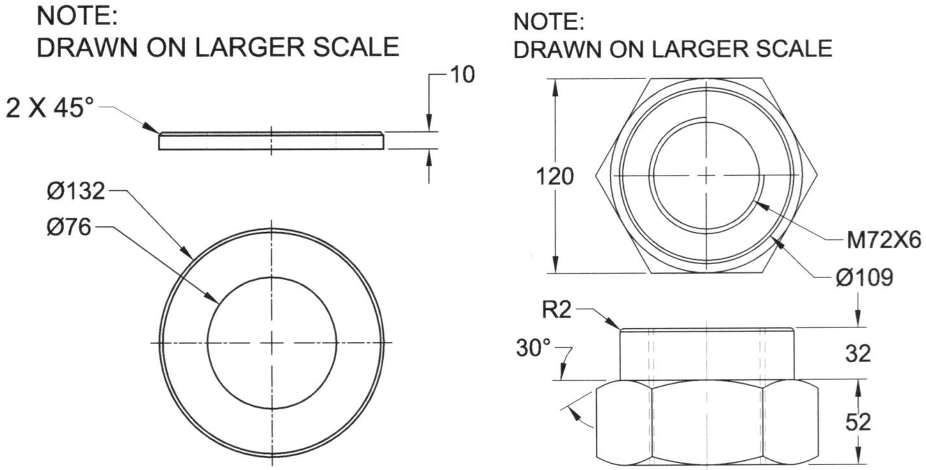

Figure 10-7 Dimensions of the Washer and Self Locking Nut

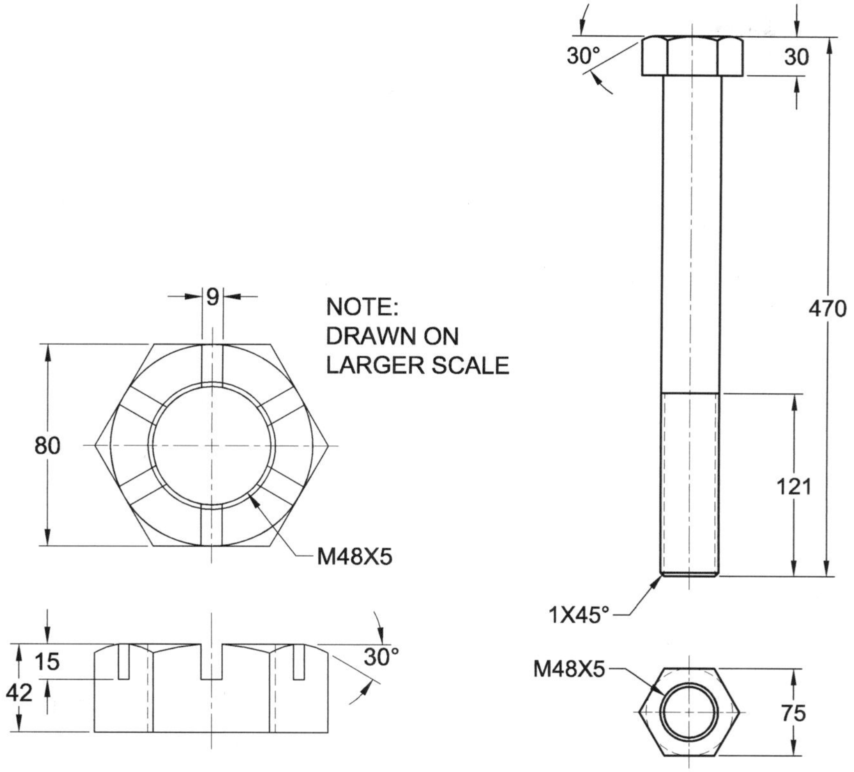

Figure 10-8 *Dimensions of the Castle Nut and Bolt*

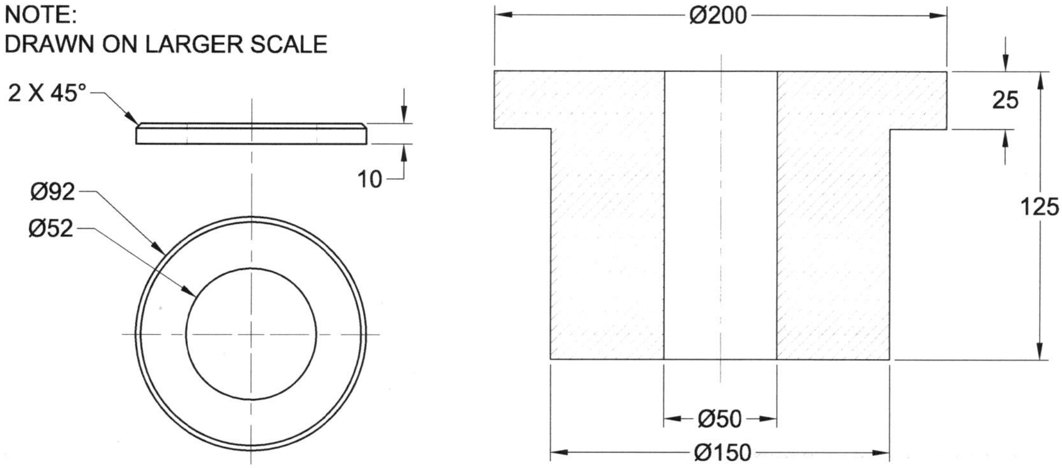

Figure 10-9 *Dimensions of the Washer and Bushing*

Note

You can download the part files of the assembly from http://cadsofttech.com. The complete path for downloading the files is as follows: Textbooks > CAD/CAM > Creo Parametric > Creo Parametric 5.0 for Novices.

The following steps are required to complete this tutorial:

a. Create all components of the assembly as separate part files in the **Part** mode.
b. Create a new file in the **Assembly** mode and then assemble the U-Support with the default datum planes.
c. Assemble the Bushing with the U-Support and then repeat the Bushing in the assembly, refer to Figures 10-10 through 10-12.
d. Assemble the Pivot with the Bushing, refer to Figures 10-13 and 10-14.
e. Suppress the two Bushings and the Pivot.
f. Assemble the Bracket with the assembly, refer to Figures 10-15 and 10-16.
g. Assemble the Washer, Hexagonal Bolt, Castle nut, and Self locking nut in the assembly, refer to Figures 10-17 through 10-19.
h. Unsuppress the suppressed components.
i. After assembling all components, create the exploded state of the assembly, refer to Figure 10-20.
j. Generate offset lines in the exploded state, refer to Figure 10-21.
k. Save the assembly and then exit the **Assembly** mode.

Before you start creating components, set the working directory to *C:\Creo-5.0\c10\Shockassembly*.

Starting the Components of the Assembly

The given assembly is created by using the bottom-up approach. As mentioned earlier, in the assemblies created using the bottom-up approach, the components are created as separate files and are placed in the assembly file. Therefore, first you need to create the components of the assembly.

1. Create all components of the assembly as separate part files and save them in the current working directory.

2. Close the part files, if opened.

Creating a New Assembly File

You need to open a new assembly file to assemble the components.

1. Choose the **New** button from the **Data** group; the **New** dialog box is displayed.

2. Select the **Assembly** radio button in the **Type** area of the **New** dialog box. In the **Sub-type** area of this dialog box, the **Design** radio button is selected by default. Enter the name of the assembly in the **File name** edit box as **SHOCKASSEMBLY**.

3. Clear the **Use Default Template** check box and choose the **OK** button; the **New File Options** dialog box is displayed.

4. Select the **mmks_asm_design** from the **Template** area and choose the **OK** button from the **New** dialog box; the assembly environment is invoked.

Assembling the U-Support with the Default Datum Planes

1. Choose the **Assemble** tool from the **Component** group or press the A key from the keyboard; the **Open** dialog box is displayed.

2. Select the **U-Support** component from the **Open** dialog box and choose the **Open** button; the **Component Placement** dashboard is displayed and you are prompted to select a reference for auto type constraining.

   In the **Component Placement** dashboard, the **Automatic** constraint type is selected by default in the **Constraint Type** drop-down list.

3. Choose the **Show component in a separate window while specifying constraints** button from the **Component Placement** dashboard to display the **U-Support** component in a separate window.

4. Now, select the **FRONT** plane from the drawing area or from the **Model Tree** in the **COMPONENT WINDOW** and select the **ASM_FRONT** plane from the drawing area or from the **Model Tree** in the assembly window. Now, these two datum planes are aligned.

5. Similarly, select the other two datum planes and align them. After all the three planes are aligned, the message **Fully Constrained** is displayed in the **Status** area of the dashboard.

6. Choose the **OK** button to complete the assembly of the first part.

Tip
Viewing the components in separate windows helps the user locate the references for the assembly constraints easily and also gives the user an independent viewing control over the two windows. You can spin, pan, and zoom the components in their respective windows.

Note
1. You can also assemble the first component in an assembly by selecting the ***Default*** *option from the* ***Constraint Type*** *drop-down list in the* ***Component Placement*** *dashboard.*

2. In an assembly, features are not displayed by default in the ***Model Tree****, so you may need to display them in the* ***Model Tree****. To do so, choose on the* ***Settings*** *button of the* ***Model Tree****; a flyout will be displayed. Choose the* ***Tree Filters*** *option from the flyout; the* ***Model Tree Items*** *dialog box will be displayed. Now, select the* ***Features*** *check box from the* ***Display*** *area of the dialog box and choose the* ***OK*** *button.*

Assembling the Bushing with the U-Support

The next component you need to assemble is the **Bushing**.

1. Choose the **Assemble** tool or press the A key from the keyboard to display the **Open** dialog box.

2. Select the **Bushing** component and then choose the **Open** button from the **Open** dialog box; the **Component Placement** dashboard is displayed. Choose the **Show component in**

separate window while specifying constraints button from the **Component Placement** dashboard to display the **Bushing** component in a separate window, if it is not chosen by default.

3. Select the **Coincident** option from the **Constraint Type** drop-down list. Select the axis of the **Bushing** component and the axis of the **U-Support** component, as shown in Figure 10-10.

4. Next, select the planar face of the **Bushing** component and the inner planar face of the **U-Support** component, as shown in Figure 10-10. Select the **Coincident** option from the **Constraint Type** drop-down list. Choose the **OK** button; the model similar to the one shown in Figure 10-11 is displayed in the drawing area.

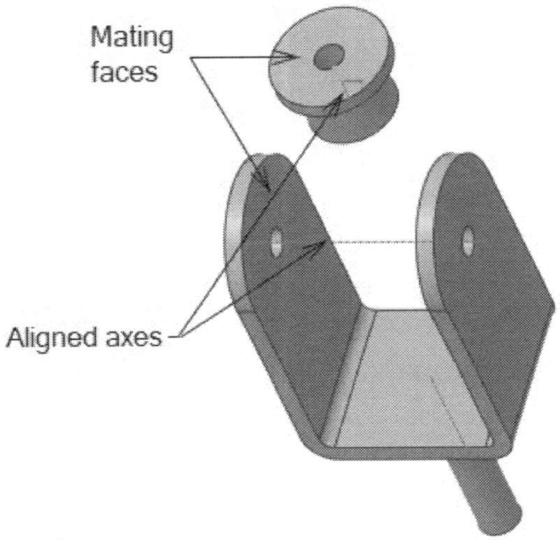

*Figure 10-10 Locations to apply the **Coincident** constraint*

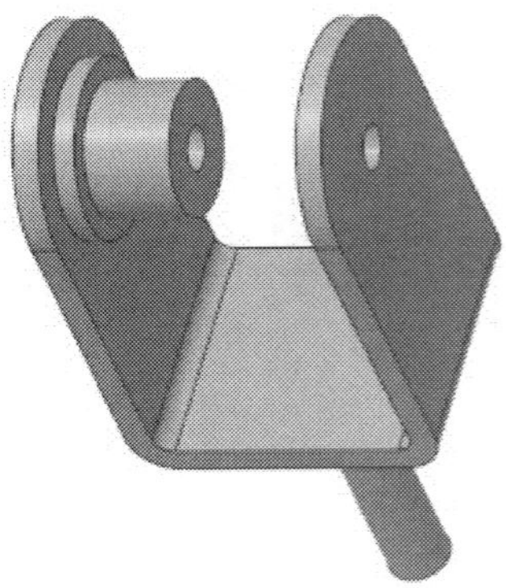

Figure 10-11 Components after assembling

Note

*You may need to choose the **Change orientation of constraint** button from the **Component Placement** dashboard to assemble the **Bushing** components in the required orientation.*

5. The next instance of the **Bushing** can be placed directly by using the **Repeat** tool. To repeat the **Bushing** component in the assembly, select it from the **Model Tree** or from the drawing area and then choose the **Repeat** tool from the **Component** group. On doing so, the **Repeat Component** dialog box is displayed and two **Coincident** constraints are displayed in the **Variable assembly references** area.

6. Select the second **Coincident** constraint from this area and then choose the **Add** button.

7. Select the inner right face of the **U-Support** component as the mating face. The copy of the **Bushing** component is assembled, as shown in Figure 10-12. Choose the **OK** button from the **Repeat Component** dialog box.

Assembling the Pivot with the Bushing

The next component that you need to assemble is the **Pivot**.

1. Choose the **Assemble** tool to display the **Open** dialog box.

2. In this dialog box, select the **Pivot** component and then choose the **Open** button to display the **Component Placement** dashboard.

3. Choose the **Placement** tab to display the panel. Select the axis of the **Pivot** and then the axis of the hole of the **Bushing**, refer to Figure 10-10, and apply the **Coincident** constraint on them.

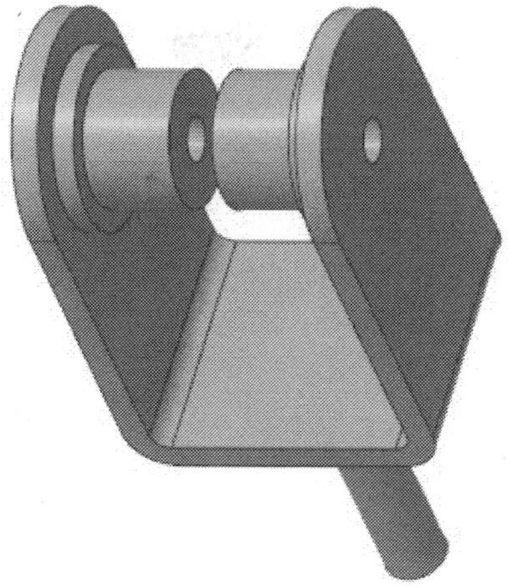

Figure 10-12 *Assembled Bushings and the U-Support*

4. Next, click on the **New Constraint** option in the **Placement** slide-down panel. Select the **Parallel** option from the **Constraint Type** drop-down list.

5. Select the top planar face of the **Pivot** and the planar face of the **U-Support**, as shown in Figure 10-13.

6. Click on the **New Constraint** option in the **Placement** slide-down panel and select the planar face of the **Pivot** and the planar face of the **Bushing**, as shown in Figure 10-13; the **Distance** constraint is automatically applied between them and is displayed by default in the **Constraint Type** drop-down list in the **Placement** tab. Select the **Coincident** option from the drop-down list so that the selected faces are made coincident.

7. Choose the **OK** button to complete the assembly. The model similar to the one shown in Figure 10-14 is displayed in the drawing area.

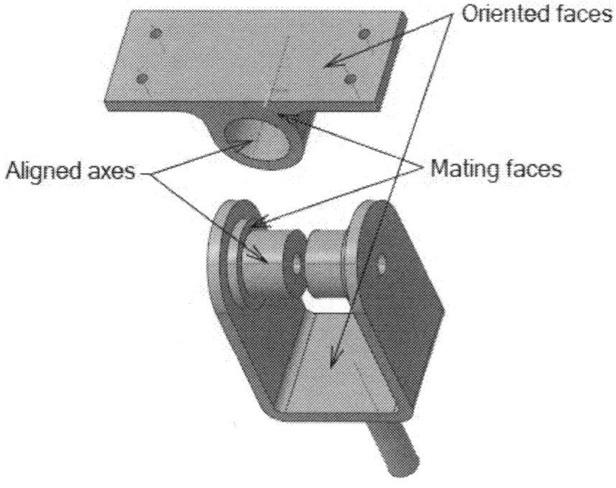

Figure 10-13 *Locations to apply constraints*

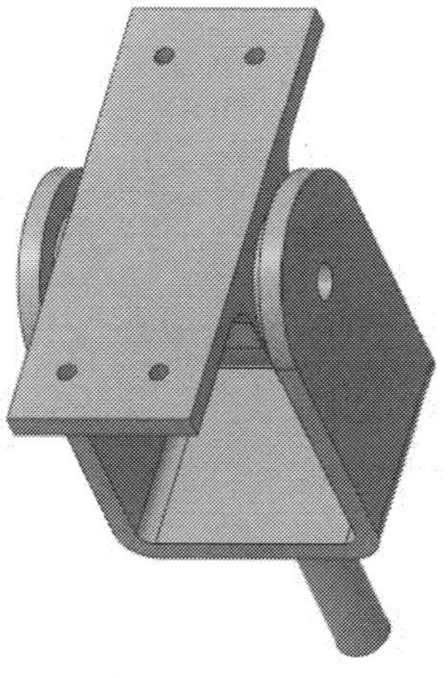

Figure 10-14 *Assembled view of the* ***Pivot, U-Support,*** *and* ***Bushing***

Suppressing Components

Next, you need to suppress both the **Bushing** and the **Pivot**. This is because when the components in the assembly increase, some components may be hidden behind the others. As a result, it gets difficult to select them. Also, more the number of components, more are the datum planes. Therefore, it becomes difficult to make selections on the components. However, when you suppress a component, its datum planes are also suppressed and the complications in the drawing area are reduced.

1. Spin the model and select the first **Bushing** from the drawing area or from the **Model Tree** and choose the **Suppress** option from the mini popup toolbar to suppress the component; the **Suppress** message box is displayed. Choose the **OK** button from the message box; the **Pivot** gets suppressed because it was dependent on the first bushing.

2. Similarly, suppress the other **Bushing** to make the assembly of the other components easy. Choose the **OK** button from the **Suppress** window to exit.

Note

Very often you need to use the datum planes and datum axes to assemble the components. Therefore, you need to turn their display on or off from the ***Graphics*** *toolbar, whenever required.*

Assembling the Bracket with the Assembly

The next component you need to assemble is the **Bracket**.

1. Choose the **Assemble** tool from the **Component** group to display the **Open** dialog box.

2. Select the **Bracket** component to open; the **Component Placement** dashboard is displayed.

3. Choose the **Placement** tab to display the panel and select the **Coincident** option from the **Constraint Type** drop-down list.

4. Select the axis of the **Bracket** and the axis of the **U-Support**, as shown in Figure 10-15.

5. Click on the **New constraint** option in the **Placement** panel and select the **Coincident** option from the **Constraint Type** drop-down list. You can also invoke the shortcut menu and choose the **New Constraint** option from it.

6. Now, select the faces to mate, refer to Figure 10-15; the preview of the assembly is displayed.

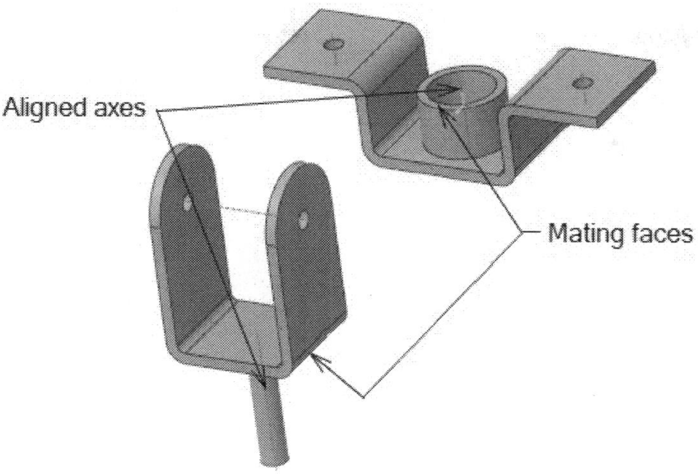

Figure 10-15 Constraints and the location to apply them

7. Choose the **OK** button from the **Component Placement** dashboard to complete the assembly. The model similar to the one shown in Figure 10-16 is displayed in the drawing area.

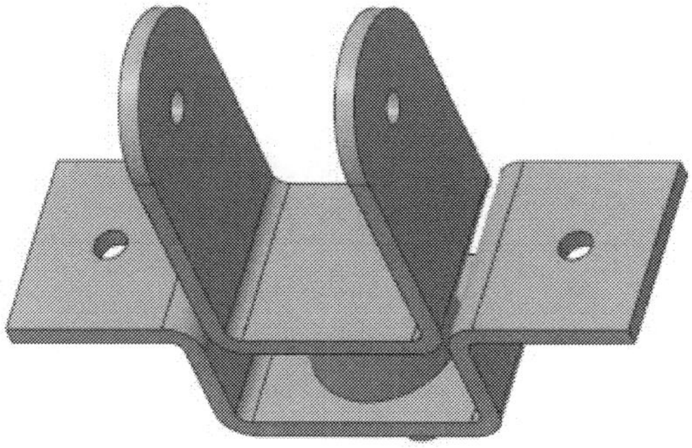

*Figure 10-16 Assembled view of the **Bracket** with **U-Support***

Assembling the Washer with the Assembly

The next component that you need to assemble is the **Washer**.

1. Choose the **Assemble** tool to display the **Open** dialog box.

2. Open the **Washer** component to display the **Component Placement** dashboard.

3. Choose the **Placement** tab from the dashboard to display the panel and select the **Coincident** option from the **Constraint Type** drop-down list to apply the align constraint.

4. Select the axis of the **Washer** and then select the axis of the **U-Support** from the drawing area to align the axes of the two components.

5. Now, click on the **New Constraint** option in the **Placement** panel and select the **Coincident** option from the **Constraint Type** drop-down list.

6. Select the top planar face of the **Washer** and then the bottom planar face of the **Bracket** to assemble the **Washer**; the message **Fully constrained** is displayed in the **Status** area of the dashboard.

7. Choose the **OK** button from the **Component Placement** dashboard to complete the assembly. The model similar to the one shown in Figure 10-17 is displayed in the drawing area.

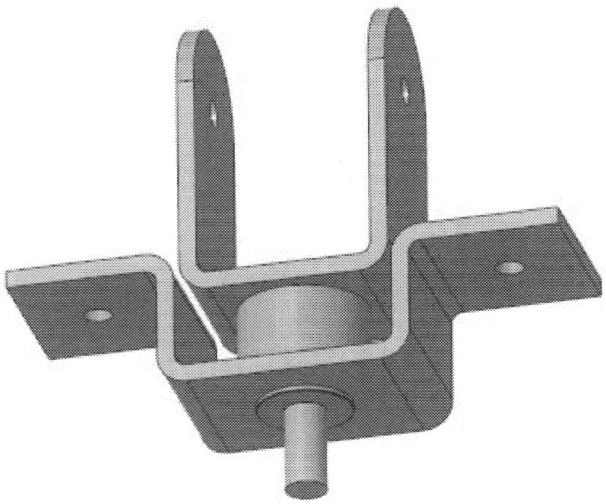

Figure 10-17** Assembled view of the **Washer** with **U-Support

Inserting the Hexagonal Bolt in the Assembly

The next component that you need to assemble is the **Hexagonal bolt**.

1. Choose the **Assemble** tool to display the **Open** dialog box.

2. Open the **Hexagonal bolt** component to display the **Component Placement** dashboard.

3. Choose the **Placement** tab to invoke a panel and select the **Coincident** option from the **Constraint Type** drop-down list to add the align constraint.

4. Select the axis or cylindrical face of the **Hexagonal bolt** and then the axis or cylindrical face of the hole on the **U-Support**. Now, the two axes are aligned. If required, you can choose the **Flip Constraint** button to flip the orientation of the **Hexagonal bolt**.

5. Click on the **New Constraint** option in the **Placement** panel and then select the **Coincident** option from the **Constraint Type** drop-down list.

6. Select the bottom planar face of the head of the **Hexagonal bolt** and then select the left outer planar face of the **U-Support**; the preview of the assembly is displayed.

7. Choose the **OK** button from the **Component Placement** dashboard to complete the assembly. The model similar to the one shown in Figure 10-18 is displayed in the drawing area.

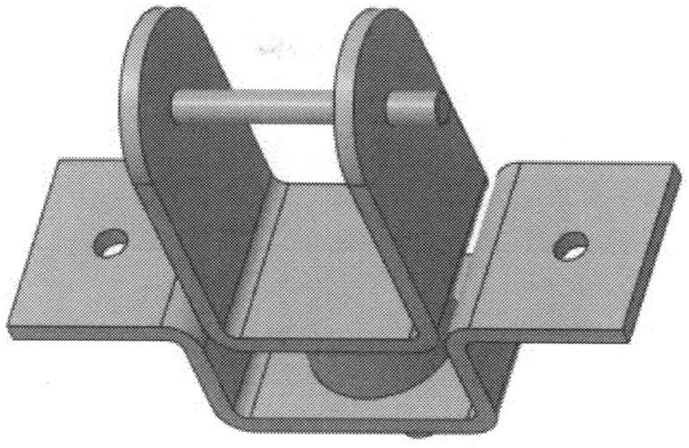

Figure 10-18 Assembly after assembling the ***Hexagonal bolt***

Assembling other Components

In this section, you need to assemble the **Washer**, **Castle nut** and the **Self locking nut** with the assembly

Unsuppressing the Components

All components are assembled at their required positions. Now, you need to unsuppress the suppressed components.

1. Choose the **Resume All** option from the **Resume** flyout of the **Operations** group; the suppressed components appear in the assembly where they were assembled. The completed assembly is shown in Figure 10-19.

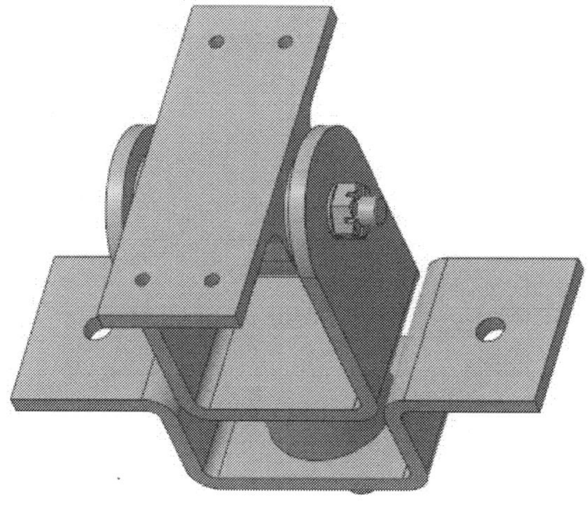

Figure 10-19 Completed assembly with all components

Creating the Exploded State of the Assembly

To view all the components in an assembly clearly, you need to create the exploded state of the assembly.

1. Invoke the **Edit position** tool from the **Model Display** group; the **Explode Tool** dashboard is displayed. Also, the default exploded state of the assembly is displayed.

   Now, you need to edit the position of the component in the exploded view.

2. Choose the **References** tab in the **Explode Tool** dashboard and click in the **Movement Reference** collector on the dashboard; you are prompted to select a reference to define the direction of movement.

3. Select the vertical axis of the **U-Support** from the drawing area; you are prompted to select the component to be moved.

4. Select the **Self locking** nut from the drawing area; a direction handle is displayed. Move the cursor to the required direction handle and drag to place it at the required position. Similarly, select and move the **Washer** and the **Bracket** to the required positions.

   Now, you need to move the components in the top half of the assembly. These components are moved with reference to the axis of the hole on the **U-Support**. Therefore, you need to select this axis as the motion reference.

5. Click in the **Movement Reference** collector on the dashboard; you are prompted to select the reference. Select the axis of the hole on the **U-Support** from the drawing area as the reference; you are now prompted to select the components to move.

6. Select and move the **Hexagonal bolt**, **Pivot**, **Bushings** and **Castle nut** to the required locations in the drawing area.

   The model similar to the one shown in Figure 10-20 is displayed in the drawing area.

7. Choose the **OK** button from the **Explode Tool** dashboard.

   Now, you need to save this exploded view.

8. To save this view, invoke the **View Manager** dialog box and then choose the **Explode** tab.

9. Choose the **New** button; the **Modified State Save** message box is displayed confirming to save the default exploded view. Choose the **Yes** button from the message box; the **Save Display Elements** dialog box is displayed.

10. Next, enter the **EXP1** in the **Explode** edit box and choose the **OK** button from the dialog box and close the **View Manager** dialog box.

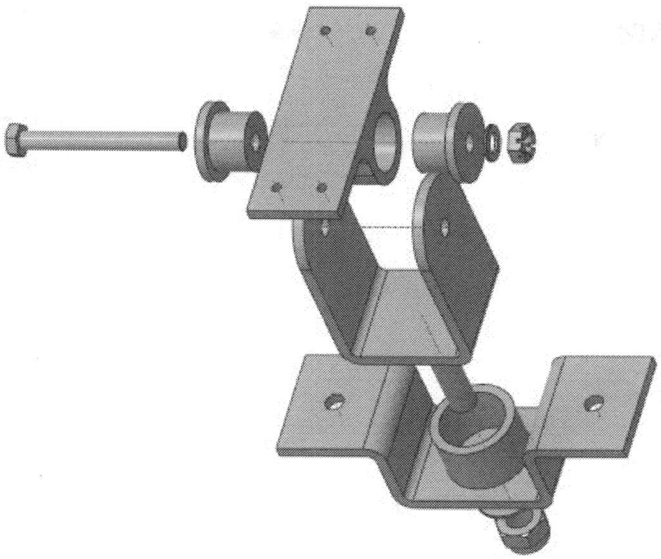

Figure 10-20 Exploded state of the assembly

Tip

*You can also view the default exploded state of an assembly by choosing the **Explode View** button from the **Model Display** group in the **View** tab.*

Creating Offset Lines

1. Invoke the **Edit position** tool from the **Model Display** group and choose the **Create cosmetic offset lines to illustrate movement of exploded component** button from the **Explode Tool** dashboard; the **Cosmetic Offset Line** dialog box is displayed.

2. Click in the **Reference 1** collector and then select the axis of the **U-Support**.

3. Next, click in the **Reference 2** collector and then select the axis of the **Self locking nut**. Next, choose the **Apply** button from the **Cosmetic Offset Line** dialog box; the blue colored explode line is created.

4. Similarly, create the offset lines by selecting the axis of the hole on the **U-Support**, followed by the **Bushing** and then the **Castle nut**. Similarly, create offset lines by selecting the axis of the hole on the **U-Support**, followed by the **Bushing** and then the **Hexagonal bolt**. The model similar to the one shown in Figure 10-21 is displayed in the drawing area.

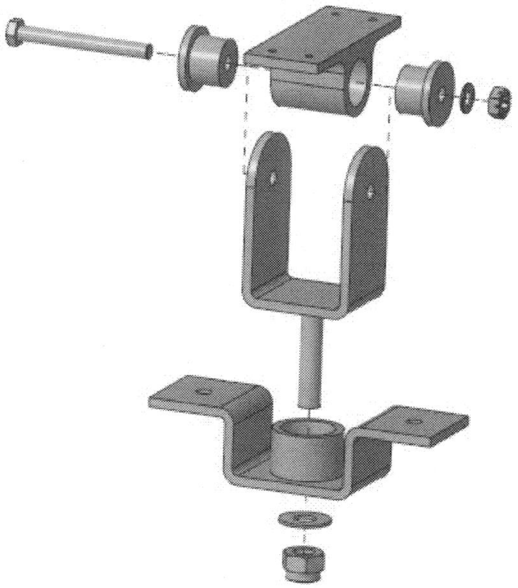

Figure 10-21 Offset lines displayed in the exploded state

5. Next, exit the **Cosmetic Offset Line** dialog box by choosing the **Close** button.

6. Choose the **OK** button from the **Explode Tool** dashboard.

Saving the Assembly File

1. Choose the **Save** button from the **File** menu and save the model.

Closing the Window

1. Choose the **Close** button from the **Quick Access** toolbar to close the assembly file.

Tutorial 2

In this tutorial, you will create all components of the Pedestal Bearing assembly and then assemble them, as shown in Figure 10-22. You will also create the exploded state of the assembly displaying the offset lines, as shown in Figure 10-23. The dimensions of the components are shown in Figures 10-24 through 10-26. **(Expected time: 2 hrs)**

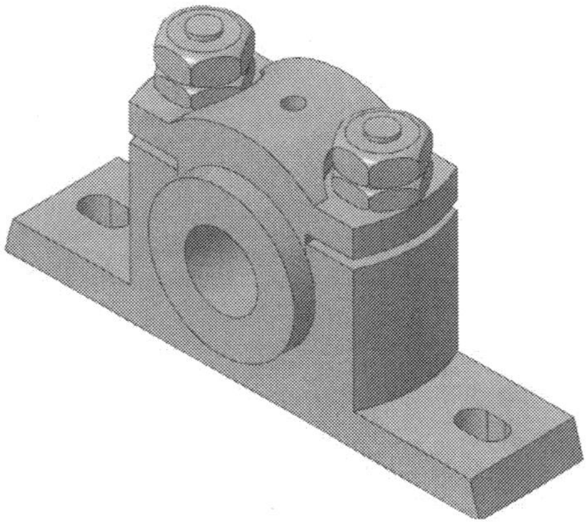

Figure 10-22 *Assembly of the Pedestal Bearing*

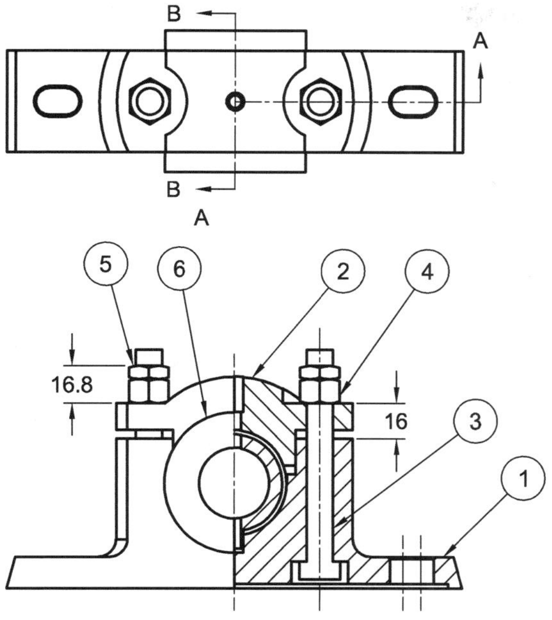

ITEM NO.	PART NO.	DESCRIPTION	QTY.
1	Casting		1
2	Cap		1
3	Bolt		2
4	Nut		2
5	Lock Nut		2
6	Brasses		1

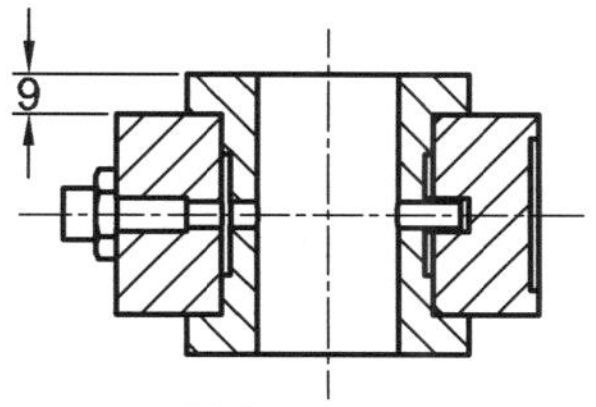

Figure 10-23 *The Pedestal Bearing assembly*

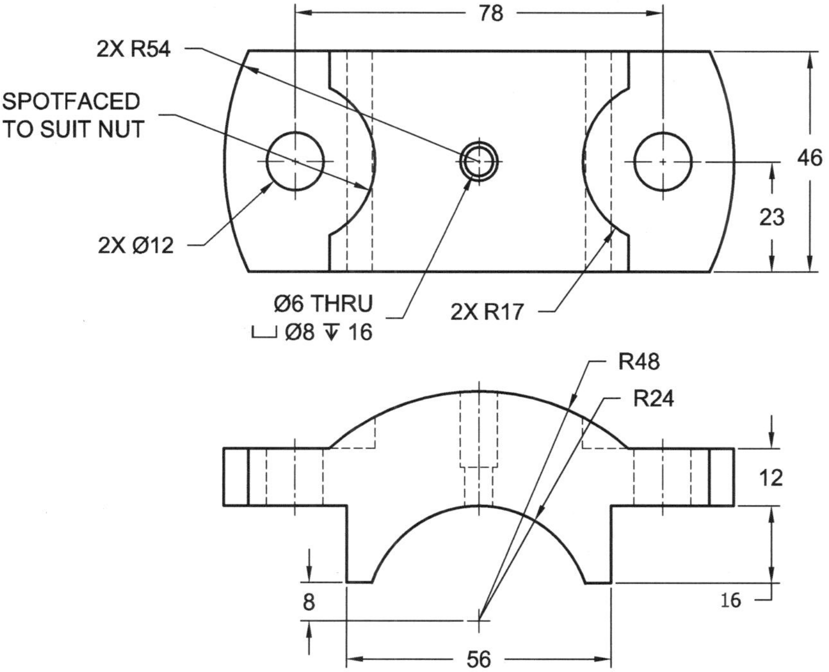

Figure 10-24 *Dimensions of the Cap*

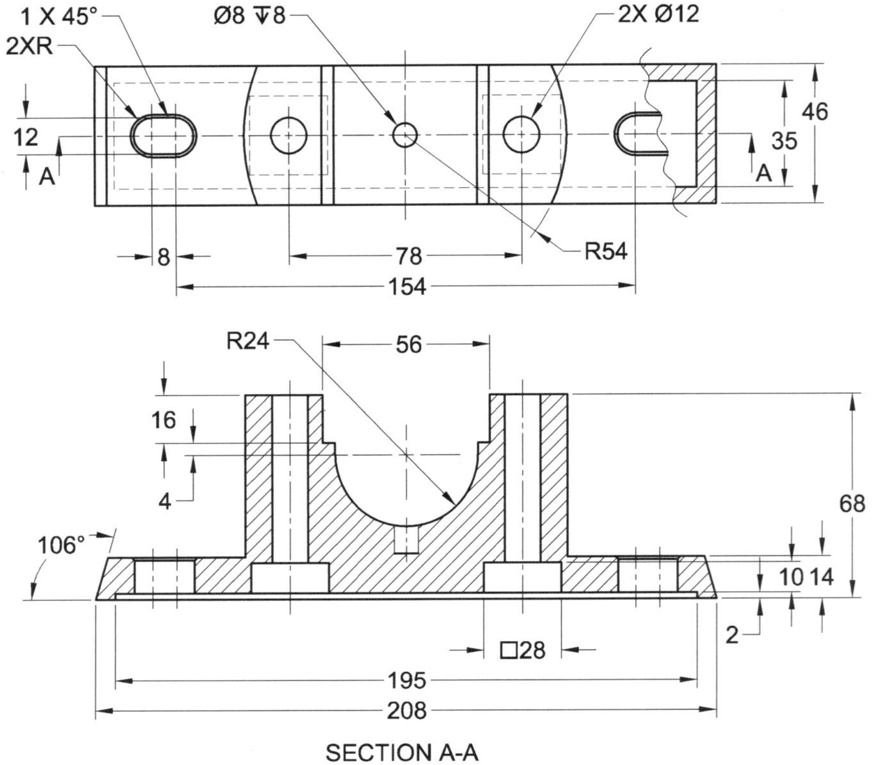

Figure 10-25 Dimensions of the Casting

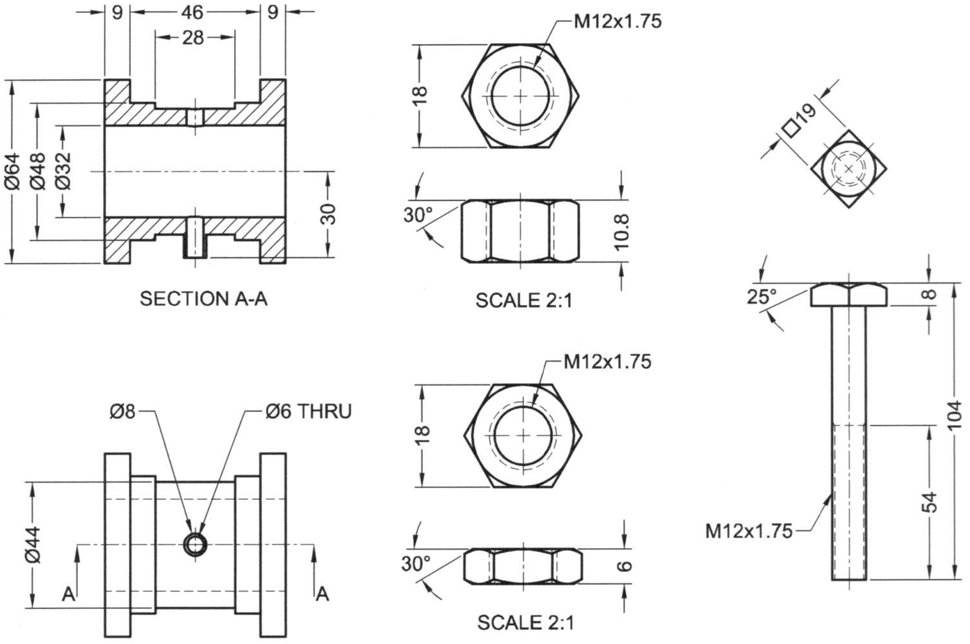

Figure 10-26 Dimensions of the Brasses, Nut, and Locknut

The following steps are required to complete this tutorial:

a. Create all components of the assembly as separate part files in the **Part** mode.
b. Create a new file in the **Assembly** mode and then assemble the **Casting** with the default assembly datum planes.
c. Assemble the **Cap** with the **Casting**, refer to Figures 10-27 and 10-28.
d. Suppress the **Cap** from the assembly.
e. Assemble the **Brasses** with the **Casting**, refer to Figures 10-29 and 10-30.
f. Insert the **Square headed bolt**, **Nut**, and **Lock nut** in the assembly and assemble them, refer to Figure 10-31.
g. Resume the suppressed components.
h. Create the exploded state of the assembly, refer to Figure 10-32.
i. Save the assembly and close the file.

Set the working directory to *C:\Creo-5.0\c10\Pedestalbearing*.

Creating Components for the Assembly

To create the assembly, all components must be created first in the **Part** mode. Also, you need to use the bottom-up approach to create the assembly.

1. Create all components of the assembly as separate part files and then save them in the current working directory.

2. Close the part files, if opened.

Creating a New Assembly File

As mentioned earlier, all components created are assembled in an assembly file that has an extension *.asm*. Therefore, you need to open a new *.asm* file.

1. Choose the **New** button from the **File** menu to display the **New** dialog box.

2. Select the **Assembly** radio button in the **Type** area of the **New** dialog box. In the **Sub-type** area of the **New** dialog box, the **Design** radio button is selected by default. Enter the name of the assembly in the **File name** edit box as **PEDESTALBEARING**.

3. Choose the **OK** button to enter into the assembly modeling environment.

Assembling the Casting with the Default Datum Planes

In the new assembly file, the three default assembly datum planes are displayed in the drawing area and the **Model Tree** is displayed at the left of the drawing area. If the display of the **Model Tree** was turned off in the previous tutorial, it would not appear. Now, you can start assembling components.

1. Choose the **Assemble** tool to display the **Open** dialog box.

2. Select **Casting** from the **Open** dialog box and choose the **Open** button; the **Component Placement** dashboard is displayed.

Note

It is recommended to choose the ***Show component in a separate window while specifying constraints*** *button to display the component to be assembled in a new window.*

3. Select the **FRONT** datum plane of the model and then select the **ASM_FRONT** plane from the assembly to align them. Similarly, align the other two default planes with the respective assembly datum planes.

Note

If the datum planes are not displayed in the drawing area, you need to turn on their display by choosing the ***Plane Display*** *tool from the* ***Datum Display Filters*** *drop-down in the* ***Graphics*** *toolbar. After assembling the first component, you may need to turn off the display of planes. This is because the datum planes clutter the drawing area and then the selection of the references for applying assembly constraints becomes difficult. You can turn the datum planes' display on when they are required.*

Assembling the Cap with the Casting

At first, you need to assemble the **Cap** with the **Casting**.

1. Choose the **Assemble** tool to display the **Open** dialog box.
2. Select the **Cap** file and then choose the **Open** button from the **Open** dialog box to display the **Component Placement** dashboard.
3. Choose the **Placement** tab to display the panel and select the **Coincident** option from the **Constraint Type** drop-down list.
4. Select the axis of the **Cap** and then the axis of the **Casting**, as shown in Figure 10-27.

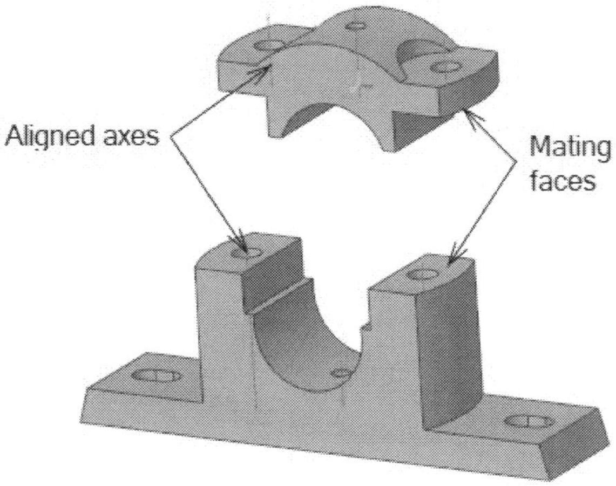

Figure 10-27 *References of the constraints used for assembling the components*

5. Choose the **New Constraint** option from the **Placement** tab and then select the **Distance** option from the **Constraint Type** drop-down list in the **Placement** panel.

6. Select the two mating faces and enter **4** in the edit box in the **Component Placement** dashboard.

7. Choose the **OK** button from the **Component Placement** dashboard to complete the creation of the assembly. The model similar to the one shown in Figure 10-28 is displayed in the drawing area.

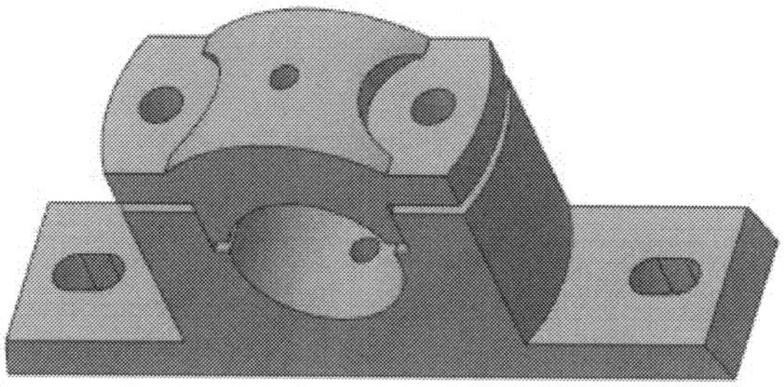

Figure 10-28 *Model after assembling the* ***Cap*** *with the* ***Casting***

Suppressing the Cap from the Assembly

You need to suppress the **Cap** because the next component has to be assembled with the **Casting**. When you suppress the **Cap** from the assembly, it becomes easier to assemble the new component. The **Cap** will be unsuppressed later.

1. Select the **Cap** from the drawing area and right-click to display the shortcut menu.

2. Choose the **Suppress** option from the shortcut menu; you are prompted to confirm the suppression. Choose the **OK** button.

Assembling Brasses with the Casting

Next, you need to assemble the **Brasses** with the **Casting**.

1. Choose the **Assemble** tool to display the **Open** dialog box.

2. Select **Brasses** and choose the **Open** button from the dialog box; the **Component Placement** dashboard is displayed.

3. Choose the **Placement** tab to display the panel and select the **Coincident** option from the **Constraint Type** drop-down list.

4. Select the axis of the **Brasses** and then the axis of the **Casting**, as shown in Figure 10-29.

Note

You can choose the ***Show component in separate window while specifying constraints*** *button from the dashboard to display the component in a separate window, which makes it easier to select the faces and then apply the constraints.*

5. Click on the **New Constraint** option in the **Placement** panel and select the faces of the **Brasses** and the **Casting**, refer to Figure 10-29.

As you select the faces, the **Oriented** option from the **Constraint Type** drop-down list will get selected.

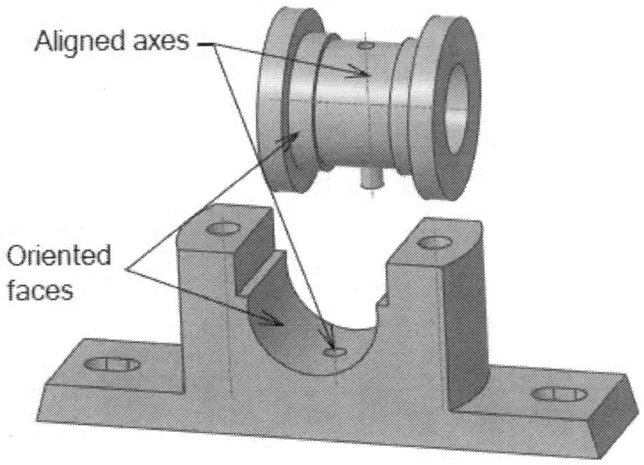

Figure 10-29 *References of the constraints used for assembling the components*

6. Choose the **OK** button from the **Component Placement** dashboard; the assembly model similar to the one shown in Figure 10-30 is displayed in the drawing area.

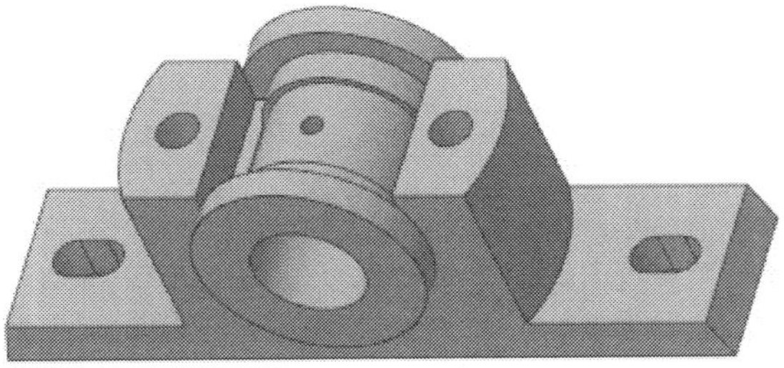

Figure 10-30 *Assembling the* ***Brasses*** *with the assembly*

7. Now, unsuppress the **Cap** by choosing **Resume > Resume All** from the **Operations** group. Similarly, assemble the remaining components. The final assembly is shown in Figure 10-31. You can choose the **Repeat** option to copy the repeated items.

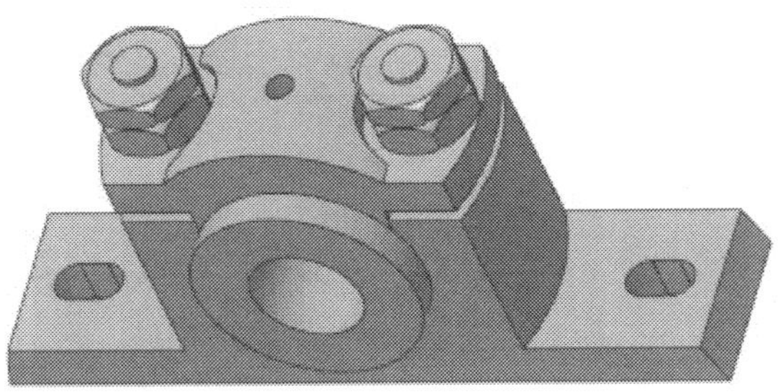

Figure 10-31 *The final Pedestal Bearing assembly*

Creating the Exploded State of the Assembly

To view all components in an assembly clearly, you need to create its exploded state.

1. Choose the **Edit Position** button from the **Model Display** tab; the **Explode Tool** dashboard and the default exploded state of the assembly are displayed.

   Now, you need to edit the position of the component in the exploded view.

2. Choose the **References** tab from the dashboard and then click in the **Movement Reference** collector; you are prompted to select a reference to define the direction of movement.

3. Select the vertical axis of the **Cap** from the drawing area; you are prompted to select the component to be moved.

4. Select the **Casting** from the drawing area; a direction handle is displayed. Move the cursor to the required direction handle and drag to place it at the required position. Similarly, place the **Cap**, two **Square headed bolts**, two **Lock nuts** and then two **Nuts** at the required location.

   Now, you need to move the **Brasses** to the required location in the drawing area.

5. Click in the **Movement Reference** collector again; you are prompted to select the reference. Select the axis of the **Brasses** from the drawing area as the reference; you are now prompted to select the component to be moved.

6. Select and move the **Brasses** to the required location in the drawing area.

   The model similar to the one shown in Figure 10-32 is displayed in the drawing area.

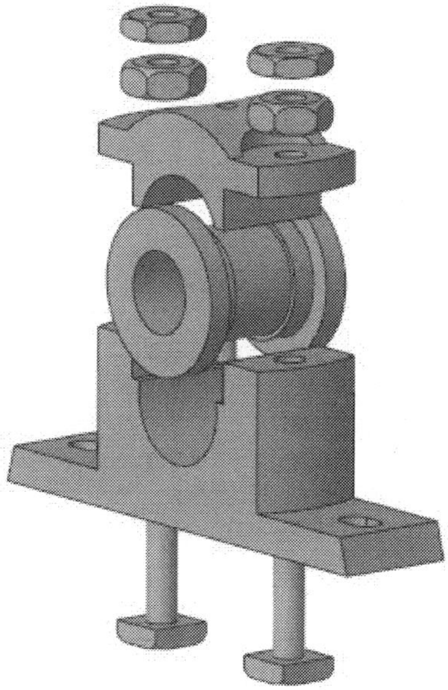

Figure 10-32 *The Exploded Pedestal Bearing assembly*

7. Now, you need to save this exploded view. To save this view, invoke the **View Manager** dialog box and choose the **Explode** tab.

8. Choose the **New** button; the **Modified State Save** message box is displayed confirming to save the default exploded view. Choose the **Yes** button from the message box; the **Save Display Elements** dialog box is displayed.

9. Next, enter the **EXP1** in the **Explode** edit box and choose the **OK** button from the dialog box and close the **View Manager** dialog box.

Creating the Offset Lines

1. Choose the **Edit Position** button from the **Model Display** tab and choose the **Create cosmetic offset lines to illustrate movement of exploded component** button from the **Explode Tool** dashboard; the **Cosmetic Offset Line** dialog box is displayed.

2. Click in the **Reference 1** collector and select the axis of the right hole on the **Casting**.

3. Next, click in the **Reference 2** collector and then select the axis of the **Square Headed bolt** on the right half of the **Casting** to display the explode line. Similarly, select the axes from the other components to create the offset lines between them.

   The model similar to the one shown in Figure 10-33 is displayed in the drawing area.

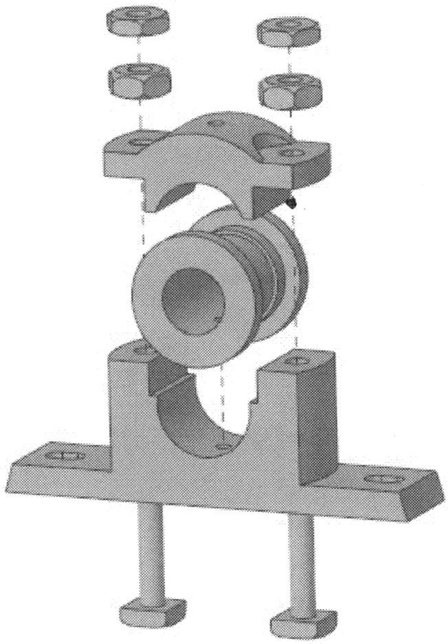

Figure 10-33 *Exploded state of the assembly displaying the offset lines*

4. Close the **Cosmetic Offset Line** dialog box and choose the **OK** button.

Saving the Assembly

1. Choose the **Save** button from the **File** menu and then save the assembly.

Closing the Window

Now, you have saved the assembly so the window can be closed.

1. Choose the **Close** button from the **Quick Access** toolbar to close the file.

EXERCISE

Exercise 1

In this exercise, you will create all components of the Crosshead assembly and then assemble them, as shown in Figure 10-34. Also, you will create an exploded state of the assembly, shown in Figure 10-35, which displays the offset lines. The dimensions of the components are shown in Figures 10-36 through 10-41. **(Expected time: 2 hrs)**

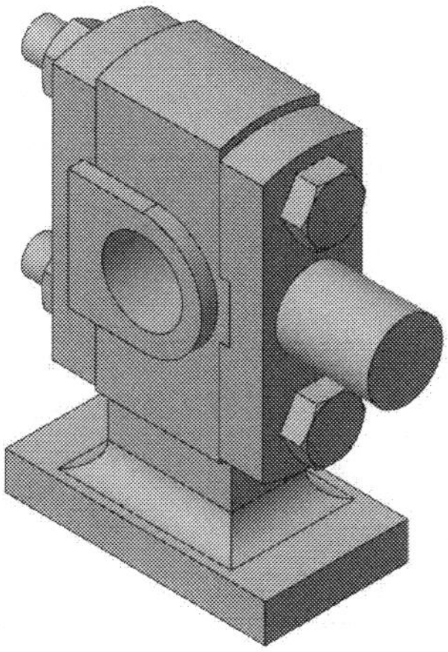

Figure 10-34 *The Crosshead assembly*

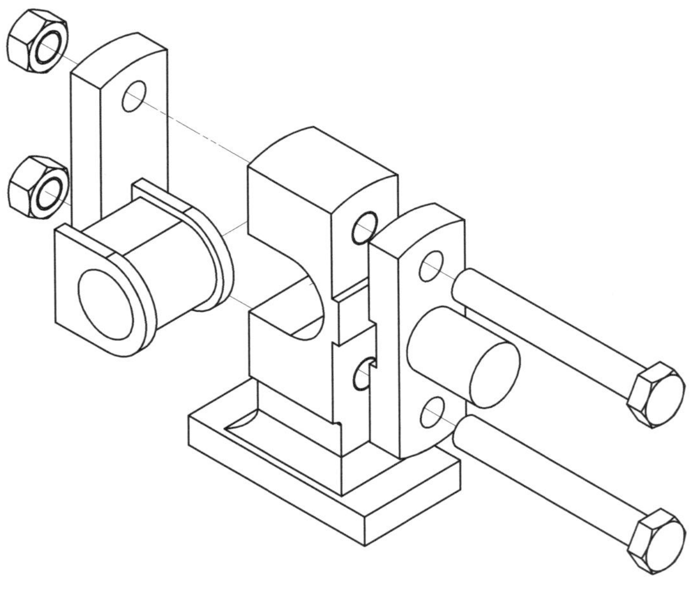

Figure 10-35 *The exploded state of the Crosshead assembly*

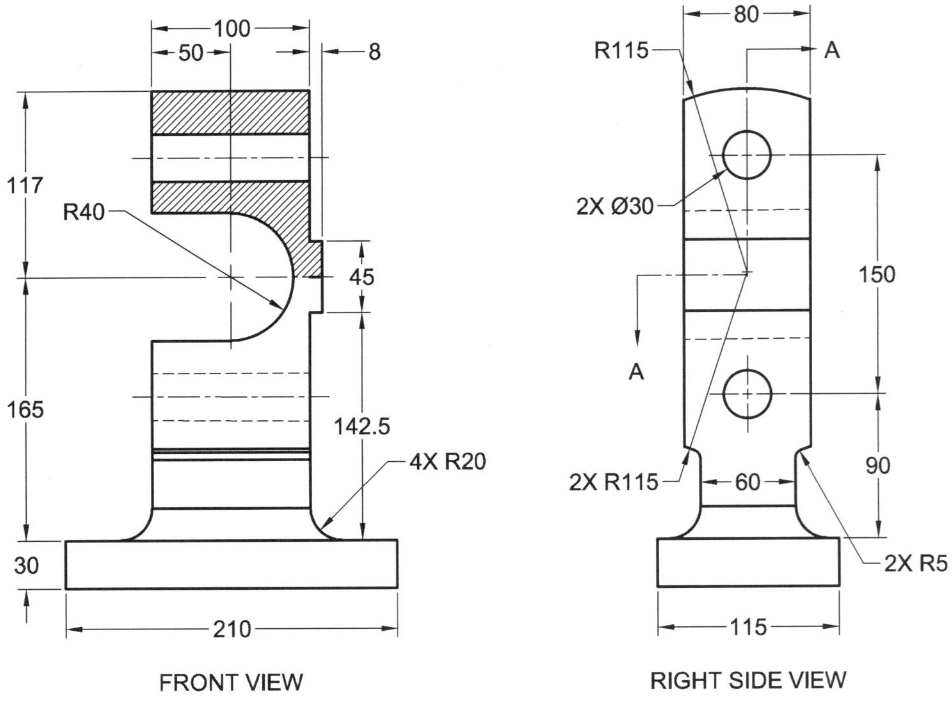

Figure 10-36 *Front view and right-side view of the Body*

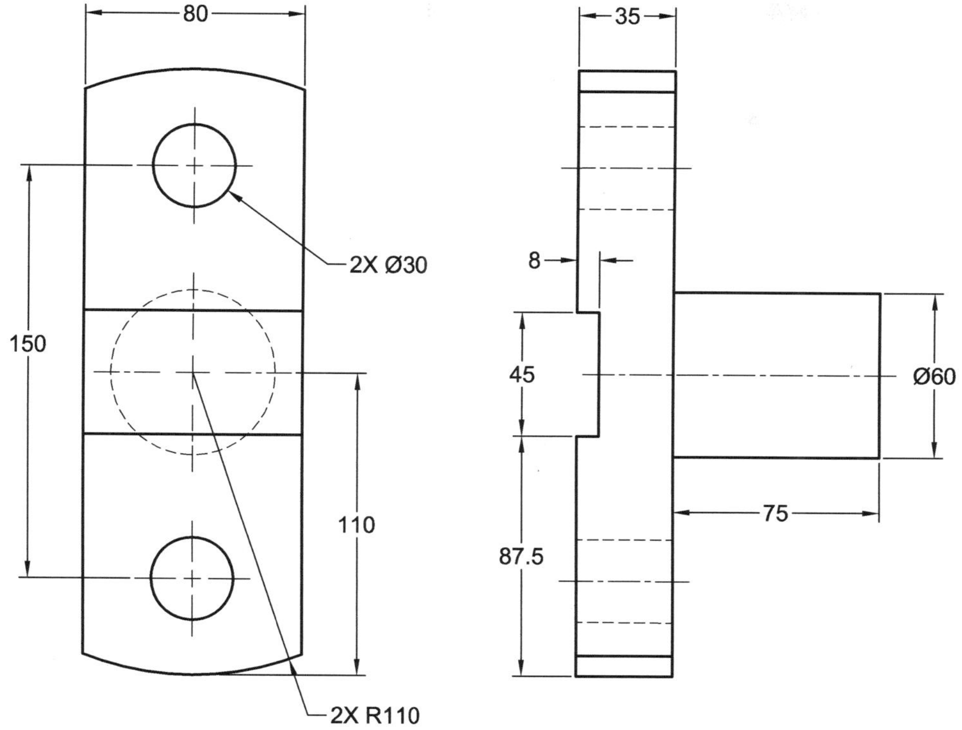

Figure 10-37 *Dimensions of the Piston Rod*

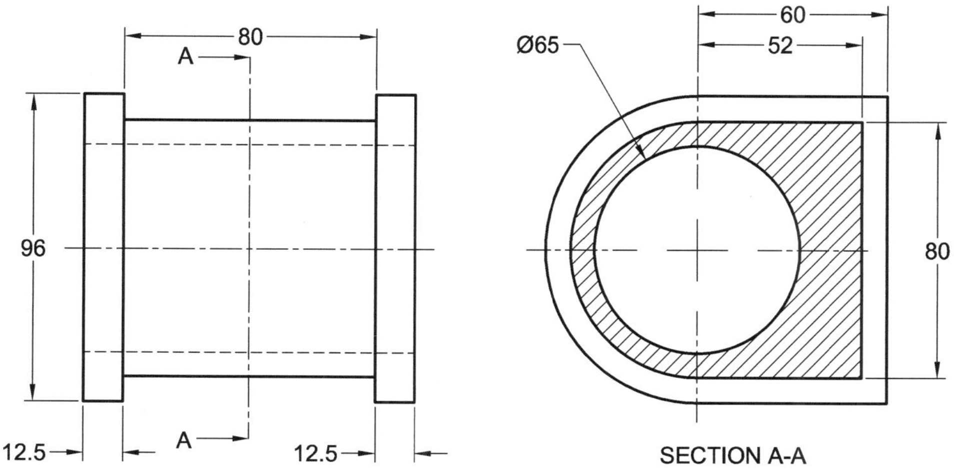

Figure 10-38 *Dimensions of the Brass*

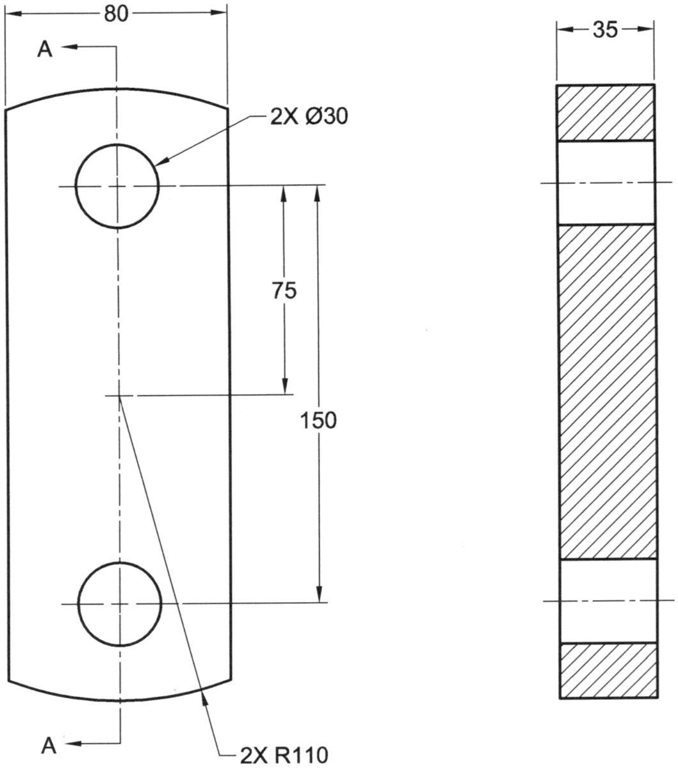

Figure 10-39 *Dimensions of the Keep Plate*

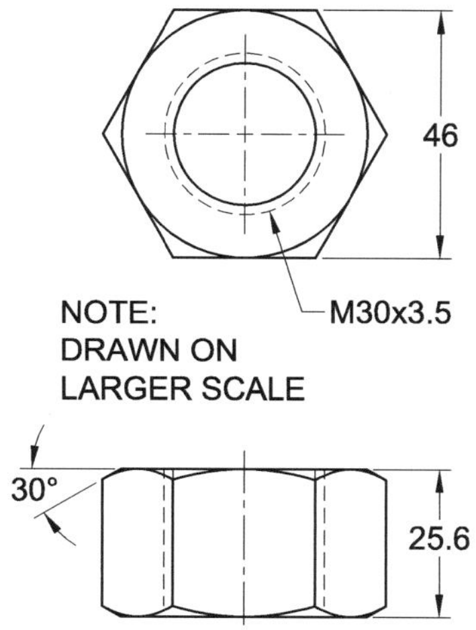

Figure 10-40 *Dimensions of the Nut*

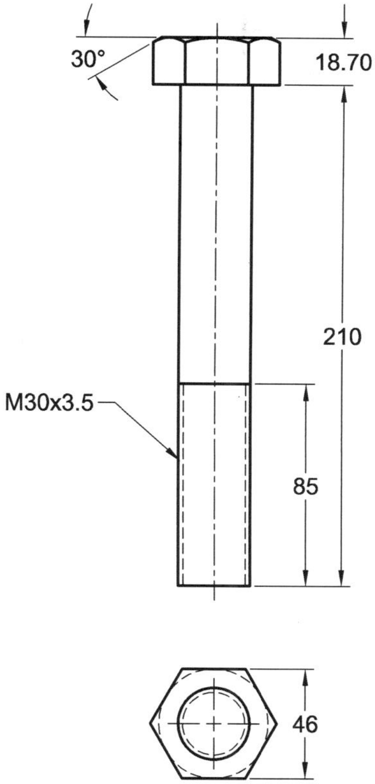

Figure 10-41 *Dimensions of the Bolt*

This page is intentionally left blank

Chapter 11

Generating, Editing, and Modifying the Drawing Views

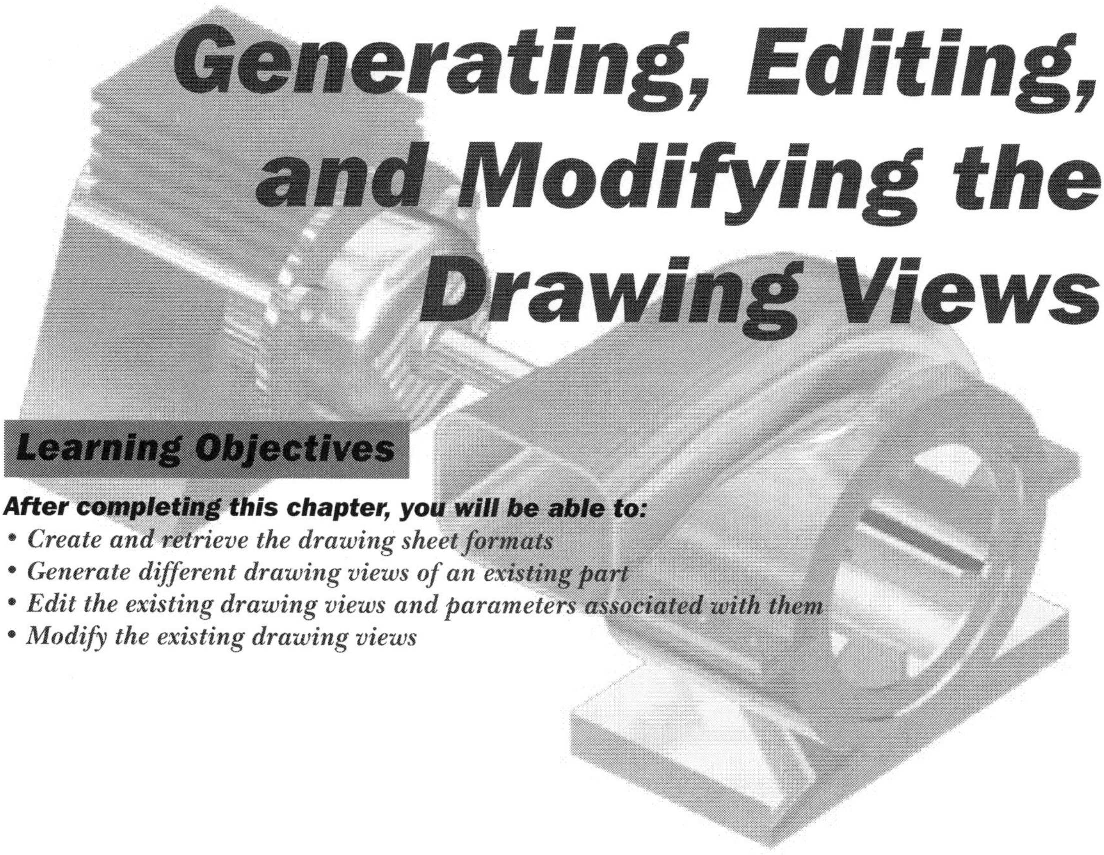

Learning Objectives

After completing this chapter, you will be able to:

- *Create and retrieve the drawing sheet formats*
- *Generate different drawing views of an existing part*
- *Edit the existing drawing views and parameters associated with them*
- *Modify the existing drawing views*

THE DRAWING MODE

Drawing views are generated in the **Drawing** mode. One of the major advantages of working with this software package is its bidirectional associative nature. This property ensures that any modifications made in the model in the **Part** mode are reflected in the drawing views of the model and vice versa. In Creo Parametric, there are two types of drafting methods: Interactive drafting and Generative drafting. In this chapter, you will learn Generative drafting.

GENERATING DRAWING VIEWS

In Creo Parametric, the first view that you need to generate is the general view. This view mostly acts as the parent view for the remaining views.

Generating the General View

Ribbon: Layout > Model Views > General View

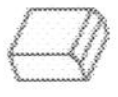
General View

The General view is the first view that is generated on the drawing sheet. This view can be the top, front, right, left, bottom, back, trimetric, isometric view, or any user-defined view of the model.

EDITING THE DRAWING VIEWS

In Creo Parametric, you can edit a drawing view as well as the items in the drawing views. All options of modifying the drawing views or items related to it can be chosen from the shortcut menu that will be displayed when you right-click and hold the right mouse button after selecting the view or the item. However, these options can also be chosen from the groups available in the **Ribbon**. Creo Parametric allows you to perform the following types of editing operations on the drawing views.

TUTORIALS

Tutorial 1

In this tutorial, you will generate the drawing views of the part created in Tutorial 1 of Chapter 9. The part is shown in Figure 11-1. The drawing views that need to be generated are shown in Figure 11-2. **(Expected time: 45 min)**

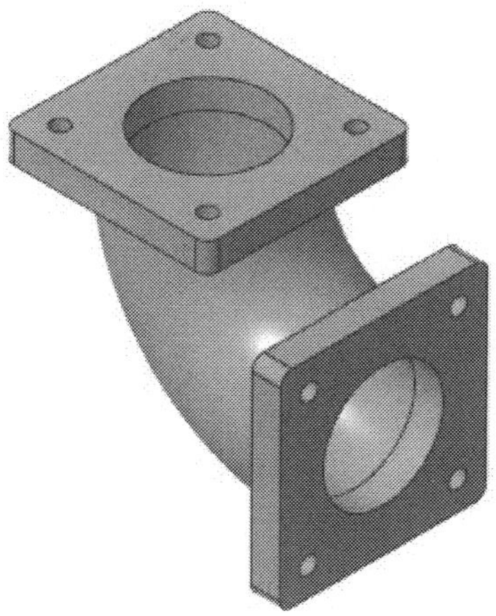

Figure 11-1 *Model for generating drawing views*

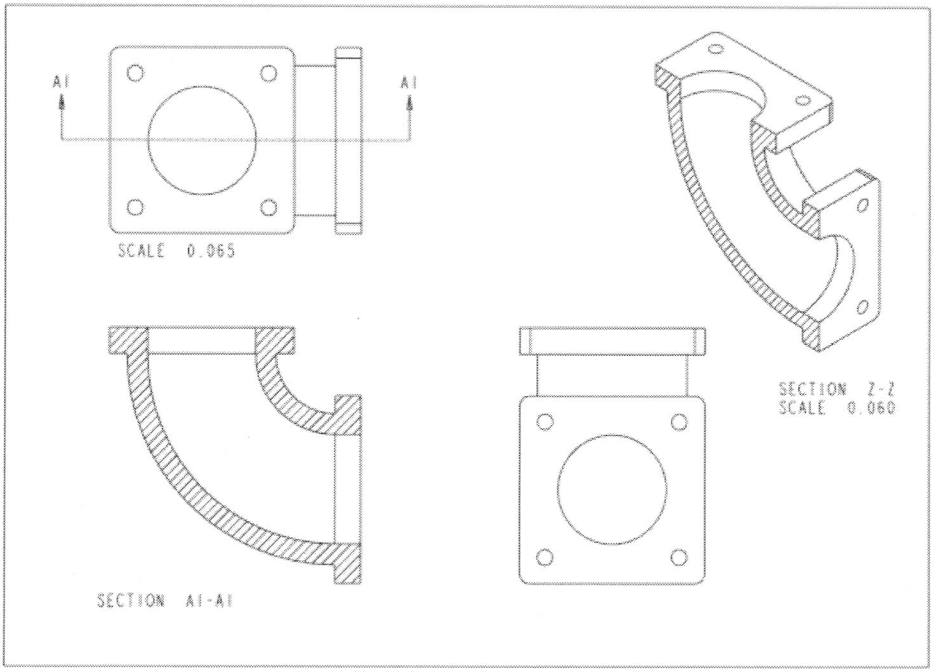

Figure 11-2 *Drawing views to be generated*

Before you start generating drawing views, copy the file *c09tut1.prt* from the *c09* folder in the current directory.

The following steps are required to complete this tutorial:

a. Start a new drawing file and select the size of the drawing sheet.
b. Generate the top view, refer to Figure 11-3.

c. Generate the sectioned front view by defining the **FRONT** datum plane as the section plane, refer to Figure 11-3.
d. Generate the right view of the sectioned front view, refer to Figure 11-4.
e. Generate the isometric sectioned view. The section will be defined by drawing a line on the **TOP** datum plane, refer to Figures 11-5 and 11-6.

Before you start generating the drawing views, set the working directory to *C:\Creo5.0\C11*. The part *(.prt)* file and the drawing *(.drw)* file should lie in the same directory or folder.

Starting a New Drawing File

To generate the drawing views, you first need to start a new drawing file.

1. Choose the **New** button from the **Data** group in the **Ribbon** to display the **New** dialog box.

2. Select the **Drawing** radio button and then enter **C11tut1** as the name of the file.

3. Choose the **OK** button from the **New** dialog box to display the **New Drawing** dialog box.

4. Choose the **Browse** button to select *C09tut1.prt* from the *C09* folder for generating drawing views and then choose the **Open** button.

5. Select the **Empty** radio button from the **Specify Template** area.

6. Choose the **Landscape** button from the **Orientation** area.

7. Select **A4** from the **Standard Size** drop-down list in the **Size** area. Choose the **OK** button from the **New Drawing** dialog box to proceed to the **Drawing** mode.

Generating the Top View

First, you need to generate the top view. All other views, except the sectioned isometric view, will be the child views of the top view. You need to generate the top view first because the required sectioned front view can be generated only from the top view. The right view can also be generated independently, but then this view will not help generate any other required view.

1. Choose the **General View** tool from the **Model Views** group in the **Ribbon**; the **Select Combined State** dialog box is displayed. Choose the **OK** button from this dialog box; you are prompted to specify a center point for the placement of the view.

2. Specify the center point for the placement of the top view close to the upper left corner of the drawing sheet; the **Drawing View** dialog is displayed.

3. Choose the **TOP** option from the **Model view names** list box and then choose the **Apply** button.

4. Choose the **Scale** option from the **Categories** list box. Next, choose the **Custom scale** radio button and enter **0.065** in the edit box.

5. Choose the **OK** button to exit the **Drawing View** dialog box.

   If necessary, move the view, refer to Figure 11-3, and then choose the **No Hidden** option from the **Display Style** drop-down in the **Graphics** toolbar. You may also need to repaint the screen.

Generating the Sectioned Front View

The sectioned front view of the model is generated from the top view. Before proceeding use the **Plane Display** button from the **Datum Display Filters** drop-down in the **Graphics** toolbar to turn on the display of datum planes and repaint the screen.

1. Choose the **Projection View** tool from the **Model Views** group of the **Layout** tab in the **Ribbon**; you are prompted to specify a center point for placement of the projection view.
2. Specify the center point for the placement of the front view below the top view, as shown in Figure 11-3.
3. Select the newly generated view and invoke the shortcut menu. Choose the **Properties** option to display the **Drawing View** dialog box.
4. Choose the **Sections** option from the **Categories** list box to display the section related options in the dialog box.
5. Select the **2D cross-section** radio button and choose the **Add cross-section to view** button; the window area below it gets activated. Select the **Create New** option from the **Name** drop-down list, if it is not selected; the **XSEC CREATE** menu is displayed.
6. Choose **Planar > Single > Done** from the **XSEC CREATE** menu.
7. Enter **A1** as the name of the cross-section in the message input window and then press ENTER to display the **SETUP PLANE** submenu; you are prompted to select a planar surface or a datum plane.
8. Select the **FRONT** datum plane (the plane that cuts the part horizontally through the center of the cylindrical feature in the top view) from the drawing area.
9. Scroll the bar in the **Drawing View** dialog box to the right. Click on the field below the **Arrow Display** column; you will be prompted to pick a view for arrows where the section is perpendicular. Select the top view to display the arrows.
10. Choose the **Apply** button and then exit the dialog box.

Modifying the Hatching

The offset distance between the hatching lines in the front sectioned view is large. Therefore you need to reduce the distance between the hatching lines.

1. Select the **X-Section** filter from the **Filter** drop-down list in the Status Bar. Select the hatching lines from the sectioned front view in the drawing sheet; the hatching lines turn

green. Hold the right mouse button to invoke the shortcut menu. Choose the **Properties** option from the shortcut menu; the **MOD XHATCH** menu is displayed.

2. In the **MOD XHATCH** menu, choose the **Spacing** option to display the **MODIFY MODE** submenu.

   The spacing between the hatching lines has to be reduced.

3. Choose the **Half** option twice and then choose the **Done** option in the **MOD XHATCH** menu. Now, the hatching appears more dense.

4. Click once in the drawing area to remove the X-hatch from the current selection set. The sheet after placing these two views should look similar to the one shown in Figure 11-3.

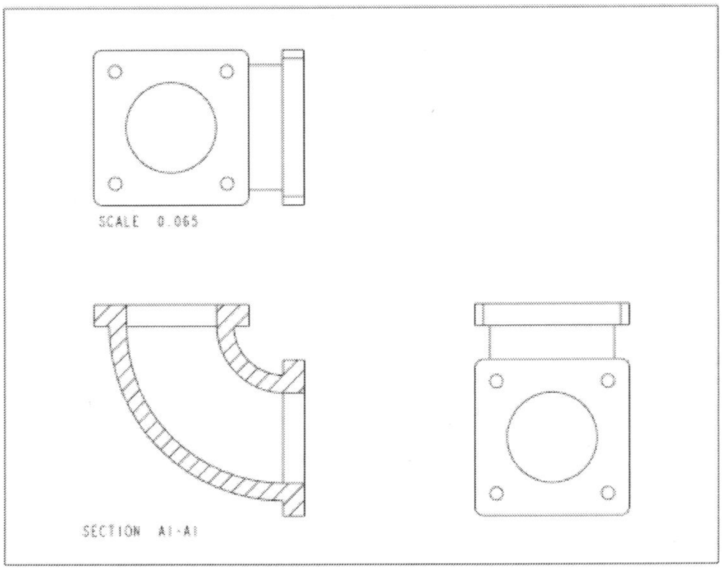

Figure 11-3 *Drawing sheet after generating the top view and the sectioned front view with the display of datum planes turned on*

5. Set the selection filter back to **General**.

Generating the Right View

The right view is the projection of the sectioned front view. Before proceeding further, turn off the display of the datum planes.

1. Choose **Layout > Model Views > Projection View** from the **Ribbon**; you are prompted to select the projection parent view.

2. Select the sectioned front view as the parent view.

3. Specify the center point for the placement of the drawing view on the right of the sectioned front view. The right view of the model is placed, as shown in Figure 11-4.

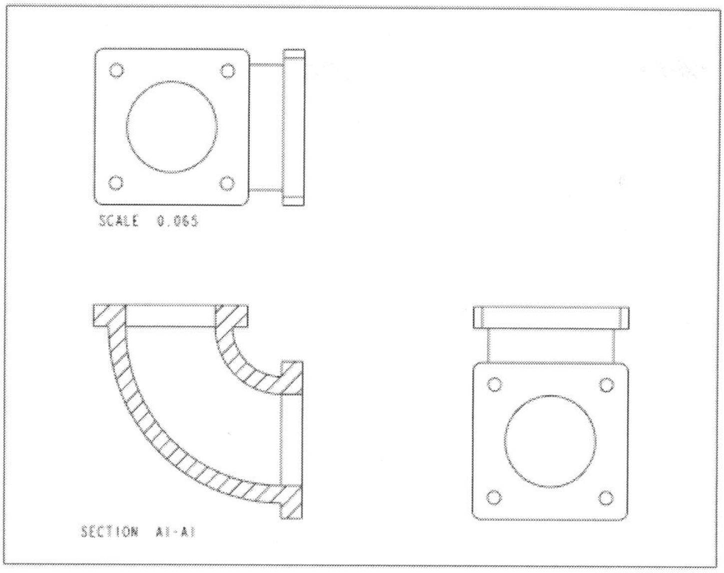

Figure 11-4 The drawing sheet after generating the top, sectioned front, and right views

Generating the Isometric Section View

The isometric section view is an independent view and it will be generated by using the **General** option.

1. Choose the **General View** tool from the **Model Views** group in the **Ribbon**; the **Select Combined State** dialog box is displayed. Choose the **OK** button from this dialog box; you are prompted to specify a center point for the placement of the view.

2. Specify the center point on the upper right corner of the drawing sheet for placing the view; the **Drawing View** dialog is displayed.

   The default view is a trimetric view but you need to display the isometric view. Therefore, you need to change the orientation.

3. Choose the **Isometric** option from the **Default orientation** drop-down list; the isometric view of the model is displayed on the drawing sheet.

4. Choose the **Scale** option from the **Categories** list box. Select the **Custom scale** radio button and enter **0.05** as the scale factor in the edit box displayed. Next, choose the **Apply** button.

5. Choose the **Sections** option from the **Categories** list box to display the section options in the dialog box.

6. Select the **2D cross-section** radio button and then choose the **Add cross-section to view** button; the collector below it gets activated. Select **Create New** from the name drop-down list; the **XSEC CREATE** menu manager is displayed.

7. Choose the **Offset > Both Sides > Single > Done** from the **XSEC CREATE** menu manager; the message input window is displayed.

8. Enter **Z** as the name of the section in the message input window and press ENTER; a separate window is displayed with the model.

9. Select the **TOP** datum plane from the subwindow. You may need to turn on the display of the datum planes. Choose the **Okay** option from the **DIRECTION** submenu; the **SKET VIEW** submenu is displayed.

10. Choose the **Right** option from the **SKET VIEW** submenu and select the **RIGHT** datum plane from the subwindow.

11. Choose **Sketch > Line > Line** from the menu bar and draw a line, as shown in Figure 11-5. This line creates a section plane.

12. Choose **Sketch > Constraint > Coincident** from the menu bar and align both ends of the line to the edges.

13. Choose **Sketch > Done** from the menu bar.

14. Choose the **Apply** button to place the view and then exit the dialog box.

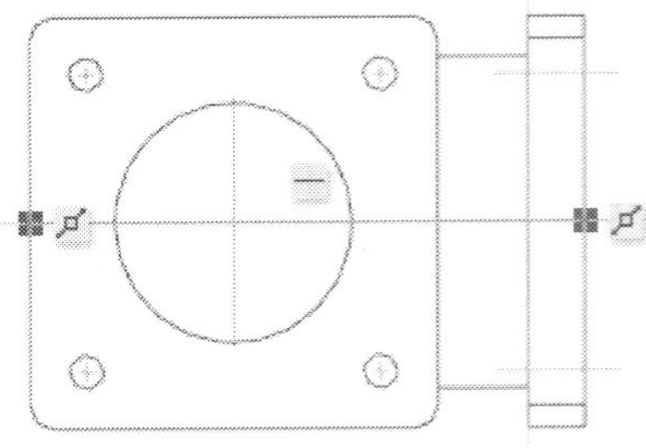

Figure 11-5 *The sketch for the section line with the constraint*

Note

If the drawing view placed on the sheet overlaps the boundary of the drawing sheet, then you can move the drawing view. To do so, select the drawing view; the selected drawing view is enclosed in a green box. Now, press the left mouse button inside the box and drag it to place it at the desired location.

Tip

*You can lock or unlock a drawing view using the **Lock View Movement** button from the **Document** group of the **Layout** tab.*

Modifying the Hatching

The spacing between the hatching lines in the sectioned isometric view is large. Therefore, you need to reduce the distance between them. Use the same procedure that was discussed earlier in this tutorial to modify the spacing between the hatch lines. The sheet after placing all views should look similar to the one shown in Figure 11-6.

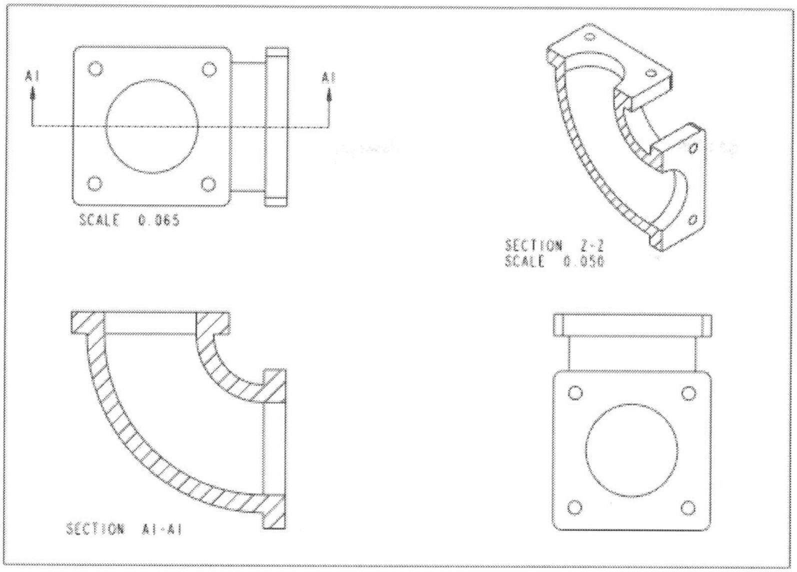

Figure 11-6 *Different drawing views generated*

Saving the Drawing File

In this section, you need to save the drawing file.

1. Choose the **Save** button from the **Quick Access** toolbar; the **Save Object** dialog box is displayed with the name of the drawing file that you have entered earlier.

2. Press ENTER to save the file.

Closing the Drawing File

After you have saved the drawing file, close the drawing file.

1. Choose **File > Close** from the menu bar; the drawing window is closed.

EXERCISE

Exercise 1

Generate the drawing views of the model shown in Figure 11-7. The drawing views to be generated are shown in Figure 11-8. To perform this tutorial, you need to download the zipped file named as *c011_creo_5.0_input* from the **Input Files** section of the CADSOFT website *http://www.cadsofttech.com*. The complete path for downloading the file is:

Textbooks > CAD/CAM > Creo Parametric > Creo Parametric 5.0 for Novices > Input Files

(Expected time: 45 min)

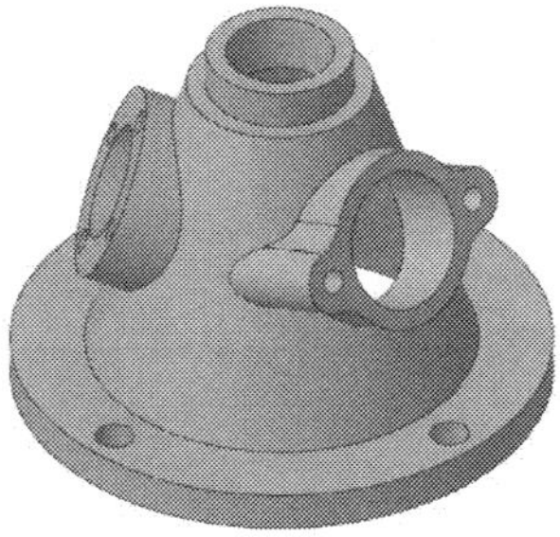

Figure 11-7 *Part for generating the drawing views*

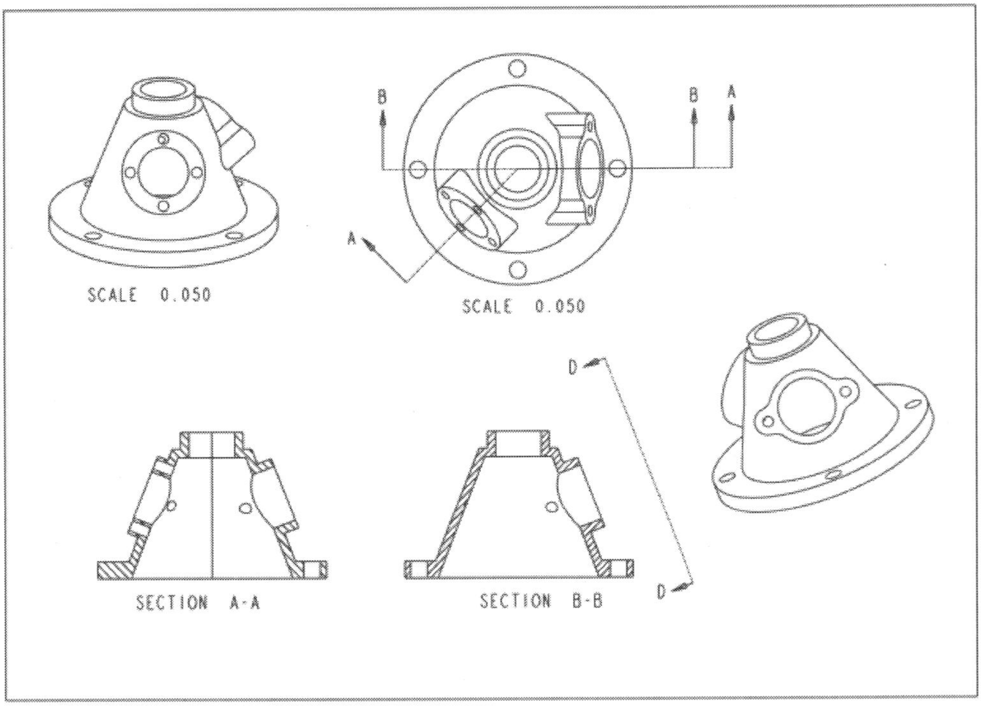

Figure 11-8 *Drawing views to be generated in Exercise 1*

Chapter 12

Dimensioning the Drawing Views

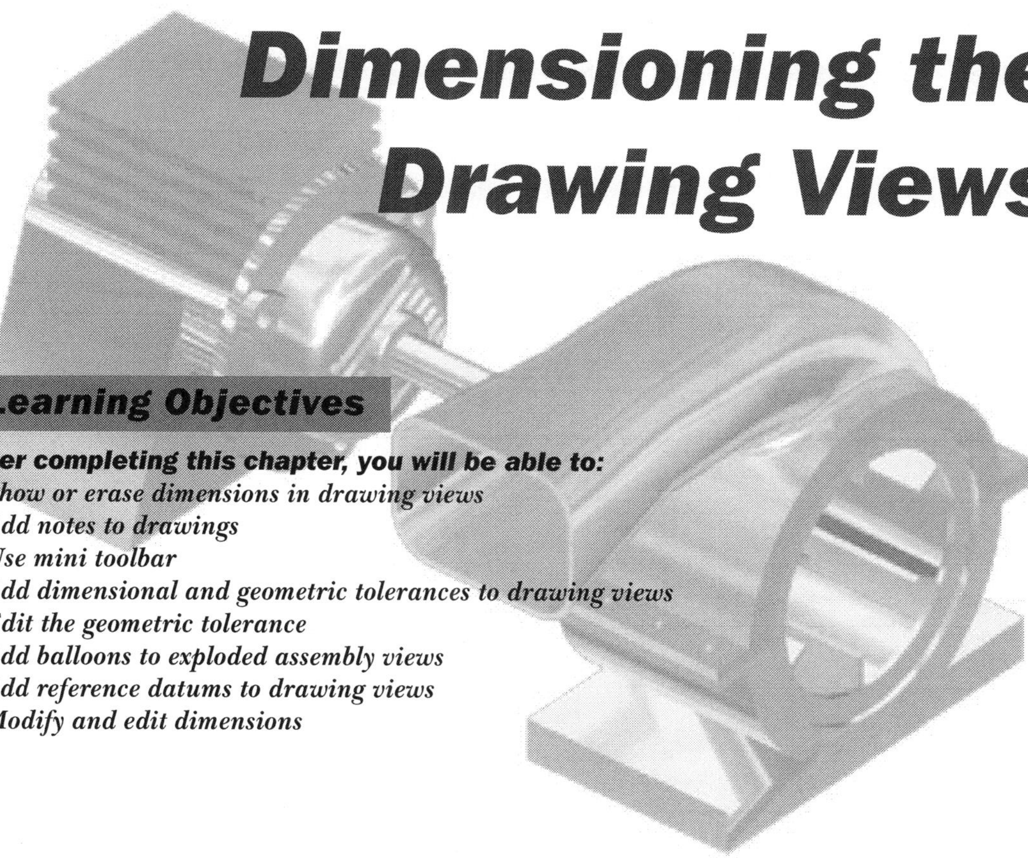

Learning Objectives

After completing this chapter, you will be able to:

- *Show or erase dimensions in drawing views*
- *Add notes to drawings*
- *Use mini toolbar*
- *Add dimensional and geometric tolerances to drawing views*
- *Edit the geometric tolerance*
- *Add balloons to exploded assembly views*
- *Add reference datums to drawing views*
- *Modify and edit dimensions*

DIMENSIONING THE DRAWING VIEWS

Once you have generated the drawing views, you need to generate dimensions, add notes, symbols, balloons, and so on in the drawing views. These dimensions are assigned to each entity of sketches in the model or the features associated with the model. These dimensions are associative in nature and they can be used to modify or drive the dimensions of a part; therefore these dimensions are called driving dimensions.

ADDING TOLERANCES TO THE DRAWING VIEWS

You can add geometric tolerances in the dimensions. Tolerance is defined as the difference between the maximum and minimum variations in the dimensions of the selected component. It is almost impossible to manufacture a component to the exact dimensions. In such cases, the tolerance value is added to dimensions to make sure that some variation that occurs during manufacturing can be taken care of. However, when you actually send a part for manufacturing, there are other parameters along with the dimension tolerances that may vary and require some tolerances. Depending upon these factors, the tolerances are divided into two types: dimensional tolerances and geometric tolerances.

ADDING DATUM FEATURE SYMBOL TO THE DRAWING VIEWS

To specify geometric tolerance on a geometry, you must specify datum references. In the drawing mode of Creo Parametric, the **Datum Feature Symbol** tool is used to specify the location of datum references.

TUTORIALS

Tutorial 1

In this tutorial, you will generate drawing views of the part shown in Figure 12-1. The views to be generated are shown in Figure 12-2. **(Expected time: 45 min)**

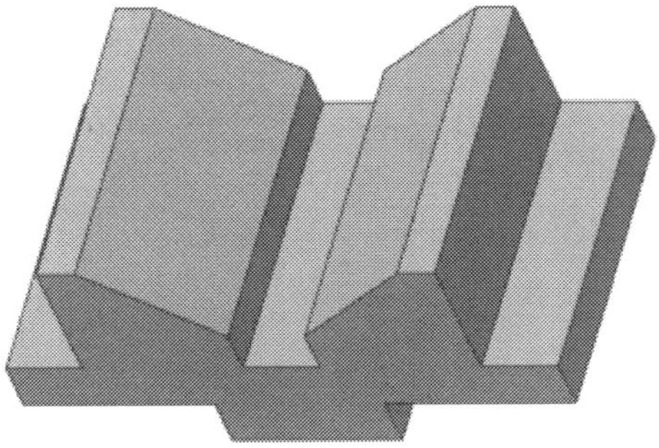

Figure 12-1 *Model for generating drawing views*

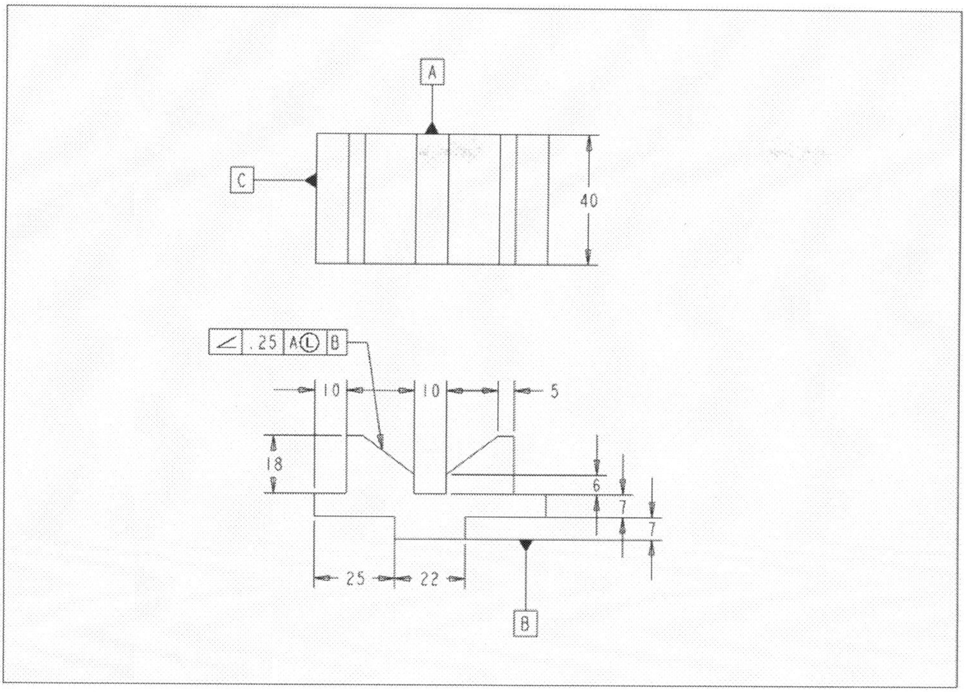

Figure 12-2 Drawing views to be created along with the dimensions

The following steps are required to complete this tutorial:

a. Create the model in the **Part** mode using the given dimensions and save the file.
b. Open a new drawing file in the **Drawing** mode and generate the top and front views of the model on the drawing sheet, refer to Figure 12-3.
c. Generate the dimensions in the views and clean them using the **Clean Dimensions** dialog box, refer to Figure 12-4.
d. Add geometric tolerance to the required entities, refer to Figure 12-5.

Before you start creating the model and the drawing views, set the working directory to *C:\Creo-5.0\C12*. Make sure that the *C12* folder exists inside the *Creo-5.0* folder.

Creating the Model

First, create the given model in the **Part** mode. Its views will be generated later.

1. Create a new **Part** file and name it as *C12tut1*.

2. Create the model in the **Part** mode and save this model.

Starting a New Drawing File

To generate drawing views, you need to start a new drawing file.

1. Choose the **New** button from the **File** menu; the **New** dialog box is displayed.

2. Select the **Drawing** radio button from the **Type** area and enter the name of the drawing as *C12tut1* in the **File name** edit box.

3. Choose the **OK** button from the **New** dialog box; the **New Drawing** dialog box is displayed.

4. If the name of the model is not displayed in the **Default Model** area, choose the **Browse** button and select the model. Select the **Empty** radio button from the **Specify Template** area, the **Landscape** button from the **Orientation** area, and the **A4** option from the **Standard Size** drop-down list.

5. Choose the **OK** button from the **New Drawing** dialog box to enter the **Drawing** mode. The **Drawing** mode is invoked and a new A4 size drawing sheet is displayed.

Generating the Drawing Views

You need to generate the top and front views of the model.

1. Choose the **General View** tool from the **Model Views** group of the **Layout** tab in the **Ribbon**; the **Select Combined State** dialog box is displayed. Choose the **OK** button from the dialog box.

2. Now, click in the sheet to specify the center point for the drawing view, refer to Figure 12-2. When you specify the point, the **Drawing View** dialog box is displayed.

3. Choose the **TOP** option from the **Model view names** list in the **View orientation** area.

4. Choose the **Scale** option from the **Categories** area and edit the scale factor for the view to **0.06**.

5. Choose the **Apply** button and close the dialog box. The default view is oriented as the top view and scaled to the specified scale factor.

6. Choose the **No Hidden** option from the **Graphics** toolbar and repaint the screen using the **Repaint** tool from the **Graphics** toolbar. The top view is shown in Figure 12-3.

7. Choose the **Projection view** tool from the **Model Views** group of the **Layout** tab in the **Ribbon**. Make sure that the angle of projection is in the third angle projection, else you will not get the desired result.

8. Specify the center point for the drawing view by clicking below the top view, as shown in Figure 12-3.

   The sheet after generating the two views is shown in Figure 12-3.

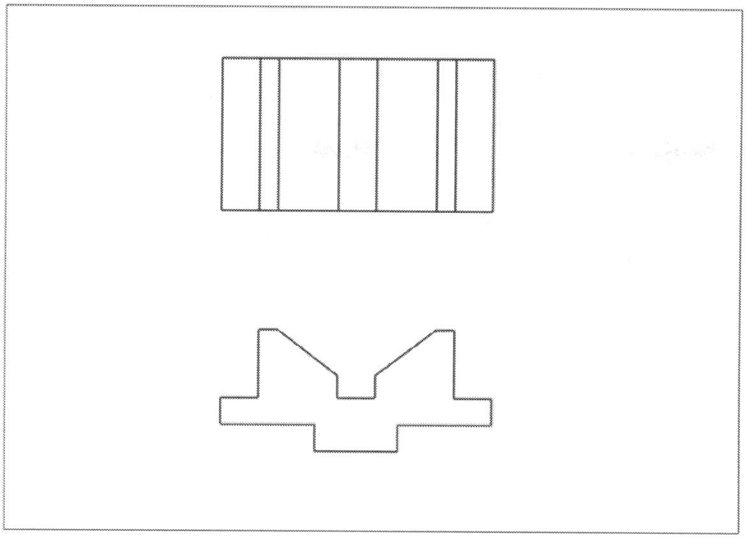

Figure 12-3 Top and front views of the model

Placing Datum Feature Symbol

1. Choose the **Datum Feature Symbol** tool from the **Annotations** group of the **Annotate** tab; the datum feature symbol gets attached to the cursor, refer to Figure 12-4.

2. Click on the top edge of the top view and drag the cursor to the desired point and press the middle mouse button to place the symbol in the drawing. The datum feature symbol is automatically named as **A** because this is the primary datum feature of the model.

3. Repeat steps 1 and 2 to place secondary and tertiary datum feature symbol B and C, respectively.

Dimensioning the Drawing Views

1. Choose **Annotate > Annotations > Show Model Annotations** from the **Ribbon**; the **Show Model Annotations** dialog box is displayed with the **Show the model dimensions** tab chosen by default. Also, you are prompted to select a view to display its annotations.

2. Select the front view (the view at the bottom); all dimensions of this view are listed in the dialog box.

3. Choose the **Select All** button available below the list box; all dimensions are displayed as strong dimensions in the selected view.

4. Choose the **Apply** button to accept and continue the settings.

5. Now, select the top view from the drawing area; its dimension is listed in the **Show Model Annotations** dialog box.

6. Select the check box on the left of the dimensions listed; the dimension of the selected item is displayed in the drawing area as a strong dimension.

7. Choose the **OK** button from the dialog box to accept the settings and exit the dialog box.

Placing the Dimensions in Order

The dimensions displayed in the drawing views are scattered and improperly placed. You need to place the dimensions in proper order using the **Clean Dimensions** dialog box.

1. Choose **Annotate > Annotations > Cleanup Dimensions** from the **Ribbon** to display the **Clean Dimensions** dialog box; you are prompted to select the dimensions to be cleaned.

2. Select all dimensions from both the views by using the window selection method; the selected dimensions are highlighted. Press the middle mouse button to activate the options in the **Placement** tab.

3. Enter **0.5** in the **Offset** and **Increment** edit boxes.

4. Clear the **Create Snap Lines** check box. Choose the **Apply** button and then choose the **Close** button to exit the **Clean Dimensions** dialog box.

5. After cleaning the dimensions, the drawing views should look similar to the ones shown in Figure 12-4. Select and move the dimensions if necessary. You can also increase the text size of the dimensions if required.

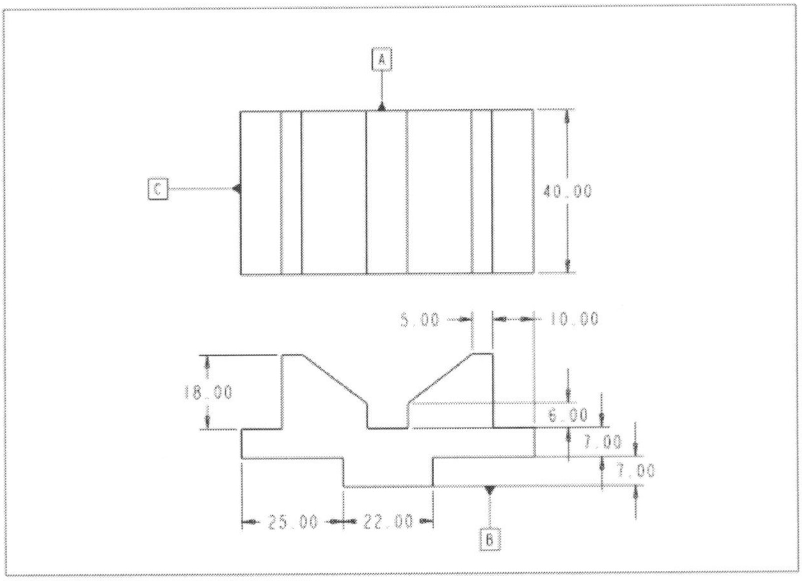

***Figure 12-4** Top and front views of the model with dimensions and datum planes*

Displaying the Geometric Tolerance

1. Choose **Annotate > Annotations > Geometric Tolerance** from the **Ribbon**; the gtol frame gets attached to the cursor and you are prompted to select edge entities, dimensions, gtols, notes, dimension, witness lines, coordinate systems axis, center axis endpoints.

2. Select the left-inclined edge from the front view and then click the middle mouse button at the desired place in the drawing area to place the gtol frame. As you place the gtol in the drawing view; the **Geometric Tolerance** dashboard gets activated.

3. Choose the **Angularity** option from the **Geometric Characteristics** drop-down list and type **0.25** in the Tolerance value edit box.

4. Specify the primary datum reference for the gtol by choosing the **Select the datum reference from the model** button from the **Tolerance & Datum** group; the **Select** dialog box is displayed. Select the **A** datum feature symbol from the drawing area and choose the **OK** button from the **Select** dialog box; the name of the datum feature symbol is displayed in the gtol frame.

5. Click in the edit box of the primary datum reference and choose the **Symbols** tool to invoke the symbols gallery. Select the least material condition symbol from the gallery; the symbol is inserted in the gtol frame.

6. Similarly select the secondary datum reference using the button available adjacent to the Secondary datum reference edit box or you can type the name of secondary reference in the edit box. Remember that the name of the datum reference is always in capital letters.

   The drawing views after adding the geometric tolerances should look similar to the ones shown in Figure 12-5. You need to move the note for the geometric tolerances to a location shown in Figure 12-6 by selecting and dragging it. Similarly, arrange all dimensions and datums on the drawing sheet, as shown in Figure 12-6.

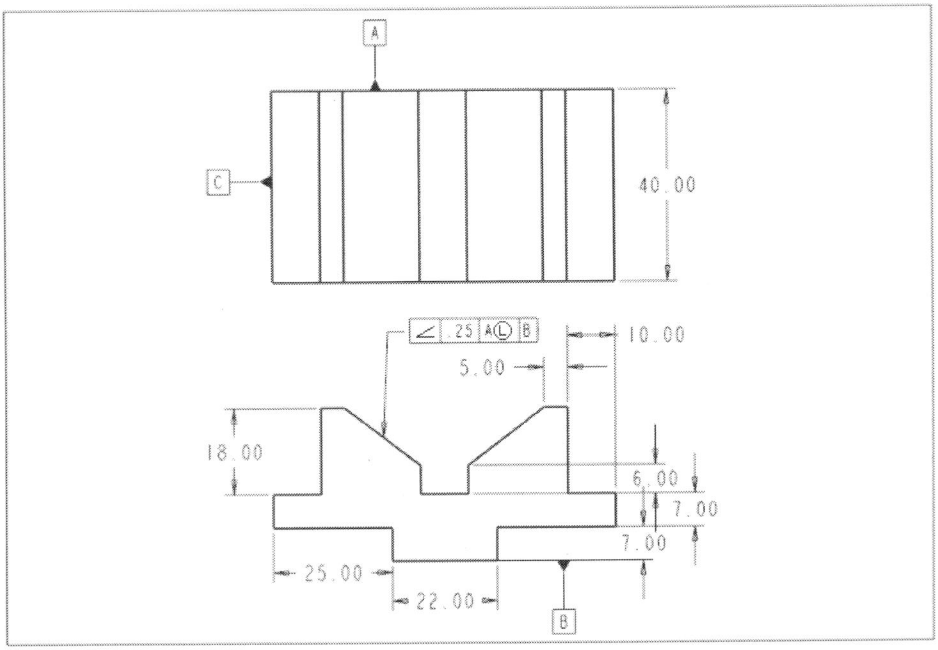

Figure 12-5 Top and front views of the model with dimensions, datum planes, and geometric tolerances

Saving the Drawing File

Now, you need to save the drawing file that you have created.

1. Choose the **Save** button from the **Quick Access** toolbar; the **Save Object** dialog box is displayed with the name of the drawing file that you had entered earlier. Press ENTER to confirm the saving of the file.

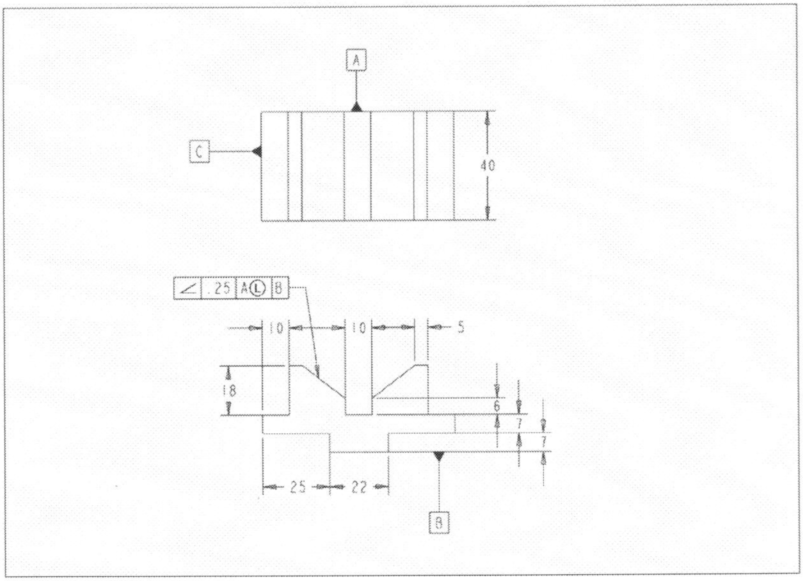

Figure 12-6 The required drawing views after arranging all dimensions and datums

Tutorial 2

In this tutorial, you will generate the drawing views of the model shown in Figure 12-7. The dimensioned drawing views are shown in Figure 12-8. **(Expected time: 45 min)**

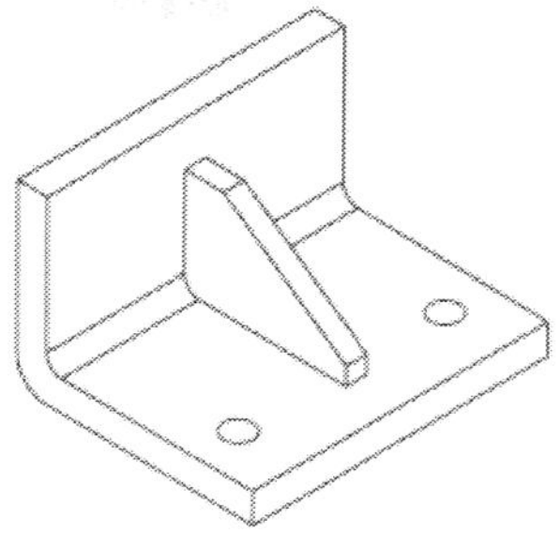

Figure 12-7 *Model for generating drawing views*

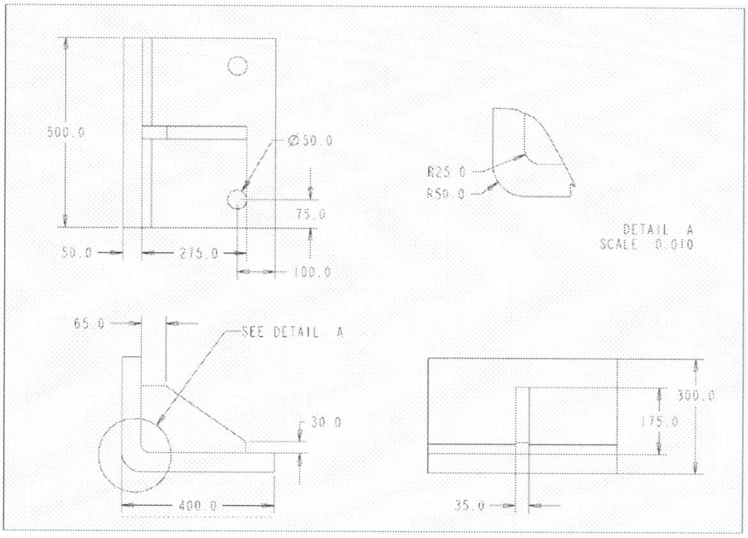

Figure 12-8 *Drawing views to be generated in Tutorial 2*

The following steps are required to complete this tutorial:

a. Create the model in the **Part** mode using the given dimensions and save the file.
b. Open a new drawing file in the **Drawing** mode and generate the top view, front, right, and detailed views of the model on the drawing sheet, refer to Figure 12-9.
c. Dimension the views and clean the dimensions using the **Clean Dimensions** dialog box, refer to Figure 12-10.

Make sure the working directory is set to *C:\Creo-5.0\C12*.

Creating the Model

First, in the **Part** mode, you need to create the model whose drawing views are to be generated.

1. Create the model in the **Part** mode using the given dimensions.

2. Save this model created in the **Part** mode with the name *C12tut2*.

Starting a New Drawing File

To generate the drawing views, you first need to start a new drawing file.

1. Choose the **New** button from the **File** menu; the **New** dialog box is displayed.

2. Select the **Drawing** radio button from the **Type** area and enter the name of the drawing as *C12tut2* in the **File name** edit box.

3. Choose the **OK** button from the **New** dialog box; the **New Drawing** dialog box is displayed. In this dialog box, the name of the model is displayed in the **Default Model** area.

4. If the name of the model is not displayed in the **Default Model** area, choose the **Browse** button and select the model.

5. Select the **Empty** radio button from the **Specify Template** area, choose the **Landscape** button from the **Orientation** area, and select the **A4** option from the **Standard Size** drop-down list.

6. Choose the **OK** button from the **New Drawing** dialog box; you enter the **Drawing** mode and a new drawing sheet of A4 size is displayed on the screen.

Creating the Drawing Views

First, the top view of the model needs to be generated and then the front view will be generated by projecting it from the top view. The right view will be projected from the front view. The detail view can be placed on the drawing sheet by specifying a scale for it.

1. Choose **Layout > Model Views > General View** from the **Ribbon**; the **Select Combined State** dialog box is displayed. Choose the **OK** button from the dialog box.

2. Now, click on the top-left side of the sheet to specify the center point for the drawing view; the **Drawing View** dialog box is displayed.

3. Choose the **TOP** option from the **Model view names** list in the **View orientation** area.

4. Choose the **Apply** button; the default view is oriented as the top view.

5. Choose the **Scale** option in the **Categories** display area and specify the scale factor for the view as **0.006** in the **Custom scale** edit box. Exit the dialog box and then choose the **No Hidden** option from the **Display Style** drop-down in the **Graphics** toolbar. Repaint the screen using the **Repaint** tool from the **Graphics** toolbar. For the top view, refer to Figure 12-9.

The front view of the model needs to be generated from the top view.

6. Choose **Layout > Model Views > Projection View** tool from the **Ribbon**. Next, specify the center point for the drawing view which is below the top view, refer to Figure 12-9. Make sure that the angle of projection is the third angle projection else the desired projection will not be displayed.

7. Similarly, place the right and detailed views on the drawing sheet, refer to Figure 12-9.

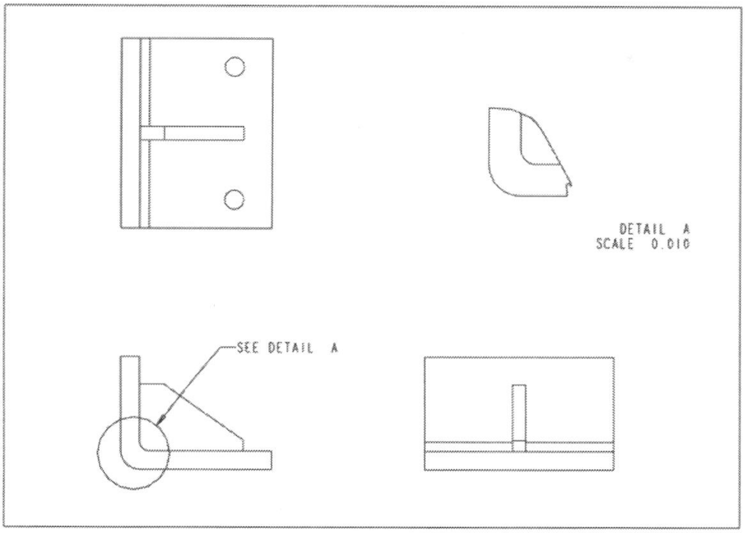

Figure 12-9 *Drawing views before adding dimensions*

Dimensioning the Drawing Views

1. Choose the **Show Model Annotations** tool from the **Annotations** group of the **Annotate** tab in the **Ribbon**; the **Show Model Annotations** dialog box is displayed with the **Show the model dimensions** tab chosen by default. Also, you are prompted to select a view to display its annotations.

2. Move the cursor toward the round feature in the detailed view and click when the round feature gets highlighted; the dimensions of the radius of round feature are listed in the dialog box.

3. Choose the dimensions of the round feature listed in the dialog box. All the radial dimensions are displayed in the selected view as strong dimensions.

4. Choose the **Apply** button to accept the settings and continue.

5. Now, select all the three views; all dimensions of the three views are listed in the dialog box.

6. Choose the **Select All** button; all dimensions are displayed on the view as strong dimensions.

7. Choose the **OK** button to accept the settings and exit the dialog box.

Placing the Dimensions in Order

The dimensions displayed in the drawing views are scattered and improperly placed. You need to use the **Clean Dimensions** dialog box to place the dimensions in order.

1. Choose the **Cleanup Dimensions** tool from the **Annotations** group of the **Annotate** tab in the **Ribbon** to display the **Clean Dimensions** dialog box; you are prompted to select the dimensions to clean.

2. Select all dimensions from the top and front views by dragging a rectangle around them. Press the middle mouse button so that the options in the **Clean Dimensions** dialog box are displayed.

   When you select the dimensions, the radius dimensions in the drawing views are not selected. This means that the radius dimensions will remain unaffected by the values that you will enter in the **Clean Dimensions** dialog box.

3. Enter **0.4** in the **Offset** edit box and **0.3** in the **Increment** edit box.

4. Clear the **Create Snap Lines** check box. Choose the **Apply** button and then choose the **Close** button to exit the **Clean Dimensions** dialog box.

   After you clean the dimensions, you will notice that the dimensions are not placed in the required order. So, you need to manually place the dimensions in proper order.

5. Select the dimension and drag it to the desired location on the drawing sheet.

   Some dimensions are repeated. For example, the diameter of the hole feature is displayed twice in the drawing view. So, you need to erase the repeated and not required dimensions.

6. Select the dimension and hold down the right mouse button on the repeated dimensions to display a shortcut menu. Choose the **Erase** option from the shortcut menu.

   In some cases, the dimensions are displayed in the views in which you do not want them to be displayed. In such cases, you can switch those dimensions to the other views. To switch the dimension to other view, select it and hold the right mouse button; a shortcut menu is displayed. Choose the **Move to View** option from it.

   After manually placing the dimensions, the drawing views should look similar to the views shown in Figure 12-10.

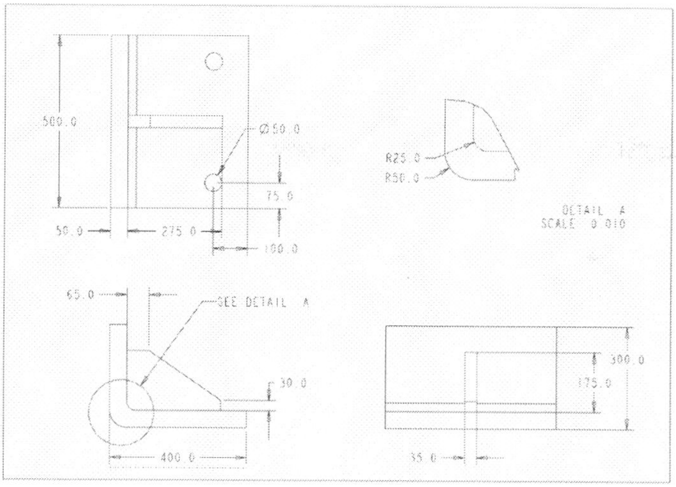

Figure 12-10 *The required drawing views*

Saving the Drawing File

1. Choose the **Save** button from the **Quick Access** toolbar; the **Save Object** dialog box is displayed with the name of the drawing file specified earlier.
2. Press ENTER to confirm the saving of the file.

EXERCISE

Exercise 1

In this exercise, you will generate the drawing views of the model created in Exercise 1 of Chapter 8, as shown in Figure 12-11. Add dimensions to the views, as shown in the Figure 12-12.

(Expected time: 45 min)

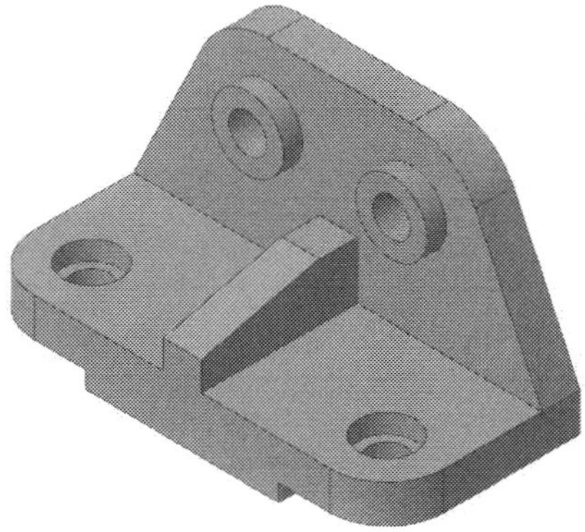

Figure 12-11 *Isometric view of the model*

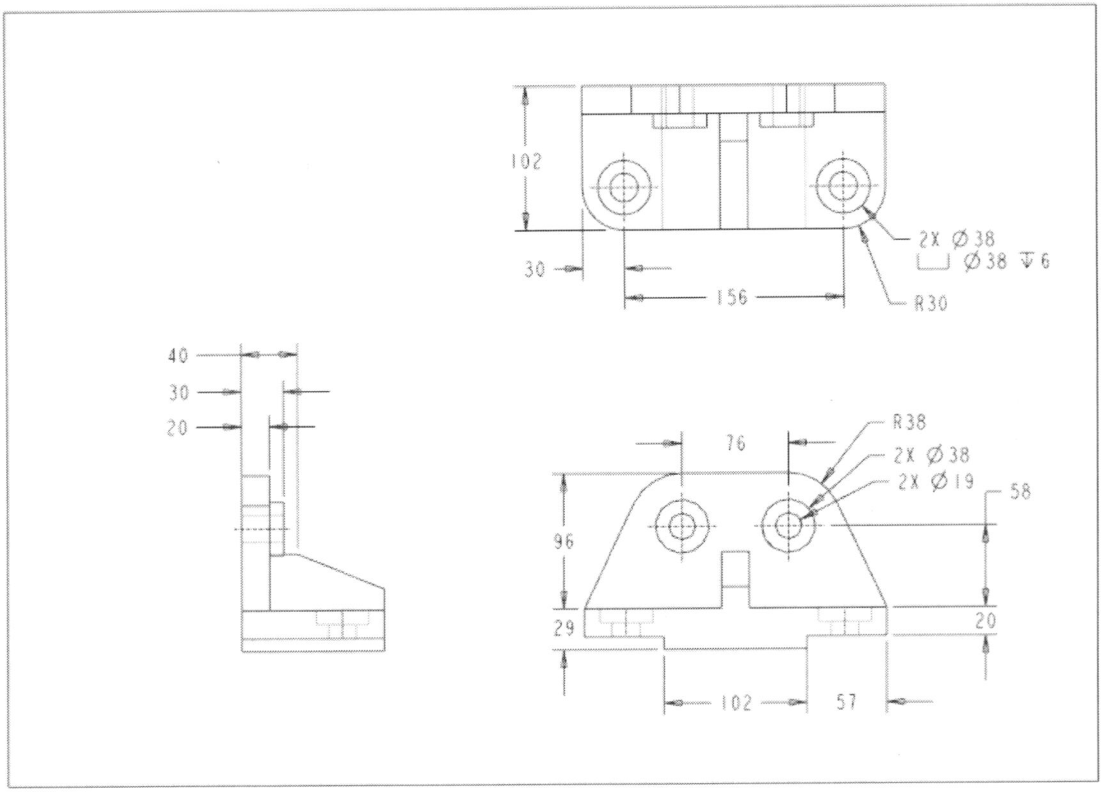

***Figure 12-12** Orthographic views of the solid model*

Note

In the given drawing views, the center lines are not generated automatically; instead they are created manually. After creating the center lines, you need to define a line style and then apply it on the center lines.

Chapter 13

Other Drawing Options

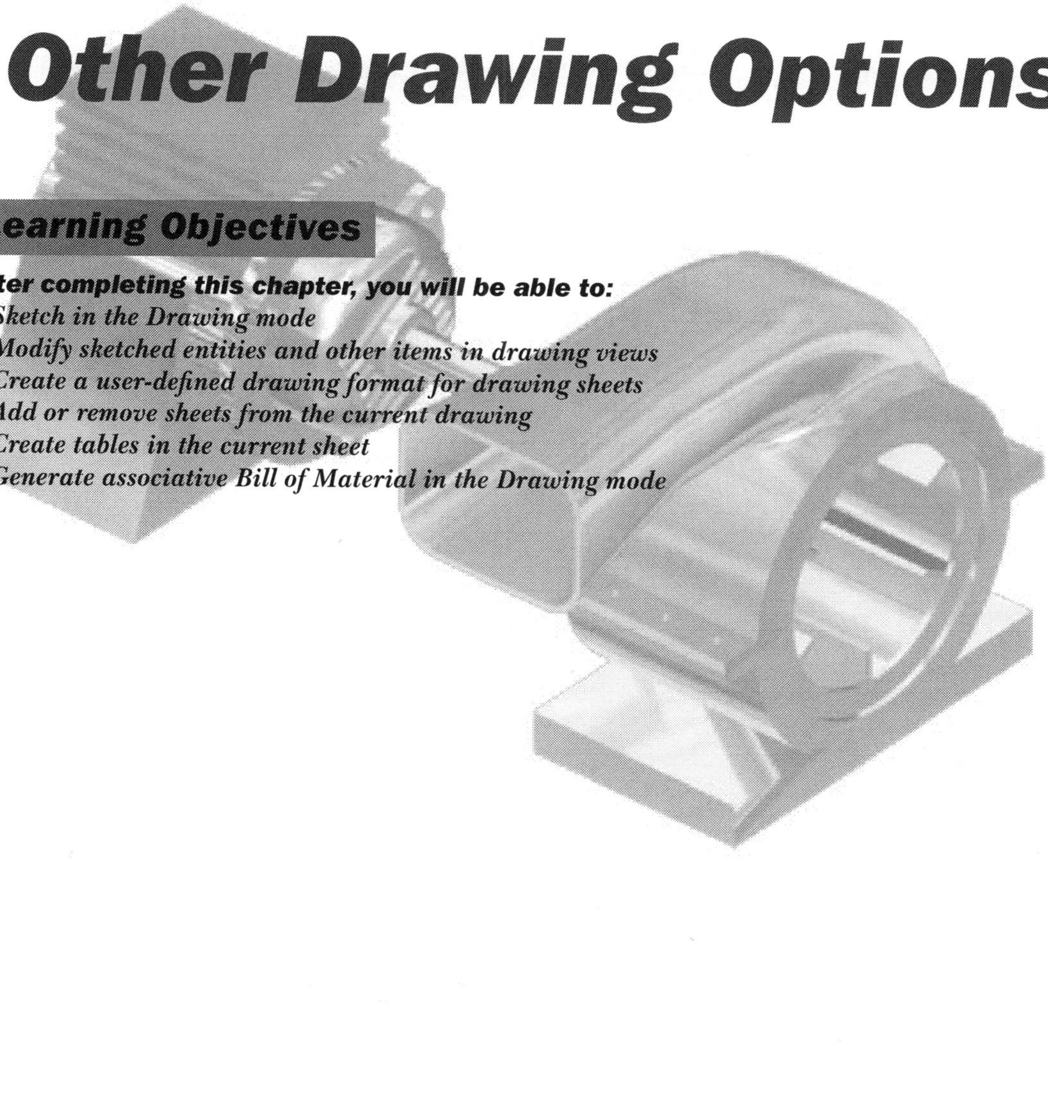

Learning Objectives

After completing this chapter, you will be able to:

- *Sketch in the Drawing mode*
- *Modify sketched entities and other items in drawing views*
- *Create a user-defined drawing format for drawing sheets*
- *Add or remove sheets from the current drawing*
- *Create tables in the current sheet*
- *Generate associative Bill of Material in the Drawing mode*

SKETCHING IN THE DRAWING MODE

Sketching in the **Drawing** mode is called drafting. As discussed earlier, there are two types of drafting techniques in Creo Parametric: Generative drafting and Interactive drafting. Any item on the drawing sheet that is not generated from a model is called a draft entity or a draft item. Drafting is extensively used for creating user-defined formats, drawing tables, and drawing title blocks in the formats. Sketching in the **Drawing** mode is almost similar to that in the other modes of Creo Parametric. Sketching can be done by using the tools in the **Sketch** tab of the **Ribbon**.

Grouping Entities

Ribbon: Sketch > Group > Draft Group

A draft entity can be grouped with a note, geometric tolerance (gtol), dimension, or with another draft entity. In Creo Parametric, the note, gtol, and dimensions are called detail items. After grouping the draft entities, the operation applied on any one of them is also applied to the grouped entity.

USER-DEFINED DRAWING FORMATS

Creo Parametric provides you with some standard drawing formats for generating drawing views. These standard formats have standard sheet sizes, tables, and title blocks. However, sometimes you may need to create a user-defined drawing format that is specifically designed as per your requirements, including sheet size, tables, and title block.

GENERATING THE BOM AND BALLOONS IN DRAWINGS

The Bill of Material (BOM) is a representation of the components and their parameters that are used in the assembly. In the **Drawing** mode, the BOM is generated and is associative in nature. Therefore, any modification in the assembly, such as addition or removal of components, is automatically reflected in the BOM.

Generating Balloons

Ribbon: Table > Balloons > Create Balloons drop-down > Create Balloons - All

Balloons are generated by using the repeat region. Before generating balloons, you need to set the repeat region for the generation of balloons.

TUTORIALS

Tutorial 1

In this tutorial, you will create a format of size A and add the title block to the format. Then you will retrieve the format in the **Drawing** mode and create the table for BOM. Next, you will generate the exploded isometric view of the **Pedestal Bearing** assembly created in Tutorial 2 of Chapter 10. Also, you will add balloons to the drawing view, as shown in Figure 13-1.

(Expected time: 45 min)

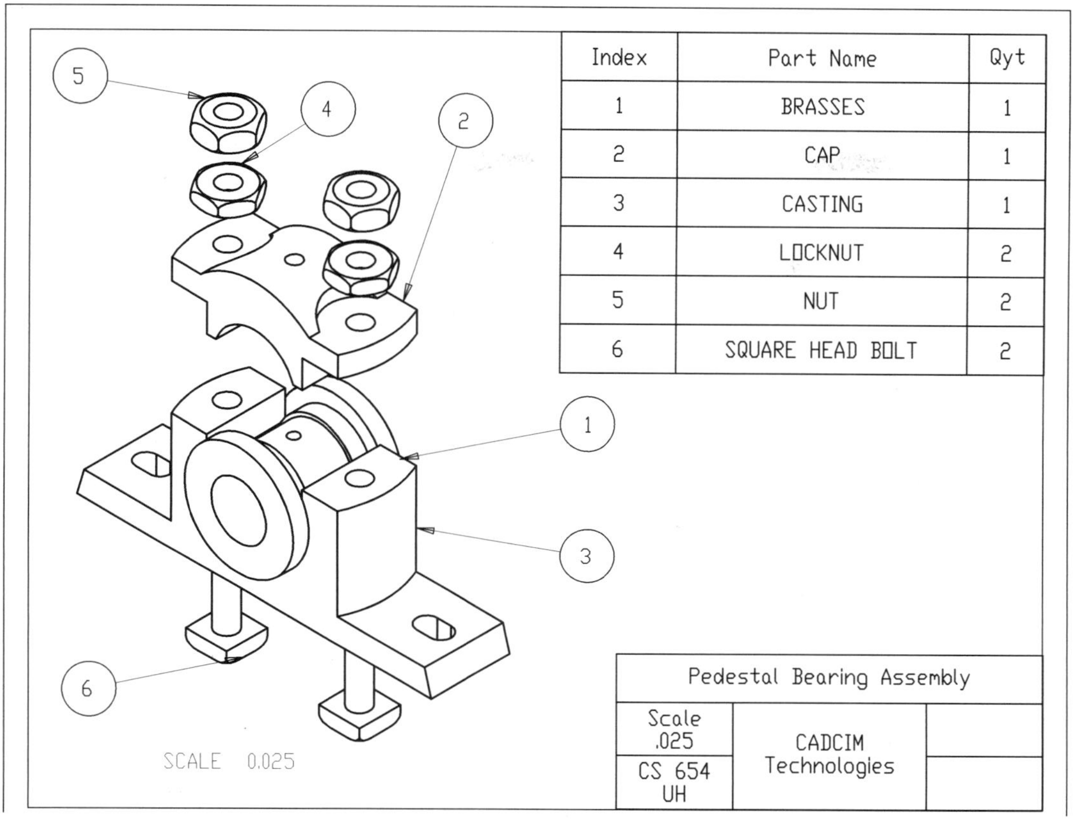

***Figure 13-1** The drawing view of the assembly showing the BOM and balloons*

The following steps are required to complete this tutorial:

a. Start a new file in the **Format** mode. Create the format of the drawing and then add the title block in the format, refer to Figures 13-2 and 13-3.
b. Save the format file and then close it.
c. Start a new drawing file in the **Drawing** mode. Select **Pedestal Bearing** as the model and retrieve the format that you have created.
d. Create the table and enter the headers.
e. Define the repeat region and then assign the report symbols, refer to Figure 13-6.
f. Generate the exploded isometric view of the assembly, refer to Figure 13-7.
g. Add balloons to the drawing, refer to Figure 13-9.

Copy the *Pedestal Bearing* folder from the *c10* folder to the *c13* folder. As this is the first tutorial of the chapter, therefore you need to create the *c13* folder inside the *Creo-5.0* folder. Set the working directory to *C:/Creo-5.0/c13/Pedestal Bearing*.

Starting the Format File

As evident from Figure 13-1, the exploded view of the Pedestal Bearing is generated on a drawing sheet with a customized format. Therefore, you need to create a format before generating the drawing view. This format will be retrieved later.

1. Choose the **New** button from the **Quick Access** toolbar; the **New** dialog box is displayed.

2. Select the **Format** radio button from the **Type** area in the **New** dialog box and enter the name as **Format1** in the **Name** edit box. Choose the **OK** button from the **New** dialog box; the **New Format** dialog box is displayed.

3. In this dialog box, the **Empty** radio button in the **Specify Template** area is selected by default and the **Landscape** button in the **Orientation** area is also chosen by default. If not so, you need to set them that ways.

4. Select **A** from the **Standard Size** drop-down list in the **Size** area and then choose the **OK** button; the **Format** mode is invoked. Note that a sheet of size **A** is displayed on the screen. This is evident from the text displayed below the sheet on the screen.

Creating the Format

1. Choose **Sketch > Sketching > Edge > Offset Edge** from the **Ribbon**; the **OFFSET OPER** menu is displayed. Choose the **Ent Chain** option from this menu and then select all the four border lines of the format by drawing a window around them. Press the middle mouse button after the selection has been made; the Message Input Window is displayed and also, an arrow pointing outward is displayed on the format boundary. This arrow displays the direction of the offset. To offset the lines in the opposite direction, specify a negative offset value.

2. Enter **-0.25** in the Message Input Window and press ENTER. Next, press the middle mouse button twice to exit the **OFFSET OPER** menu.

3. Choose **Table > Table > Table** drop-down **> Insert Table** from the **Ribbon**; the **Insert Table** dialog box is displayed.

4. Choose **Table growth direction: leftward and ascending** button from the **Direction** area to set the direction of table growth.

5. Set the value to **3** in both the **Number of Columns** and **Number of Rows** spinners in the **Table Size** area.

6. Enter **0.5** as height in the **Height (INCH)** edit box and **1.2** as width in the **Width (INCH)** edit box.

7. Choose the **OK** button from the **Insert Table** dialog box; the **Select Point** dialog box is displayed and you are prompted to select a point to place the table. Also, the preview of the table gets attached to the cursor.

8. Click on the desired location in the drawing to place the table, refer to Figure 13-2.

9. Select any cell of the second column and then choose the **Height and Width** tool from the **Rows & Columns** group in the **Table** tab; the **Height and Width** dialog box is displayed.

10. Enter the value **2** in the **Width (drawing units)** edit box in the **Columns** area and then choose the **OK** button; the title block created will be similar to the one shown in Figure 13-2. Note that the title block created is not the required one. Therefore, you need to modify it.

Figure 13-2 *Format with the title block*

11. Select the entire first row and then choose the **Merge Cells** button from the **Rows & Columns** group.

12. Similarly, select the second and third cells in the second column and choose the **Merge Cells** button.

    The table after merging rows and columns should look similar to the one shown in Figure 13-3.

13. You need to set the alignment of the text that will be later entered in the cells of the table. To do so, double-click on any of these cells; the **Format** tab is added.

14. Choose the inclined arrow from the **Style** group of the **Format** tab; the **Text Style** dialog box is displayed.

15. In the **Note/Dimension** area of this dialog box, select the **Middle** option from the **Vertical** drop-down list and select the **Center** option from the **Horizontal** drop-down list. Exit the **Text Style** dialog box by choosing the **OK** button.

16. Now, the text is aligned vertically to the middle of the cell and horizontally to the center of the cell. Similarly, set the alignment in all the cells of the table individually.

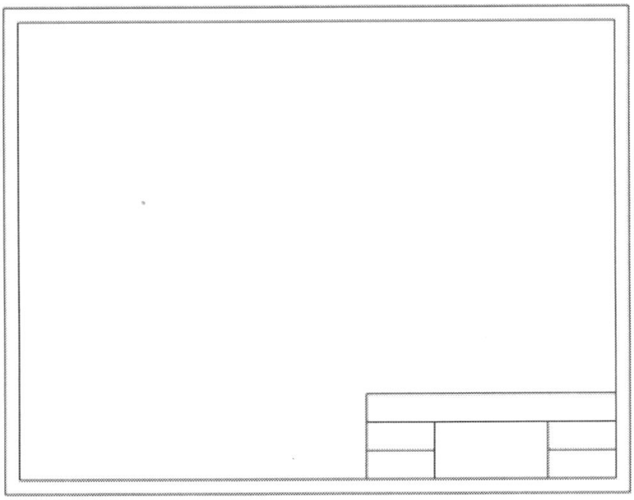

Figure 13-3 *Modified title block*

Saving the Format File

You need to save the format file that you have created so that it can be used as a template in the **Drawing** mode. Using this template, you will generate the exploded drawing view of the Pedestal Bearing. The file will be stored in the *.frm* file format.

1. Choose the **Save** button from the **Quick Access** toolbar; the **Save Object** dialog box is displayed with the name of the file that you entered earlier. Press ENTER.

2. Close the current window by choosing the **Close** button from the **Quick Access** toolbar.

Starting a New Drawing File

You need to start a new drawing file to generate the exploded drawing view of the Pedestal Bearing.

1. Choose the **New** button from the **File** menu to display the **New** dialog box. In this dialog box, select the **Drawing** radio button and specify the name of the drawing as *c13tut1*. Choose the **OK** button; the **New Drawing** dialog box is displayed.

2. In the **New Drawing** dialog box, choose the **Browse** button from the **Default Model** area and select the assembly file named *pedestalbearing.asm*.

3. Select the **Empty with format** radio button from the **Specify Template** area in the **New Drawing** dialog box.

4. Select **Format1** from the **Format** drop-down list. If **Format1** is not available in the drop-down list, choose the **Browse** button to locate **Format1**. Then, choose the **OK** button to start a new drawing file with the selected format.

Creating the Table for BOM

1. Choose **Table > Table > Table** drop-down **> Insert Table** from the **Ribbon**; the **Insert Table** dialog box is displayed.

2. Choose the **Table growth direction: leftward and descending** button from the **Direction** area.

3. Set the value to **3** in the **Number of Columns** spinner and **2** in the **Number of Rows** spinner in the **Table Size** area of the dialog box. Now, you need to enter the dimensions of the BOM table.

4. Enter **0.5** as height in the **Height (INCH)** edit box and **1.2** as width in the **Width (INCH)** edit box.

5. Choose the **OK** button from the **Insert Table** dialog box; the **Select Point** dialog box is displayed and you are prompted to select a point to place the table.

6. Click on the desired location to place the table; refer to Figure 13-2.

7. Select any cell of the second column and then choose the **Height and Width** tool from the **Rows & Columns** group in the **Table** tab; the **Height and Width** dialog box is displayed.

8. Enter the value **3** in the **Width (drawing units)** edit box in the **Column** area and then choose the **OK** button.

9. Similarly, select any cell of the third column and change its width to **0.8**.

Setting the Repeat Region

Repeat regions are the smart cells in a table that can expand depending on the amount of data inserted in a cell. You need to define the second row as the repeat region.

1. Choose the **Repeat Region** tool from the **Data** group of the **Table** tab in the **Ribbon**; the **TBL REGIONS** menu is displayed.

2. Choose the **Add** option from the **TBL REGIONS** menu; you are prompted to locate the corners of the region.

3. Select the first cell of the second row and then select the third cell of the second row to locate the corners of the region. Press the middle mouse button twice to exit the menu.

Creating the Column Headers

Now, you need to create headings in the table.

1. Double-click in the first cell of the first row; the **Format** tab will be added to the **Ribbon.** Enter **Index** and click outside twice to exit.

2. Similarly, enter **Part Name** and **Qty** in the second and third cells, respectively.

Assigning the Report Symbols in the Repeat Region

The information in the cells that are defined as repeat region is determined by the text written in the form of report symbols. These report symbols associate the information in the cells with the assembly file directly.

1. Double-click in the first cell of the second row; the **Report Symbol** dialog box is displayed, as shown in Figure 13-4. Select **rpt** from this dialog box; the **rpt** options are displayed, as shown in Figure 13-5.

*Figure 13-4 The **Report Symbol** dialog box*

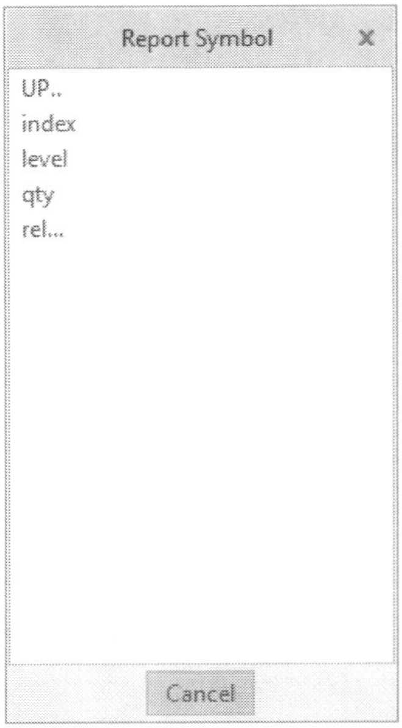

*Figure 13-5 The **Report Symbol** dialog box with the **rpt** options*

2. Select the **index** option from the dialog box; **rpt.index** is added to the cell.

3. Double-click in the second cell of the second row; the **Report Symbol** dialog box is displayed. Select **asm** from this dialog box; the **asm** options are displayed.

4. Select the **mbr** option from the dialog box; the **mbr** options are displayed.

5. Select the **name** option from the dialog box; **asm.mbr.name** is added to the cell.

6. Double-click in the third cell of the second row; the **Report Symbol** dialog box is displayed. Select **rpt** from this dialog box; the **rpt** options are displayed.

7. Select the **qty** option from the dialog box.

   If the text selected to enter in the cells is overlapping the other text or is extending beyond the cell, ignore it. This is because the BOM will be created in the repeat region where this text exists and after the BOM is created, the text of the BOM will fit inside the cells. The drawing sheet after creating the column headers and entering the report symbols is shown in Figure 13-6.

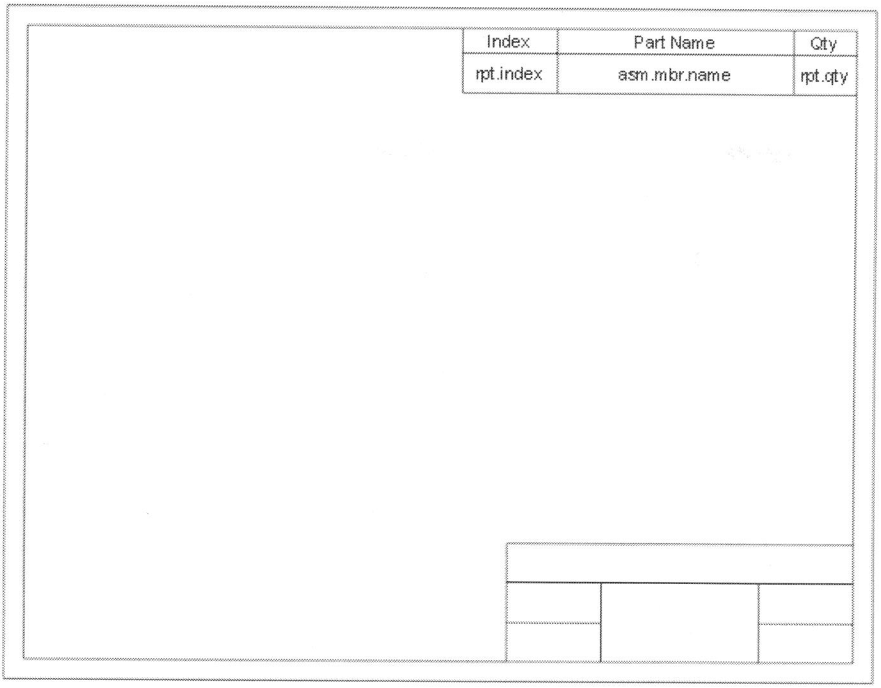

Figure 13-6 *Drawing sheet after adding the report symbols*

Generating the Exploded Drawing View

In Tutorial 2 of Chapter 10, you created the exploded view of the Pedestal Bearing in the **Assembly** mode. The name of the exploded view that you created was **EXP1**. As a result, the exploded view of the Pedestal Bearing is integrated with the assembly file that you copied from the *c10* folder. Now, you will use this exploded state to generate the exploded drawing view.

1. Choose the **General View** tool from the **Model Views** group in the **Layout** tab of the **Ribbon**; the **Select Combined State** dialog box is displayed.

2. Select the **Do not prompt for Combined State** check box and then choose the **OK** button to exit the dialog box; you are prompted to select the center point for the drawing view.

3. Select a point on the left of the drawing sheet; the **Drawing View** dialog box is displayed. Select the **Isometric** option from the **Default Orientation** drop-down list in the **Drawing View** dialog box and then choose the **Apply** button.

4. Choose the **Scale** option from the **Categories** list box. Next, select the **Custom scale** radio button from the **Scale and perspective options** area. Enter the value **.025** in the edit box and then choose the **Apply** button.

5. Select the **View States** option from the **Categories** list box. Next, select the **Explode components in view** check box and select the **EXP1** from the **Assembly explode state** drop-down list. Next, choose the **Apply** button.

6. Close the **Drawing View** dialog box to complete the placement of the required view. If the model is not placed properly inside the boundaries of the drawing sheet, unlock it and then drag the drawing view and place it, as shown in Figure 13-7. Note that the BOM automatically appears in the table by extending the repeat region and the **Qty** column is empty.

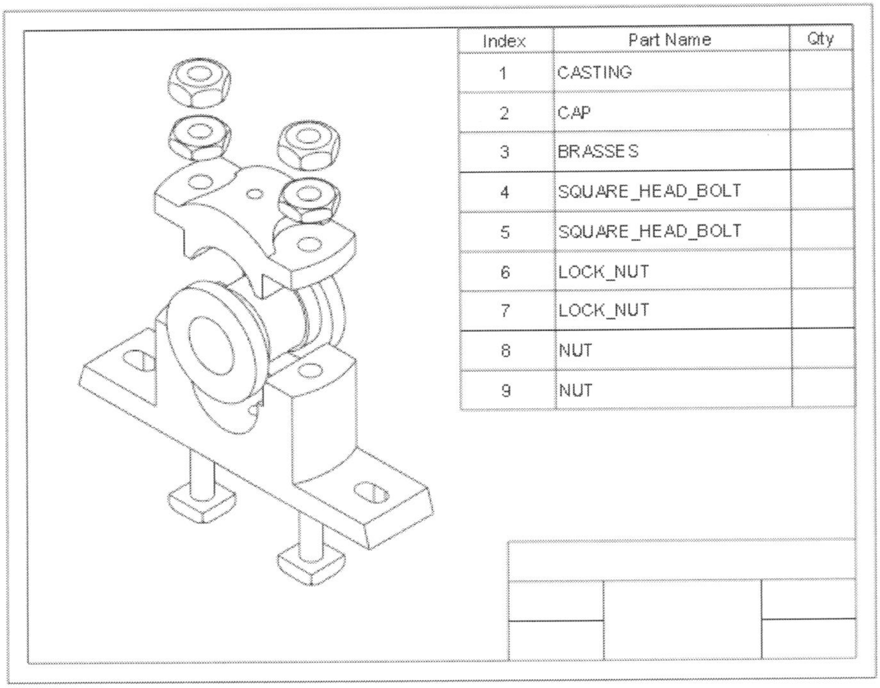

Figure 13-7 *Exploded isometric view of the assembly*

Note

1. If the BOM is not displayed, then choose ***Table > Data > Update Tables*** *from the* ***Ribbon****.*

2. You may need to select the ***No Hidden*** *option from the* ***Display Style*** *drop-down list from the* ***Graphics*** *toolbar to change the view.*

Setting the No Repeat Option

As some components are repeated in the assembly, they are also repeated in the BOM. The BOM needs to be set such that no part name is repeated.

1. Choose the **Repeat Region** tool from the **Data** group of the **Table** tab in the **Ribbon**; the **TBL REGIONS** menu is displayed.

2. Choose the **Attributes** option from the menu; you are prompted to select a region.

3. Move the cursor over the table below the first row. All rows, except the first row, are highlighted in green. The highlighted portion is called region. Select the region; the **REGION ATTR**

submenu is displayed. Choose the **No Duplicates** option and then choose the **Done/Return** button from the submenu.

4. Press the middle mouse button twice and exit the menu; the drawing sheet appears, as shown in Figure 13-8.

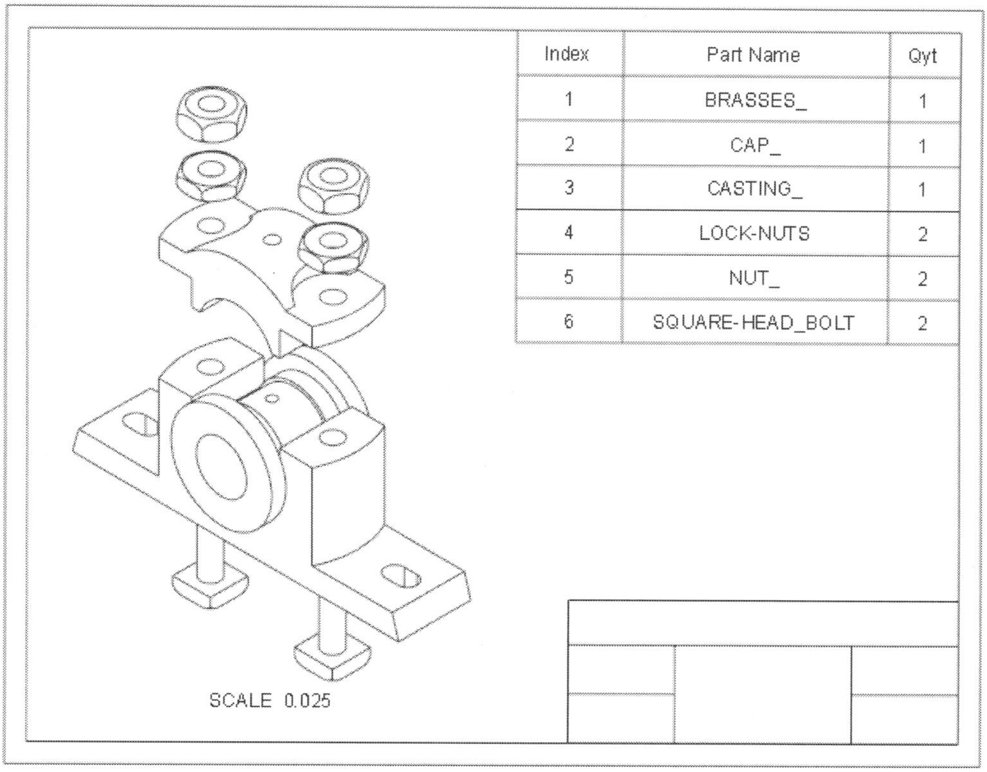

Figure 13-8 *The drawing view showing BOM with no duplicates*

5. If the information is not updated in the table, choose the **Update Tables** tool from the **Data** group of the **Table** tab from the **Ribbon** to update the information.

Generating Balloons

Balloons are generated by using the repeat region. Before generating the balloons, you need to set the repeat region for generating balloon.

1. Choose the **Create Balloons - All** option from the **Create Balloons** drop-down list available in the **Balloons** group of the **Table** tab in the **Ribbon**; balloons are attached with all the parts in the exploded view. The balloons contain the index number so that the parts in the assembly can be referred to as BOM.

2. Drag the balloons to place them appropriately in the drawing sheet, as shown in Figure 13-9. You can select a balloon and then hold the right mouse button to invoke the shortcut menu. From this menu, you can select the required options to modify the attachment points of balloons.

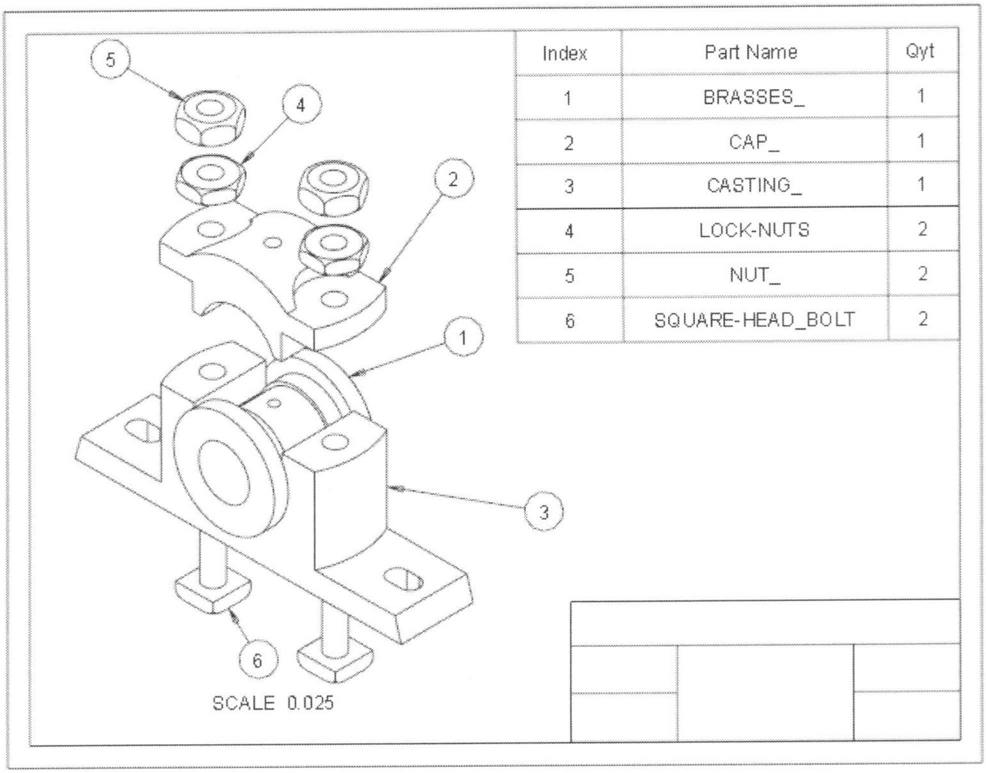

Figure 13-9 The drawing view of the assembly showing the BOM and balloons

Entering Text in the Title Block

The text alignment in various cells of the title block was defined while creating the format. As a result, when you enter the text, it will automatically be aligned to the middle left.

1. Double-click on the top cell in the table; the **Format** tab is added to the **Ribbon**.

2. Now, enter **Pedestal Bearing Assembly** in the cell. Next, choose the inclined arrow from the **Style** group of the **Format** tab; the **Text Style** dialog box is displayed. Now, select the **Default** check box at the right of the **Height** property in the character area and choose the **OK** button.

3. Double-click on the cell that is in the first column and the second row; the **Format** tab is added to the **Ribbon**.

4. Now, enter **Scale** and then press ENTER. Now, enter **0.025** and click outside twice to exit; the text **Scale: 0.025** is entered in the two lines in the title block.

5. Similarly, enter the text in all the remaining cells one by one, as shown in Figure 13-10.

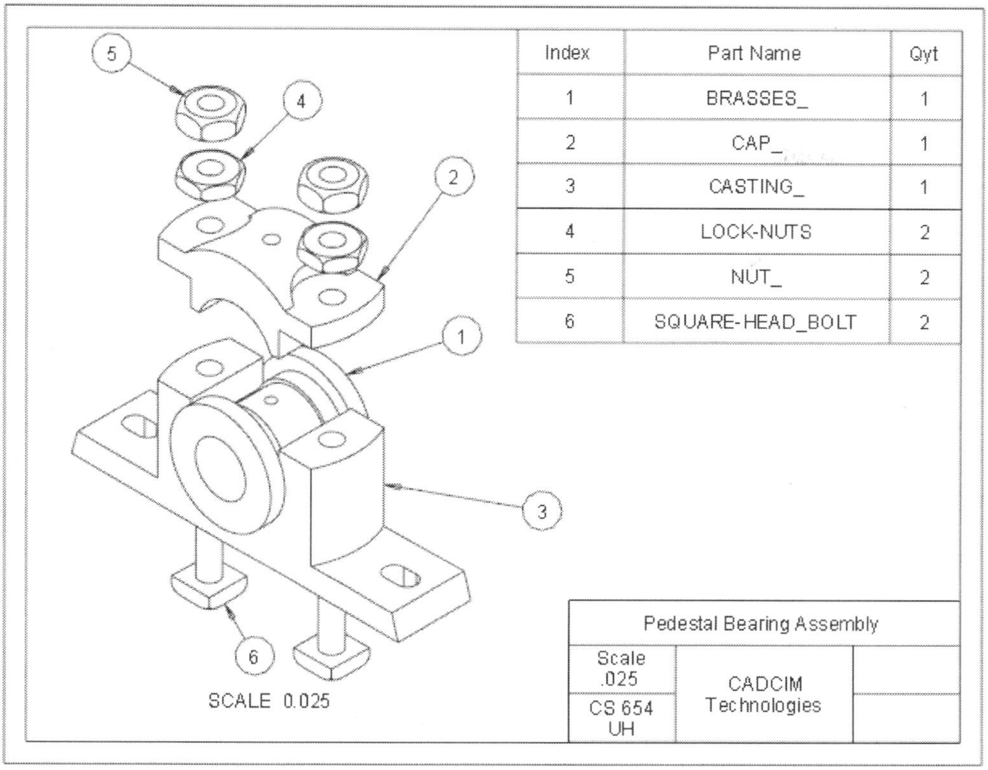

Figure 13-10 The drawing view of the assembly showing the title block, balloons, and the BOM

Saving the Drawing File

1. Choose the **Save** button from the **Quick Access** toolbar to save the drawing file; the **Save Object** dialog box is displayed. Now, save the file by pressing ENTER.

Closing the Window

The drawing file has been saved and now you can exit the **Drawing** mode.

1. Choose **File > Close** from the menu bar to close the current session.

Note

The occurrence of components in BOM depends on the order in which they were assembled in the assembly. In other words, the component placed first will be placed first in the BOM.

Tutorial 2

In this tutorial, you will create a format of size A and add the title block in the format. You will retrieve the format in the **Drawing** mode and generate the exploded isometric view of the Shock Assembly created in Tutorial 1 of Chapter 10. Also, you will add the associative Bill of Material and the Balloons to the drawing view, as shown in Figure 13-11. **(Expected time: 45 min)**

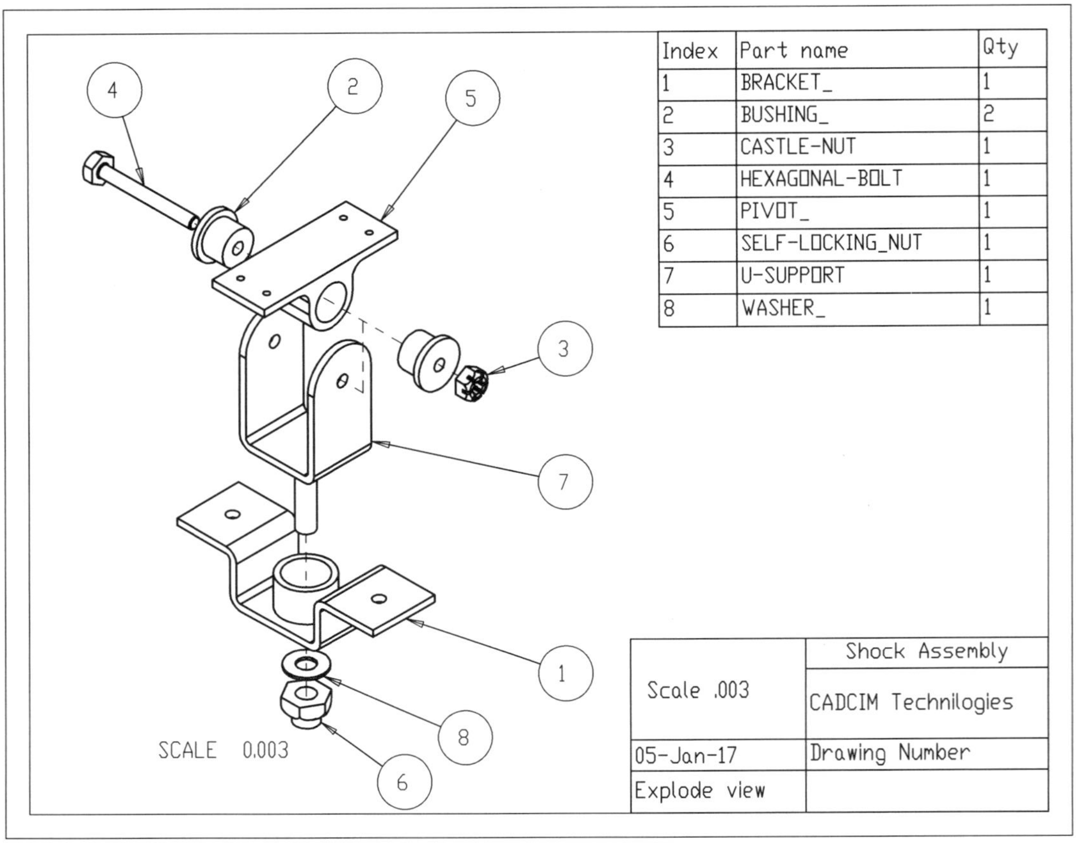

Index	Part name	Qty
1	BRACKET_	1
2	BUSHING_	2
3	CASTLE-NUT	1
4	HEXAGONAL-BOLT	1
5	PIVOT_	1
6	SELF-LOCKING_NUT	1
7	U-SUPPORT	1
8	WASHER_	1

Figure 13-11 *The drawing view of the exploded assembly with BOM*

The following steps are required to complete this tutorial:

a. Start a new file in the **Format** mode. Create the format of the drawing and add the title block in the format, refer to Figures 13-12 and 13-13.
b. Save the format file and then close it.
c. Start a new drawing file in the **Drawing** mode, select *shockassembly.asm* as the model, and retrieve the format created earlier.
d. Create the table for BOM and enter the headers.
e. Define the repeat region and then assign the report symbols, see Figure 13-14.
f. Generate the exploded isometric view of the assembly and generate balloons in the drawing view, refer to Figures 13-15 and 13-16.
g. Add text in the title block, refer to Figure 13-17.

Copy the *Shock Assembly* folder from the *c10* folder to the *c13* folder and set this folder as the working directory.

Starting the Format File

As mentioned in the tutorial description, you need to create a format of size A that will be retrieved later to create the drawing view of the Shock Assembly.

1. Choose the **New** button from the **Quick Access** toolbar to display the **New** dialog box.

2. Select the **Format** radio button from the **Type** area in the **New** dialog box and name the file as *Format2*. Choose the **OK** button from the **New** dialog box; the **New Format** dialog box is displayed.

3. Select the **Empty** radio button from the **Specify Template** area and the **Landscape** button from the **Orientation** area of the **New Format** dialog box, if they are not already selected.

4. Select **A** from the **Standard Size** drop-down list in the **Size** area and then choose the **OK** button to invoke the **Format** mode. Note that a sheet of size A is displayed on the screen. This is evident from the text displayed below the sheet on the screen.

Creating the Format

1. Choose **Sketch > Sketching> Edge > Offset Edge** from the **Ribbon**; the **OFFSET OPER** menu is displayed. Choose the **Ent Chain** option and then select all the four border lines of the format by drawing a window around them. Press the middle mouse button after the selection has been made; an arrow is displayed pointing outward. This arrow displays the direction of the offset. Since you need to offset the lines in the opposite direction, you will specify a negative offset distance.

2. Enter **-0.25** in the Message Input Window and press ENTER. Press the middle mouse button twice to exit the **OFFSET OPER** menu.

3. Choose **Table > Table** drop-down **> Insert Table** from the **Ribbon**; the **Insert Table** dialog box is displayed.

4. Choose the **Table growth direction: leftward and ascending** button from the **Direction** area.

5. Set the value to **2** in the **Number of Columns** spinner and **4** in the **Number of Rows** spinner in the **Table Size** area.

6. Enter **0.5** as height in the **Height (INCH)** edit box and **1.2** as width in the **Width (INCH)** edit box.

7. Choose the **OK** button from the **Insert Table** dialog box; the **Select Point** dialog box is displayed and you are prompted to select a point to place the table.

8. Click on the required location to place the table, refer to Figure 13-12; the table is placed at the specified place.

9. Select any cell of the second column and then choose the **Height and Width** tool from the **Rows & Columns** group in the **Table** tab; the **Height and Width** dialog box is displayed.

10. Enter the value **3.2** in the **Width (drawing units)** edit box in the **Column** area and then choose the **OK** button.

11. Select any cell in the second row and choose the **Height and Width** tool from the **Rows & Columns** group in the **Table** tab; the **Height and Width** dialog box is displayed.

12. Enter the value **1** in the **Height (drawing units)** edit box in the **Row** area of the dialog box and then choose the **OK** button. Note that you may need to clear the **Automatic height adjustment** check box to enter the **Height (drawing units)**.

13. Select any cell in the first row and choose the **Height and Width** tool from the **Rows & Columns** group in the **Table** tab; the **Height and Width** dialog box is displayed.

14. Now, clear the **Automatic height adjustment** check box in the **Row** area. Enter the value **0.7** in the **Height (drawing units)** edit box in the **Row** area and then choose the **OK** button.

    The title block created will be similar to the one shown in Figure 13-12. However, this title block is not the required one. Notice the difference between the title block shown in Figure 13-12 and the one shown in Figure 13-13. So, you need to modify the title block that you have created such that it is similar to the one shown in Figure 13-13.

Figure 13-12 *Format with the title block*

15. Select a cell in the first row and a cell in the second row of the first column to merge them. Then, choose the **Merge Cells** tool from the **Rows & Columns** group in the **Table** tab of the **Ribbon**; the cells are merged.

    The table after merging the rows and columns should look similar to the one shown in Figure 13-13. You need to set the alignment of the text that will be entered in the cells of the table later.

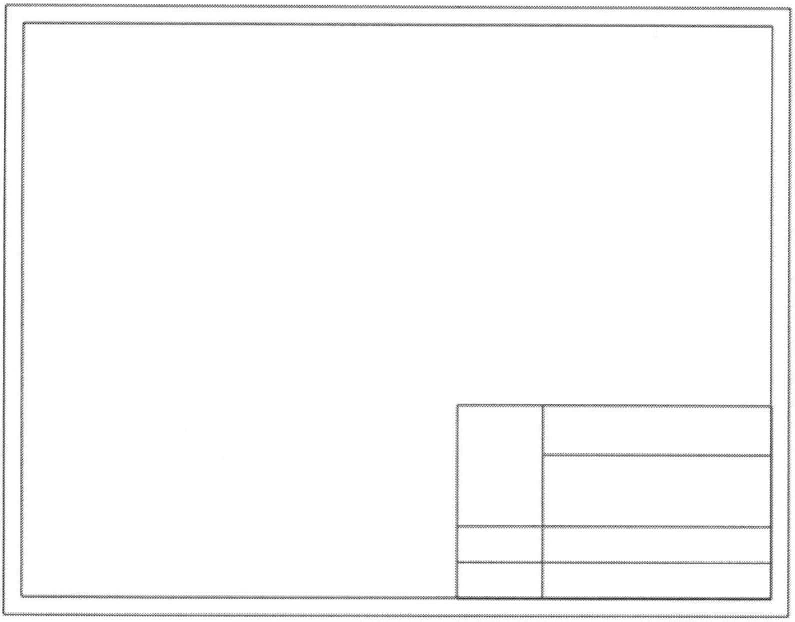

Figure 13-13 *Modified title block*

16. Double-click in a cell to invoke the **Format** tab and click on the inclined arrow from the **Style** group of the **Format** tab; the **Text Style** dialog box is displayed.

17. In the **Note/Dimension** area of this dialog box, select the **Middle** option from the **Vertical** drop-down list and the **Center** option from the **Horizontal** drop-down list. Next, choose the **OK** button to exit the dialog box.

    Now, the text is aligned vertically to the middle of the cell and horizontally to the center of the cell. Similarly, set the alignment in all the cells of the table individually.

Saving the Format File

You need to save the format file that you have created so that it can be used as a template in the **Drawing** mode. The file will be stored in the *.frm* file format.

1. Choose the **Save** button from the **Quick Access** toolbar; the **Save Object** dialog box is displayed with the name of the file entered earlier. Next, press ENTER to save it.

2. Close the current window by choosing **File > Close** from the menu bar.

Starting a New Drawing File

You need to start a new drawing file for generating the exploded drawing view of the Shock Assembly and for generating the BOM.

1. Choose the **New** button to display the **New** dialog box. Select the **Drawing** radio button from the dialog box, specify the name of the drawing as **c13tut2**, and then choose the **OK** button; the **New Drawing** dialog box is displayed.

2. Choose the **Browse** button from the **Default Model** area and select the assembly file named *shockassembly.asm*.

3. Select the **Empty with format** radio button in the **Specify Template** area of the **New Drawing** dialog box.

4. Select **Format2** from the **Format** drop-down list. If **Format2** is not available in the drop-down list, choose the **Browse** button to locate **Format2**. Next, choose the **OK** button to exit the **New Drawing** dialog box.

Creating the Table for BOM

1. Choose **Table > Table** drop-down **> Insert Table** from the **Ribbon**; the **Insert Table** dialog box is displayed.

2. Choose **Table growth direction: leftward and descending** button from the **Direction** area.

3. Set the value to **3** in the **Number of Columns** spinner and **2** in the **Number of Rows** spinner in the **Table Size** area.

4. Enter **0.5** as height in the **Height (INCH)** edit box and **1.2** as width in the **Width (INCH)** edit box.

5. Choose the **OK** button from the **Insert Table** dialog box; the **Select Point** dialog box is displayed and you are prompted to select a point to place the table.

6. Select a point close to the upper right corner of the inner rectangle; the table is placed at the specified point.

7. Select any cell in the second column and then choose the **Height and Width** tool from the **Rows & Columns** group in the **Table** tab; the **Height and Width** dialog box is displayed.

8. Enter the value **3** in the **Width (drawing units)** edit box in the **Column** area and then choose the **OK** button. Similarly, change the width of the third column to **0.8**.

Setting the Repeat Region

Repeat regions are the smart cells in a table that can expand depending on the amount of data inserted in the cell. The second row will be defined as the repeat region.

1. Choose **Repeat Region** from the **Data** group of the **Table** tab in the **Ribbon**; the **TBL REGIONS** menu is displayed.

2. Choose the **Add** option from the **TBL REGIONS** menu; you are prompted to locate the corners of the region.

3. Select the first cell of the second row and the third cell of the second row to define the region. Press the middle mouse button twice to exit the menu.

Creating the Column Headers

Next, you need to create headings in the table.

1. Double-click in the first cell of the first row; the **Format** tab is added to the **Ribbon.** Enter **Index** in the cell and click outside twice to exit.

2. Similarly, enter **Part Name** and **Qty** in the second and third cells, respectively.

Entering the Report Symbols in the Repeat Region

As mentioned earlier, the information in the cells that are defined as the repeat region is determined by the text written in the form of report symbols. These report symbols associate the information in the cells with the assembly file directly.

1. Double-click in the first cell of the second row; the **Report Symbol** dialog box is displayed. Select **rpt** from this dialog box; the **rpt** options are displayed.

2. Select the **index** option from this dialog box.

3. Next, double-click in the second cell of the second row; the **Report Symbol** dialog box is displayed. Select **asm** from this dialog box; the **asm** options are displayed.

4. Select the **mbr** option from this dialog box and then select the **name** option from the options that appear.

5. Double-click in the third cell of the second row; the **Report Symbol** dialog box is displayed. Select **rpt** from this dialog box; the **rpt** options are displayed.

6. Select the **qty** option from this dialog box. The drawing sheet after adding the report symbols is shown in Figure 13-14.

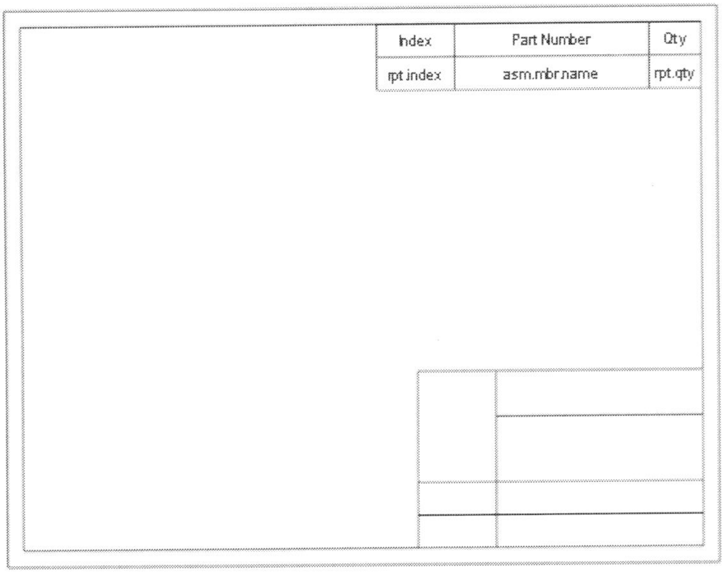

Figure 13-14 Drawing sheet after adding the report symbols

Generating the Drawing View

In Tutorial 1 of Chapter 10, you created the exploded view of the Shock Assembly in the **Assembly** mode. The name of the exploded view that you have specified was **EXP1**. As a result, the exploded view of the Shock assembly has been integrated with the assembly file that you copied from the *c10* folder. Now, you will use this exploded state to generate the exploded drawing view in the **Drawing** mode.

1. Choose the **General** tool from the **Model Views** group of the **Layout** tab in the **Ribbon**; the **Select Combined State** dialog box is displayed.

2. Select the **Do not prompt for Combined State** check box and then choose the **OK** button; the dialog box is closed and you are prompted to select the center point for the drawing view.

3. Select a point on the left of the drawing sheet; the **Drawing View** dialog box is displayed. Select the **Isometric** option from the **Default orientation** drop-down list in the **Drawing View** dialog box and then choose the **Apply** button.

4. Select the **Scale** option from the **Categories** list box. Also, select the **Custom scale** radio button from the **Scale and perspective options** area. Enter **.003** in the edit box and then choose the **Apply** button.

5. Select the **View States** option from the **Categories** list box. Select the **Explode components in view** check box. Select the **EXP1** option from the **Assembly explode state** drop-down list and then choose the **Apply** button.

6. Choose the **Close** button from the **Drawing View** dialog box; the exploded isometric view of the assembly is displayed.

If the model is not placed properly inside the boundary of the drawing sheet, unlock it and then drag the drawing view and place it, as shown in Figure 13-15. If the BOM is not displayed, choose **Table > Data > Update Tables** from the **Ribbon** to display it.

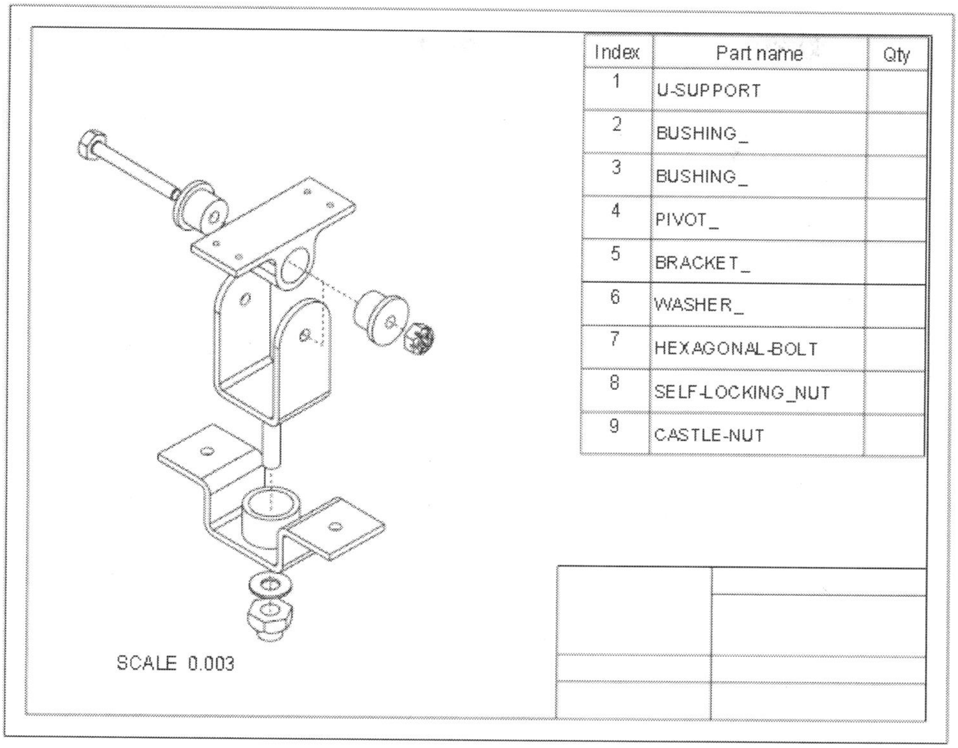

***Figure 13-15** Exploded isometric view of the assembly*

Setting the No Duplicates Option

The Shock assembly consists of two instances of Bushing. Therefore, in BOM, it is also repeated. But, the BOM needs to be set such that no part name is repeated.

1. Choose **Repeat Region** from the **Data** group of the **Table** tab in the **Ribbon**; the **TBL REGIONS** menu is displayed.

2. Choose the **Attributes** option from the menu; you are prompted to select a region.

3. Move the cursor on the table below the first row; all the rows, except the first row, are highlighted in green. Select the region; the **REGION ATTR** submenu is displayed. Choose the **No Duplicates** option and then choose the **Done/Return** option.

4. Press the middle mouse button twice to exit the menu. The drawing sheet appears, as shown in Figure 13-16.

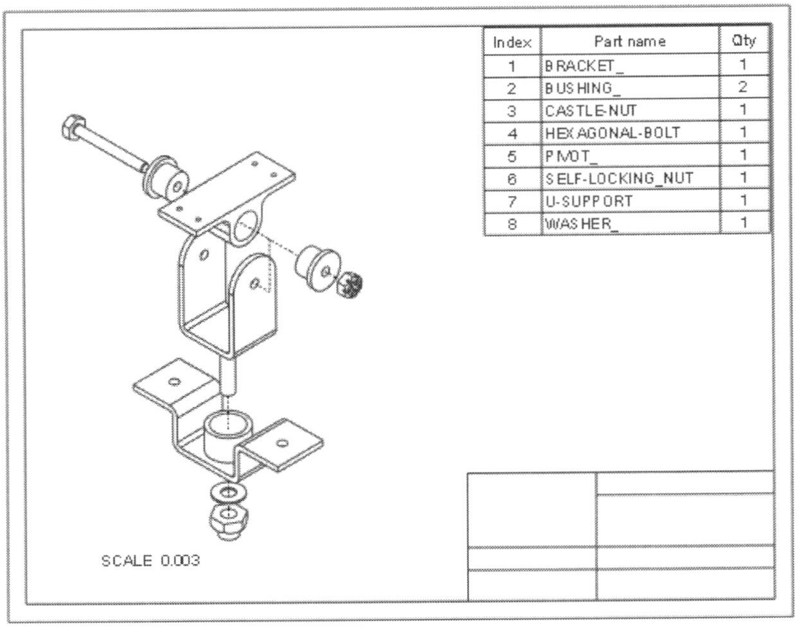

Figure 13-16 *The drawing view of the assembly showing the BOM and balloons*

5. If the information is not updated in the table, choose **Table > Data > Update Tables** from the **Ribbon** to update it.

Generating Balloons

Balloons are generated by using the repeat region. Before generating the balloons, you need to set the repeat region for the balloon generation.

1. Choose the **Create Balloons - All** option from the **Create Balloons** drop-down list available in the **Balloons** group of the **Table** tab of the **Ribbon**; balloons get attached with all the parts of assembly in the exploded view. The balloons contain the index number so that the parts in the assembly can be referred to as BOM.

2. Drag the balloons to place them appropriately on the drawing sheet, refer to Figure 13-16. You can modify the attachment points of a balloon by selecting the balloon and right-clicking. Next, choose the required option from the shortcut menu displayed to modify the attachment points.

Entering Text in the Title Block

1. Double-click in the cell formed by merging the first row and the second column; the **Format** tab is added to the **Ribbon**.

2. Enter **Shock assembly** in the cell. Next, click on the inclined arrow from the **Style** group of the **Format** tab; the **Text Style** dialog box is displayed. Now, select the **Default** check box at the right of the **Height** property and choose the **OK** button.

3. Now, select the cell from the second row and the second column. Enter **CADSOFT Technologies** in the cell and click outside twice to exit.

4. Similarly, enter the text in all the remaining cells one by one, as shown in Figure 13-17.

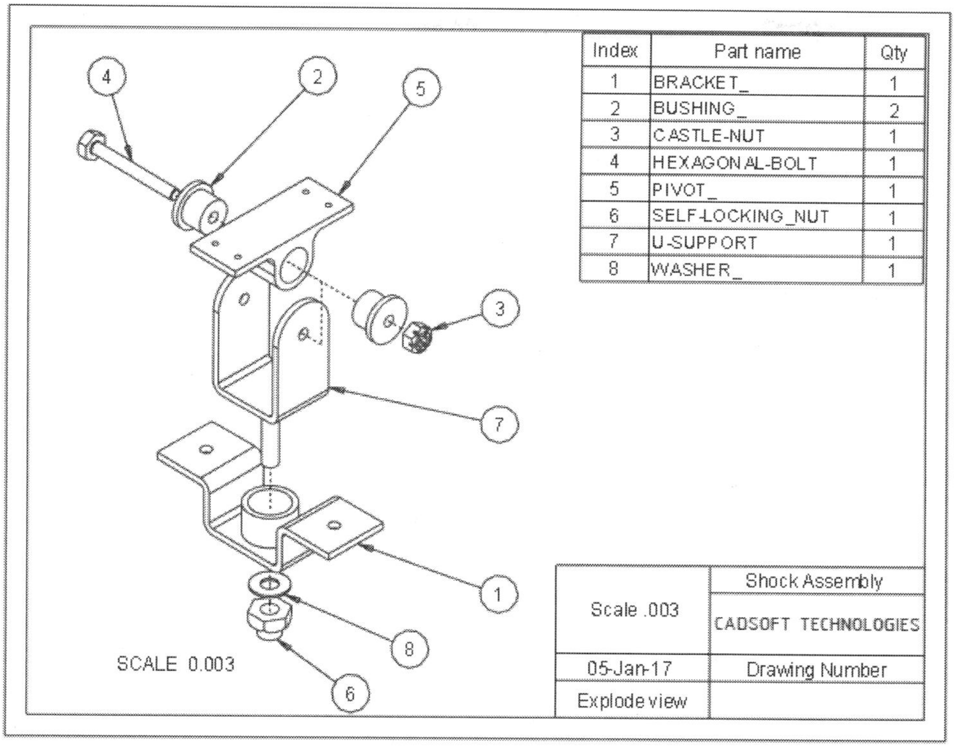

Figure 13-17 The drawing view of the assembly showing the title block, balloons, and the BOM

Saving the Drawing File

1. Choose the **Save** button from the **Quick Access** toolbar to save the drawing file; the **Save Object** dialog box is displayed. Press ENTER to save the file.

Closing the Window

The drawing file has been saved and now you can exit the **Drawing** mode.

1. Choose **File > Close** from the menu bar to close the window.

EXERCISE

Exercise 1

Create a format of size A and then retrieve it in the **Drawing** mode to place the Right view, and an exploded isometric view of the **Cross Head** assembly created in Exercise 1 of Chapter 10. Add the associative assembly BOM and the balloons to the drawing view, as shown in Figure 13-18. **(Expected time: 45 min)**

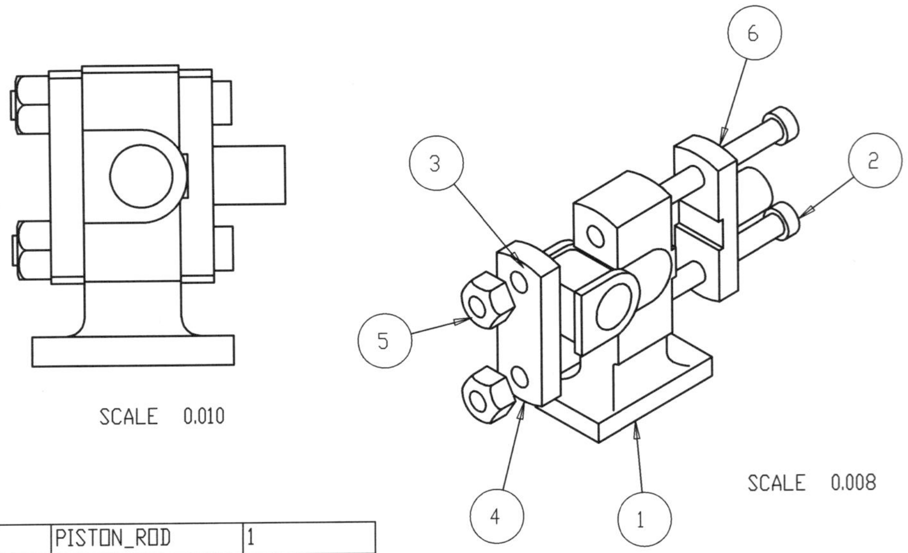

6	PISTON_ROD	1
5	NUT	2
4	KEEP_PLATE	1
3	BRASSES	1
2	BOLT	2
1	BODY	1
Index	Description	Qty

Crosshead Assembly		
Scale 0.01	CADCIM Technologies	
CS 654 UH		

Figure 13-18 Drawing for Exercise 1

Chapter 14

Working with Sheetmetal Components

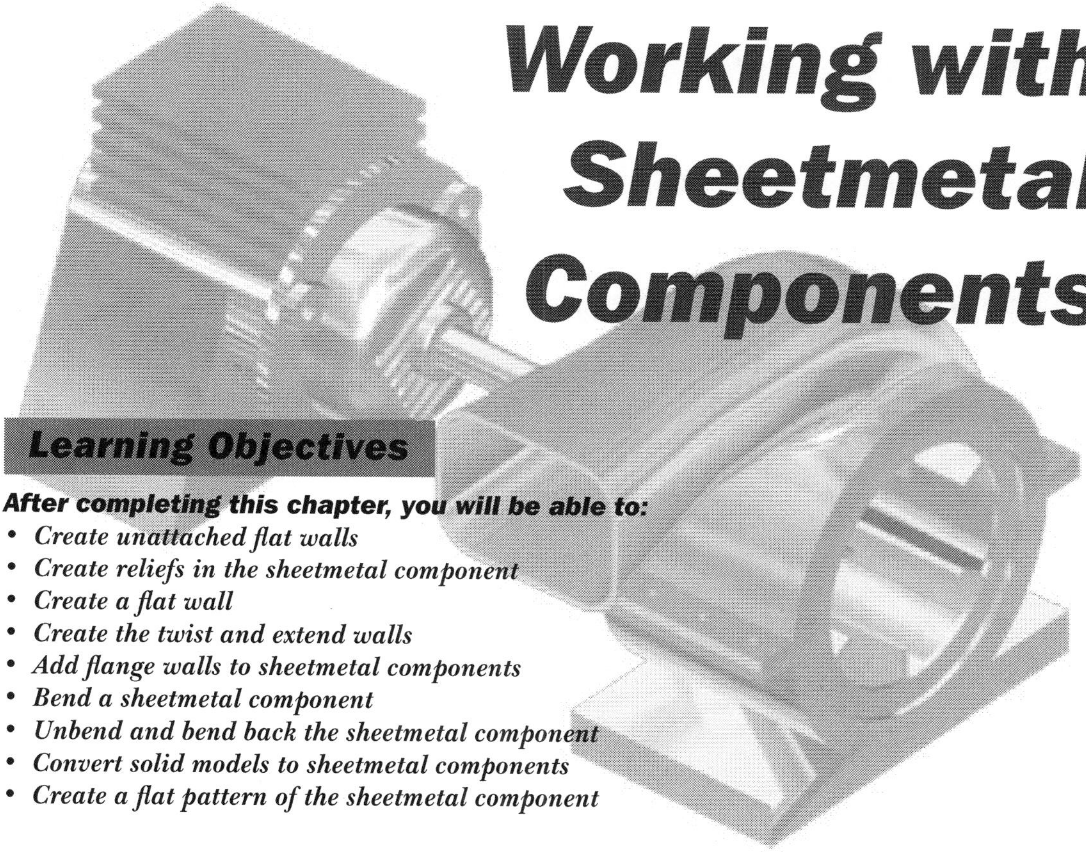

Learning Objectives

After completing this chapter, you will be able to:

- *Create unattached flat walls*
- *Create reliefs in the sheetmetal component*
- *Create a flat wall*
- *Create the twist and extend walls*
- *Add flange walls to sheetmetal components*
- *Bend a sheetmetal component*
- *Unbend and bend back the sheetmetal component*
- *Convert solid models to sheetmetal components*
- *Create a flat pattern of the sheetmetal component*

INTRODUCTION TO SHEETMETAL

A sheetmetal component is created by bending, cutting, or deforming a thin sheet of uniform thickness. The sheetmetal designs are created in the **Sheetmetal** mode of Creo Parametric. The **Sheetmetal** mode provides you the options to create primary and secondary walls and convert a solid model into a sheetmetal component.

INTRODUCTION TO SHEETMETAL WALLS

Primary Walls

Primary walls are created as base features and are independent entities. The different type of primary walls are unattached flat wall, unattached extruded wall, revolved wall, blended wall, offset wall, and so on.

Secondary Walls

Secondary walls are the walls that are dependent on at least one primary wall. They share a parent-child relation with the primary walls. Secondary walls include all the primary walls and also walls such as flat wall, flange wall, extend wall, twist wall, and so on.

CREATING THE BEND FEATURE

Ribbon: Model > Bends > Bend drop-down > Bend

The **Bend** tool transforms the sheetmetal wall into an angular or roll shape. You need to sketch a bend line and determine the direction of the bend using the direction arrows or by orienting the sketching view. The bend line is used as the reference for calculating the developed length and creating the bend geometry. Bends can be added at any time during the design process as long as a wall feature exists. You can add bends across form features, but you cannot add them where they cross another bend. Depending on where you place the bend in your sheetmetal design, you may need to add reliefs to the bend.

CREATING THE UNBEND FEATURE

Ribbon: Model > Bends > Unbend drop-down > Unbend

The **Unbend** option flattens any curved surface on the sheetmetal part. The curved surface may be a bend feature or a primary or secondary wall created previously.

CONVERSION TO SHEETMETAL PART

Converting solid parts to sheetmetal parts enables you to modify your existing solid design with sheetmetal industry features. The conversion can serve as a shortcut in your design process because you can use existing solid designs to reach your sheetmetal design intent. Also, you can include multiple features within a single conversion feature. After you convert a part to sheetmetal, the part behaves like any other sheetmetal component.

CREATING CUTS IN THE SHEETMETAL COMPONENT

The process of creating cuts in a sheetmetal component is similar to the one followed to create cuts in a solid model. The **Extrude** tool in the **Shapes** group can be used to create cuts in a sheetmetal component.

CREATING THE FLAT PATTERN

The **Flat Pattern** tool is equivalent to the **Unbend** feature. This feature flattens any curved surface, whether it is a bend feature or a wall. But the most distinctive characteristic of the **Flat Pattern** tool is that it automatically places itself at the end of the **Model Tree** to maintain the flat model view. If new features are added to your design, the flat pattern is suppressed but is automatically resumed after the features are added. If you do not want to flip between the flat pattern and the solid views for each new feature, manually suppress and resume the flat pattern as required.

TUTORIALS

Tutorial 1

In this tutorial, you will create the sheetmetal component of the Holder Clip shown in Figure 14-1. The flat pattern of the component is shown in Figure 14-2. The dimensions are shown in Figure 14-3. The thickness of the sheet is 1 mm. After creating the sheetmetal component, create its flat pattern. **(Expected time: 45 min)**

Figure 14-1 *Model for Tutorial 1*

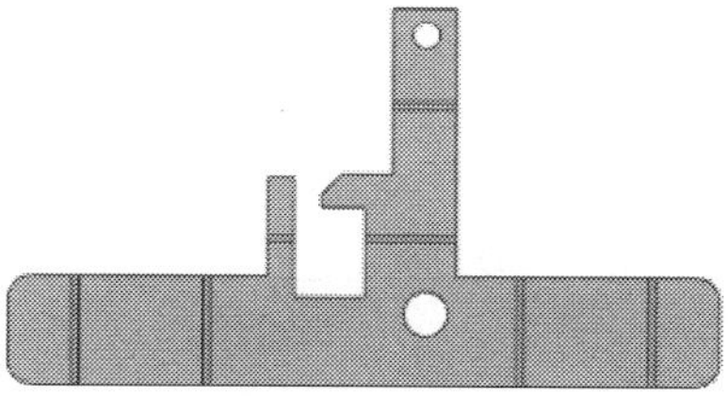

Figure 14-2 *The flat pattern of the component*

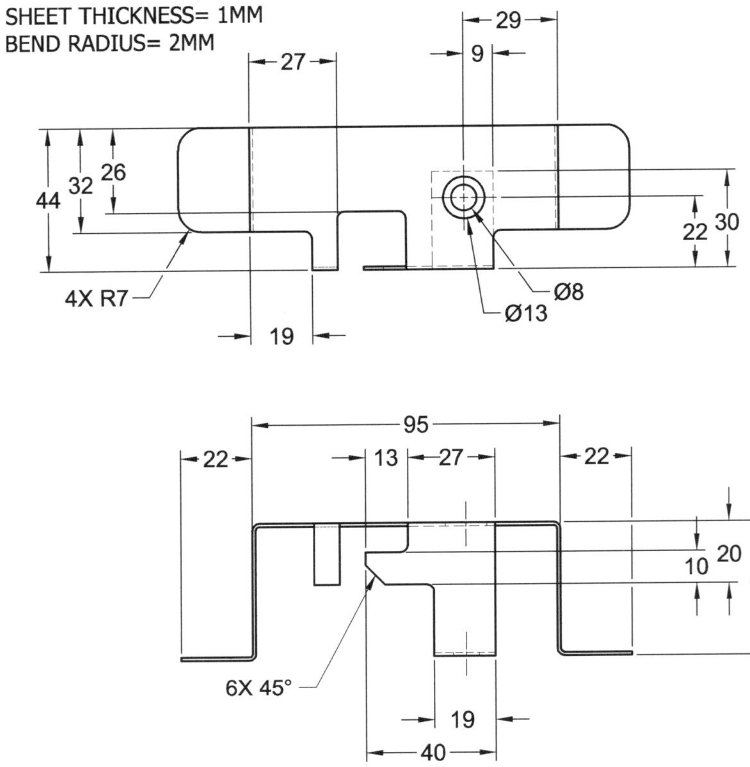

Figure 14-3 *Top and front views of the Holder Clip*

The following steps are required to complete this tutorial:

a. Create the base feature, refer to Figures 14-4 and 14-5.
b. Create the flange walls on the right and left edges of the base wall, refer to Figures 14-6 through 14-10.
c. Create the next flange wall on the front edge of the base wall, refer to Figure 14-12. Next, create a cut feature on it, refer to Figures 14-13 and 14-14.
d. Create a flat wall attached to the flange wall created in the previous step, refer to Figures 14-15 and 14-16.
e. Create the next flange wall, refer to Figures 14-17 and 14-18.
f. Create the next flange on the other front edge of the base wall, refer to Figure 14-20.
g. Create the round feature and the chamfer feature, refer to Figures 14-21 and 14-22.
h. Create the hole features on the top and bottom walls, refer to Figures 14-23 and 14-24.
i. Create the flat pattern of the model, refer to Figure 14-25.

Starting a New Object File

1. Start a new file in the **Part** mode and select **Sheetmetal** from the **Sub-type** area. Next, enter **c14tut1** in the **Name** edit box.

2. Set the template to **mmns_part_sheetmetal**.

Creating the Base Wall

1. Choose the **Planar** tool from the **Shapes** group; the **Planar** dashboard is displayed and you are prompted to select a closed sketch.

2. Choose the **References** tab from the **Planar** dashboard to display a slide-down panel and then choose the **Define** button from the slide-down panel; the **Sketch** dialog box is displayed and you are prompted to select a sketching plane.

3. Select the **TOP** datum plane from the drawing area and then choose the **Sketch** button from the **Sketch** dialog box.

4. Draw the sketch for the base wall, as shown in Figure 14-4, and then choose the **OK** button to exit the sketcher environment.

5. Enter **1** in the edit box of the **Planar** dashboard and choose the **OK** button from the dashboard to complete the creation of the feature. The model similar to the one shown in Figure 14-5 is displayed in the drawing area.

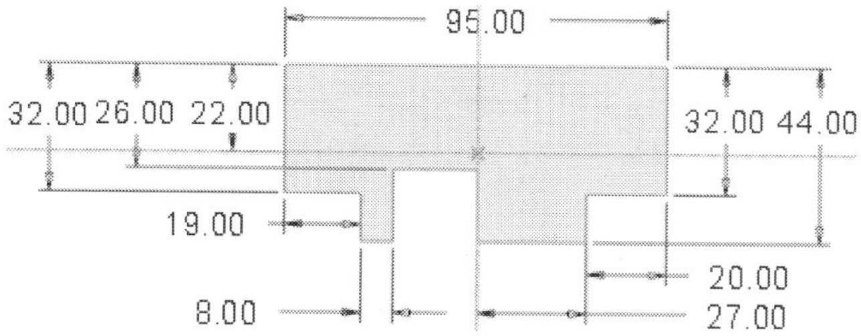

Figure 14-4 The sketch for the base wall

Figure 14-5 Isometric view of the model

Creating the First Flange Wall

1. Choose the **Flange** tool from the **Shapes** group; the **Flange** dashboard is displayed and you are prompted to select the edge to attach the wall.

2. Select the right edge of the base wall, as shown in Figure 14-6; the preview of the flange wall is displayed in the drawing area. The **I** profile is selected by default in the selection box above the **Placement** tab in the **Flange** dashboard. The dimension, indicating the bend radius at the attachment edge, is also displayed in the drawing area. The default value of radius is selected as **Thickness**, which is dimensioned from the inner surface of the bend.

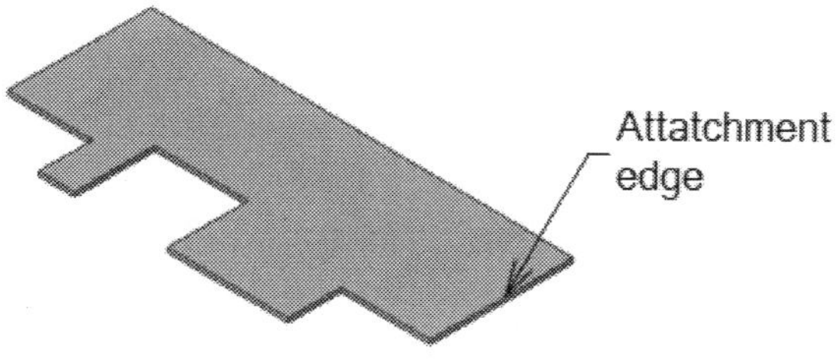

Figure 14-6 The attachment edge for the flange wall

3. Choose the **Shape** tab from the **Flange** dashboard to display the slide-down panel and then double-click on the default height value in the slide-down panel. Next, enter **41**. The angle at which the wall is created with respect to the base wall is selected as 90-degree by default. You can also drag the handles and adjust the values dynamically. Alternatively, double-click on the values displayed on the model; the edit boxes will be displayed where you can enter new values as per your requirement.

4. Choose the **Dimension the outer surface of the bend** button from the **Flange** dashboard and enter **2** in the edit box; the preview of the wall is displayed in the drawing area.

5. Choose the **OK** button from the **Flange** dashboard to complete the creation of the wall; the model similar to the one shown in Figure 14-7 is displayed in the drawing area.

Figure 14-7 *Model with the flange wall*

Creating the Next Flange Wall

1. Select the attachment edge, which is the inner edge of the previously created flange wall, as shown in Figure 14-8, and create a flange wall as discussed earlier, with **22** as its height value and **2** as its radius value. The model similar to the one shown in Figure 14-9 is displayed in the drawing area.

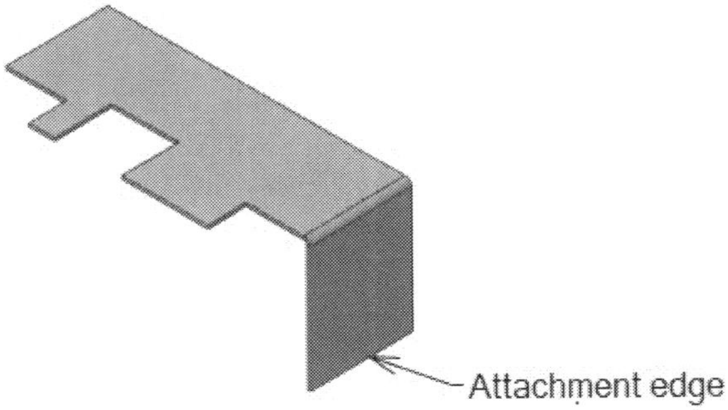

Figure 14-8 *The attachment edge chosen for the flange wall*

Figure 14-9 *Model with the flange wall*

Creating the Flange Walls on the Left Edge of the Base Wall

1. Create the flange walls on the left edge of the base wall by following the similar procedure that you used for creating the flange walls on the right edge. The model after creating the flange walls on the left edge of the base wall is shown in Figure 14-10. You can also mirror the left side flange walls about the **RIGHT** plane.

Figure 14-10 Model after creating the flange wall on the left edge

Creating the Next Flange Wall on the Front Edge of the Base Wall

1. Choose the **Flange** tool from the **Shapes** group; the **Flange** dashboard is displayed and you are prompted to select the edge to attach the wall.

2. Select the front edge of the base wall, as shown in Figure 14-11; the preview of the flange wall is displayed in the drawing area.

3. Choose the **Shape** tab to display the slide-down panel. Specify **41** as the value of height of the wall in the edit box in the slide-down panel. You can also double-click on the height value displayed on the model and enter the new value. The angle at which the wall is created with respect to the base wall is selected as 90-degree by default.

4. Choose the **Dimension the outer surface of the bend** button from the **Flange** dashboard and enter the value **2** in the edit box. The preview of the flange wall is displayed in the drawing area.

5. Choose the **OK** button from the **Flange** dashboard; the model similar to the one shown in Figure 14-12 is displayed in the drawing area.

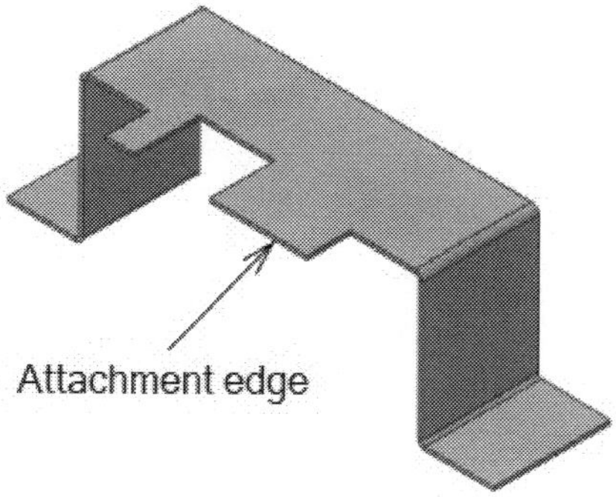

Figure 14-11 The attachment edge for the flange wall

Figure 14-12 Model with the flange wall

Creating the Cut Feature on the Front Flange Wall

1. Choose the **Extrude** tool from the **Shapes** group and select the front face of the front flange wall as the sketching plane. Make sure that the **Remove Material** button is chosen.

2. Draw the sketch for the extruded cut feature, as shown in Figure 14-13.

3. Exit the sketcher environment and extrude the sketch through all faces in the model. The sheetmetal component after creating the cut feature is shown in Figure 14-14.

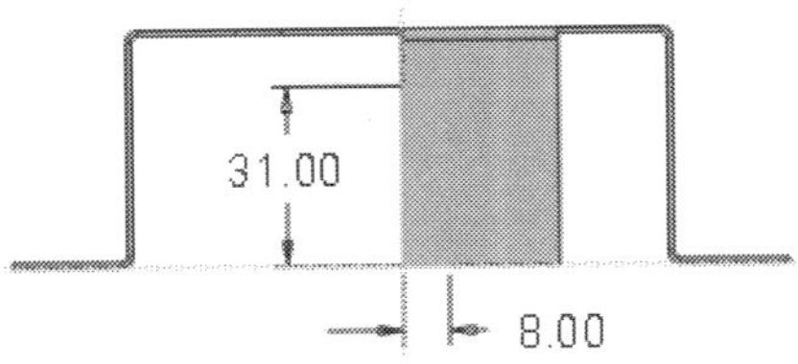

Figure 14-13 The sketch for the extruded cut feature

Figure 14-14 Model after creating the cut feature

Creating the Flat Wall

1. Choose the **Flat** tool from the **Shapes** group; the **Flat** dashboard is displayed and you are prompted to select the attachment edge to create the wall.

2. Select the attachment edge, as shown in Figure 14-15; the **Rectangle** profile of the wall is selected by default and the preview of the wall is displayed in the drawing area.

3. Change the angle value to **Flat** in the **Angle** edit box of **Flat** dashboard.

4. Choose the **Shape** tab from the **Flat** dashboard; the slide-down panel is displayed with the default dimensions of the flat wall. Specify **21** as the value for the height of the wall and specify **-21** as the offset distance from the coordinate system in the slide-down panel. The preview of the flat wall is displayed in the drawing area.

5. Choose the **OK** button from the **Flat** dashboard to complete the feature creation. The model similar to the one shown in Figure 14-16 is displayed in the drawing area.

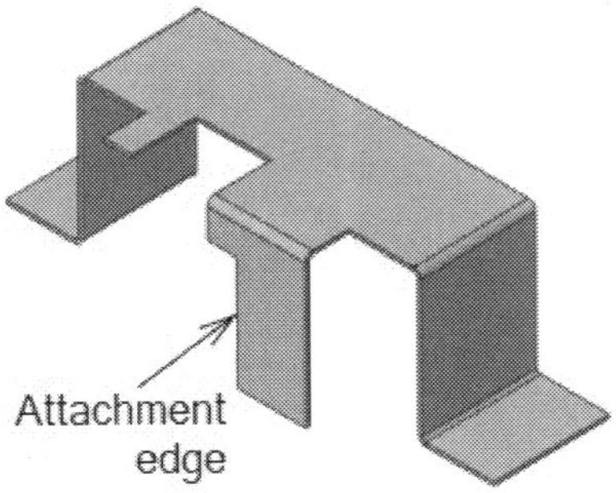

Figure 14-15 The attachment edge for the flat wall

Figure 14-16 Model after creating the flat wall

Creating the Next Flange Wall

1. Choose the **Flange** tool from the **Shapes** group; the **Flange** dashboard is displayed and you are prompted to select the attachment edge to create the wall.

2. Select the attachment edge, as shown in the Figure 14-17; the preview of the flange wall is displayed in the drawing area.

3. Choose the **Shape** tab from the **Flange Wall** dashboard to display the slide-down panel. Specify **30** as the value for the height of the wall in the edit box in the slide-down panel. The angle at which the wall is created with respect to the front flange wall is selected as 90-degree by default.

4. Enter **2** as the dimension for creating the bend at the attachment edge which is dimensioned from the outer surface of the bend.

5. Choose the **OK** button from the **Flange** dashboard to complete the creation of the feature. The model similar to the one shown in Figure 14-18 is displayed in the drawing area.

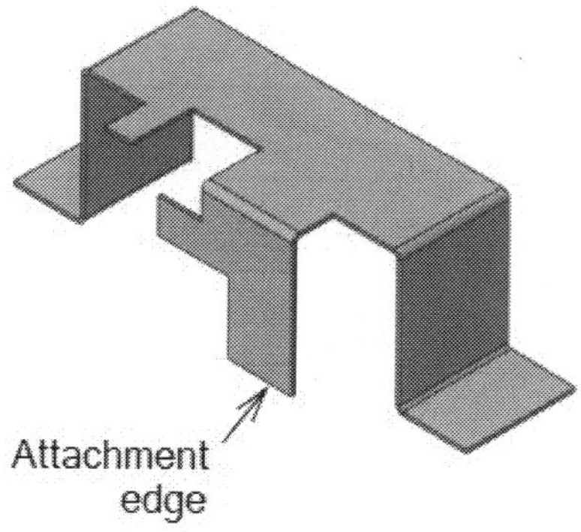

Figure 14-17 *The attachment edge for the flange wall*

Figure 14-18 *Model showing the flange wall*

Creating the Next Flange Wall on the Front Edge of the Base Wall

1. Create the next flange wall on the front edge of the base wall by selecting the attachment edge, as shown in Figure 14-19, and specifying the height of the wall as **19**. The model similar to the one shown in Figure 14-20 is displayed in the drawing area.

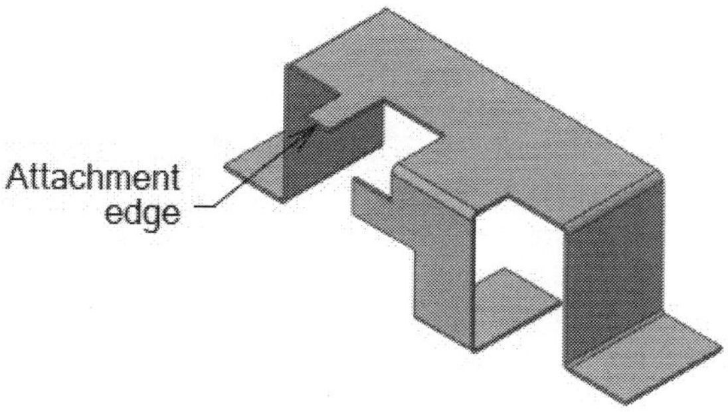

Figure 14-19 *The attachment edge for the flange wall*

Figure 14-20 *Model showing the flange wall on the front edge of the base wall*

Creating the Round and Chamfer Features

1. Create the rounds with radius 7, as shown in Figure 14-21, using the **Round** tool available in the **Engineering** group.

2. Create the edge chamfer, as shown in Figure 14-22, using the **Edge Chamfer** tool available in the **Engineering** group. For dimensions of the edge chamfer, refer to Figure 14-3.

Figure 14-21 *Model after creating the round feature*

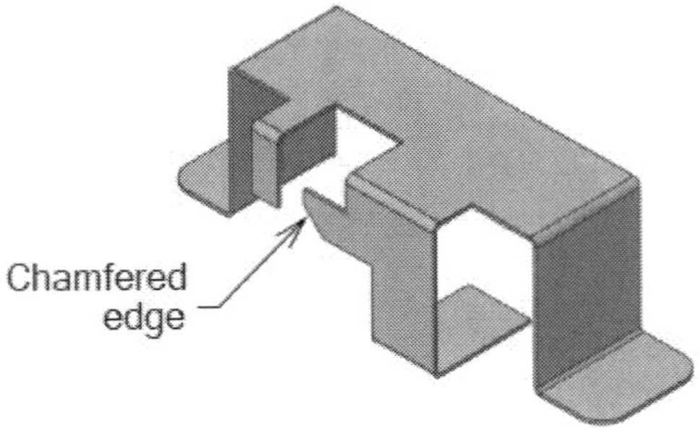

Figure 14-22 *Model after creating the chamfer feature*

Creating the Hole Features on the Top and Bottom Walls

1. Choose the **Hole** tool from the **Engineering** group of the **Ribbon**; the **Hole** dashboard is displayed.

2. Select the top surface of the base wall and specify the placement parameters for the hole having diameter 13, refer to Figure 14-3.

3. Choose the **OK** button from the **Hole** dashboard to exit it. The model after creating the hole feature on the top surface of the base wall is shown in Figure 14-23.

4. Similarly, create the hole feature on the bottom flange wall by specifying the placement parameters, refer to Figure 14-3. The model after creating the hole feature on the bottom flange wall is shown in Figure 14-24.

Figure 14-23 *Model with the hole feature on the base wall*

Figure 14-24 *Model with the hole feature on the bottom flange wall*

Creating the Flat Pattern of the Model

1. Choose the **Flat Pattern** button from the **Bends** group; the flat pattern of the model is created with the 3D notes.

2. To clear the 3D notes, choose **Annotation Display Filter > Annotation Display** from the **Graphics** toolbar. The flat pattern of the model is displayed, as shown in Figure 14-25.

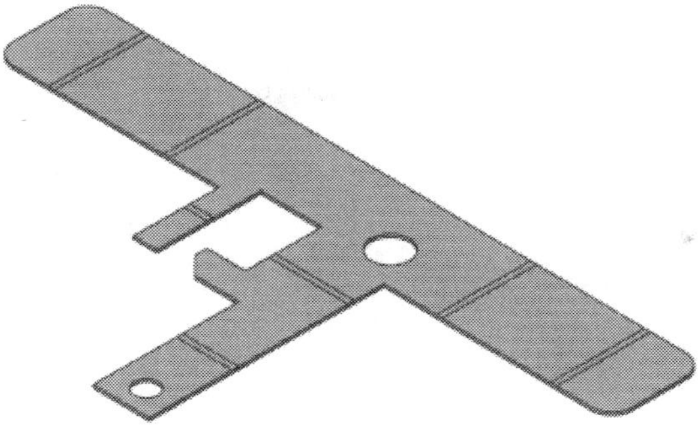

Figure 14-25 *Model after creating the flat pattern*

Saving the Model

1. Choose the **Save** button from the **Quick Access** toolbar to save the model.

Tutorial 2

In this tutorial, you will create the sheetmetal component shown in Figure 14-26. The flat pattern of the component is shown in Figure 14-27. The dimensions are shown in Figure 14-28. The thickness of the sheet is 1 mm, Bend Radius is 1 mm, Relief Type is Rectangular, and width value is equal to thickness. After creating the sheetmetal component, create its flat pattern.

(Expected time: 45 min)

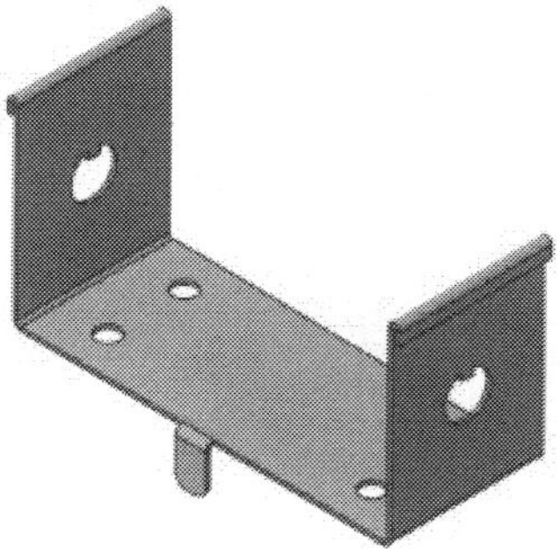

Figure 14-26 *Model for Tutorial 2*

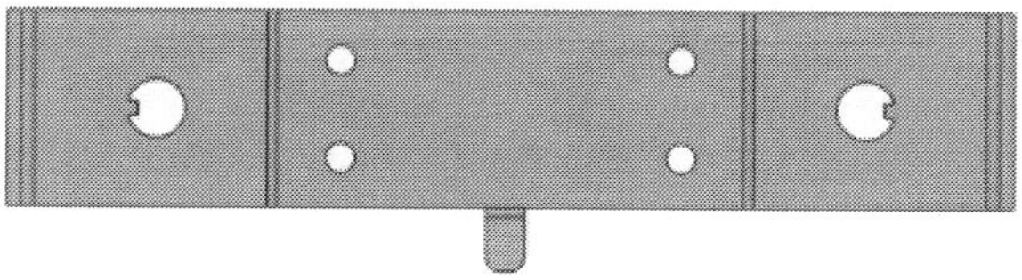

Figure 14-27 *Top view of the model*

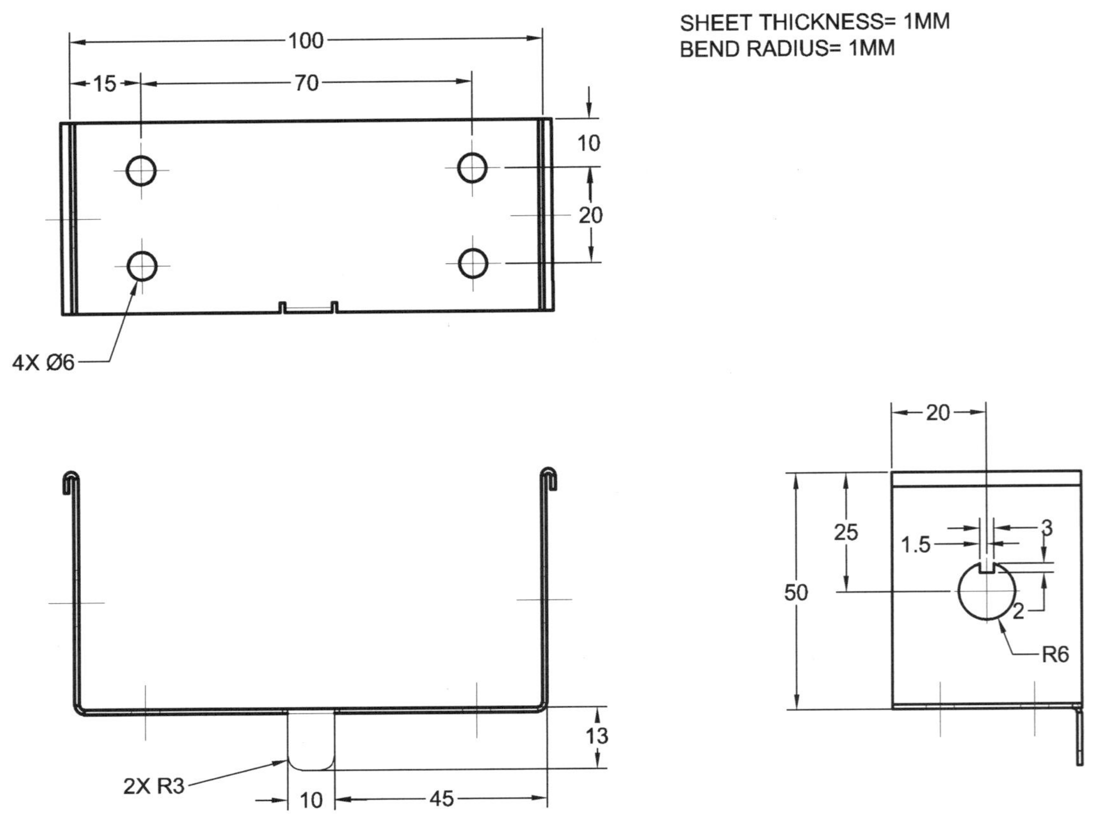

Figure 14-28 *Orthographic views of the model*

The following steps are required to complete this tutorial:

a. Create the base feature, refer to Figures 14-29 and 14-30.
b. Create the hole feature on the top surface of the base wall, refer to Figure 14-31 and pattern the hole feature, refer to Figure 14-32.
c. Create a flange wall on the right edge of the base wall, refer to Figure 14-33.
d. Create the cut feature, refer to Figure 14-34 and Figure 14-35.
e. Create the next flange wall attached to the wall created previously, refer to Figure 14-37.
f. Create the two flange walls along with the cut feature on the left side of the base wall as created previously, refer to Figure 14-38.
g. Create the flat wall on the front edge of the base wall, refer to Figure 14-39.
h. Create the flat pattern of the component, refer to Figure 14-41.

If required, set the working directory to *C:\Creo-5.0\c14*.

Starting a New Object File

1. Start a new file in the **Sheetmetal** mode and name it as *c14tut2*.

2. Set the template to **mmns_part_sheetmetal**.

Creating the Base Wall

1. Choose the **Planar** tool from the **Shapes** group; the **Planar** dashboard is displayed and you are prompted to select a closed sketch.

2. Choose the **References** tab from the **Planar** dashboard to display the slide-down panel. Choose the **Define** button from the slide-down panel; the **Sketch** dialog box is displayed and you are prompted to select a sketching plane.

3. Select the **TOP** datum plane from the drawing area and choose the **Sketch** button from the **Sketch** dialog box.

4. Draw the sketch for the base wall, as shown in Figure 14-29, and choose the **OK** button to exit the sketcher environment.

5. Enter the value **1** in the edit box in the **Planar** dashboard and choose the **OK** button from the dashboard to complete the creation of the feature. The model similar to the one shown in Figure 14-30 is displayed in the drawing area.

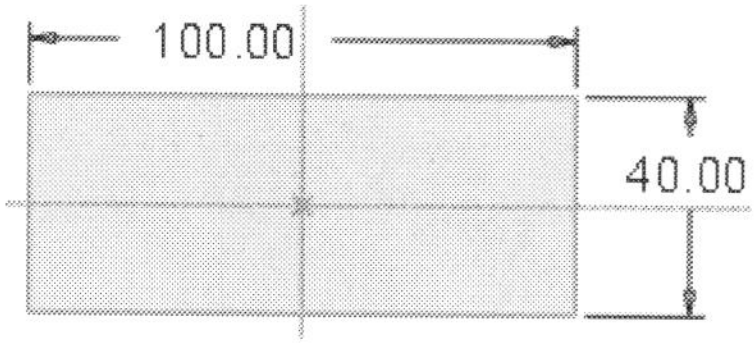

Figure 14-29 *The sketch for the base wall*

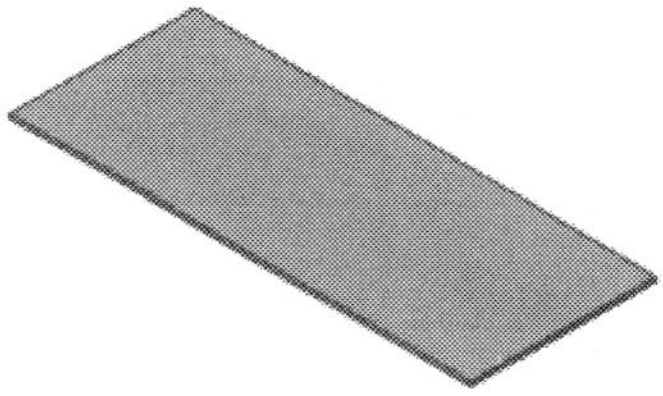

Figure 14-30 *Model of the base wall*

Creating the Hole Feature

1. Create a hole on the lower right corner of the base wall, as shown in Figure 14-31, by specifying the dimensions shown in Figure 14-28.

Creating the Pattern of the Hole Feature

1. Create the rectangular pattern of the hole, as shown in Figure 14-32.

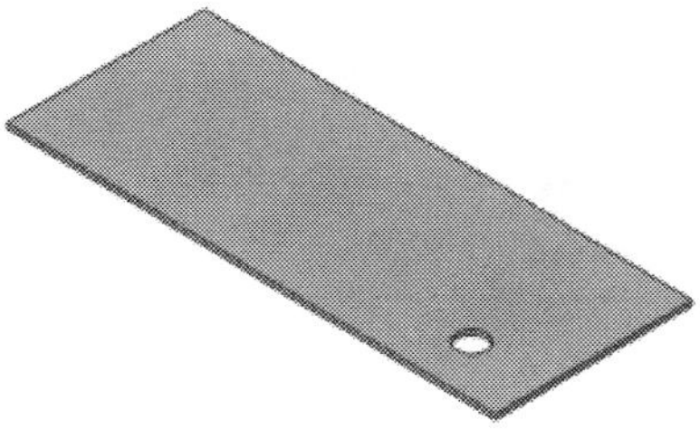

Figure 14-31 *Model after creating the hole feature*

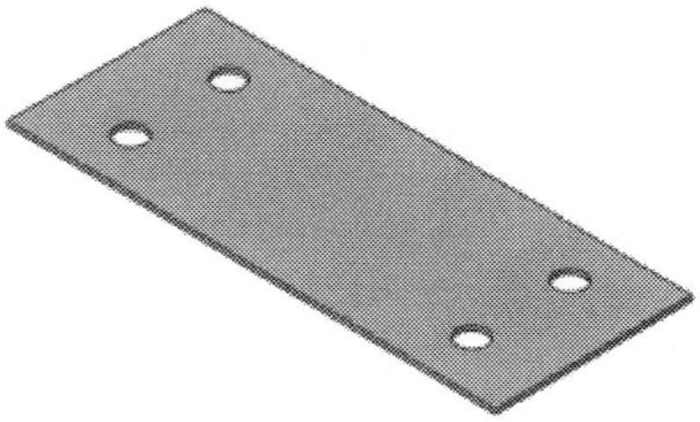

Figure 14-32 *Model after creating the pattern of the hole feature*

Creating the First Flange Wall

1. Choose the **Flange** tool from the **Shapes** group; the **Flange** dashboard is displayed and you are prompted to select the edge to attach the wall.

2. Select the right edge of the base wall; the preview of the flange wall is displayed in the drawing area.

3. Choose the **Shape** tab from the **Flange Wall** dashboard to display the slide-down panel and specify **50** as the value for the height of the wall in the edit box.

4. Choose the **Dimension the outer surface of the bend** button from the **Flange** dashboard and enter **2** as the value in the edit box; the preview of the wall is displayed in the drawing area.

5. Choose the **OK** button from the **Flange** dashboard to complete the creation of the wall. Figure 14-33 shows the model after creating the flange wall.

Figure 14-33 *Model with the flange wall*

Creating the Cut Feature

1. Choose the **Extrude** tool from the **Shapes** group and select the front face of the flange wall as the sketching plane. Also, make sure that the **Remove Material** button is chosen.

2. Draw the sketch for the cut feature on the front face of the previous flange wall, as shown in Figure 14-34.

3. Extrude the sketch through all faces of the model to create the cut feature, as shown in Figure 14-35.

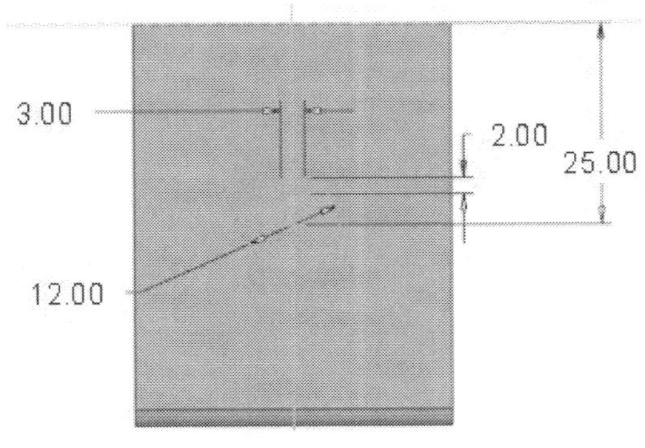

Figure 14-34 *The sketch for the cut feature*

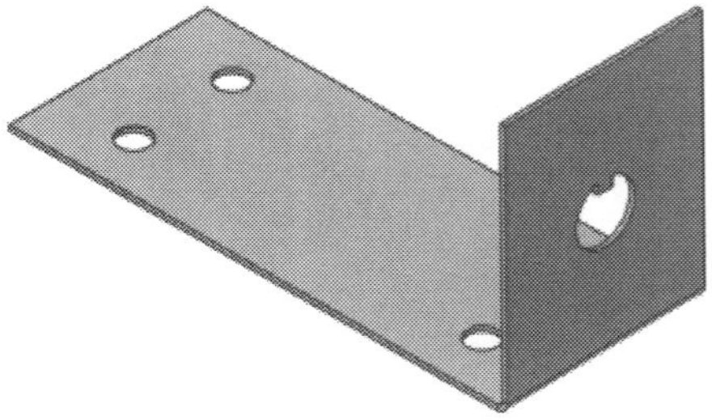

Figure 14-35 *Model with the cut feature*

Creating the Next Flange Wall

1. Choose the **Flange** tool from the **Shapes** group; the **Flange** dashboard is displayed and you are prompted to select the edge to attach the wall.

2. Select the attachment edge, as shown in Figure 14-36; the preview of the flange wall is displayed in the drawing area.

3. Select the **Open** option from the **Wall profile** drop-down list in the dashboard; the preview of the wall is displayed in the drawing area.

4. Accept the default parameters in the **Shape** slide-down panel and choose the **OK** button from the **Flange** dashboard to complete the creation of the wall.

Figure 14-37 shows the model with the flange wall.

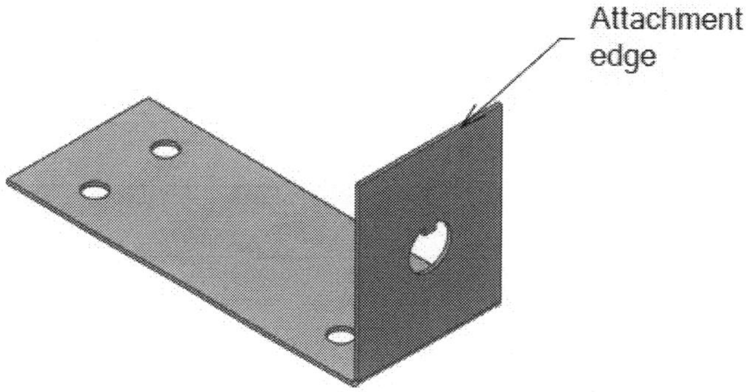

Figure 14-36 *The attachment edge for the flange wall*

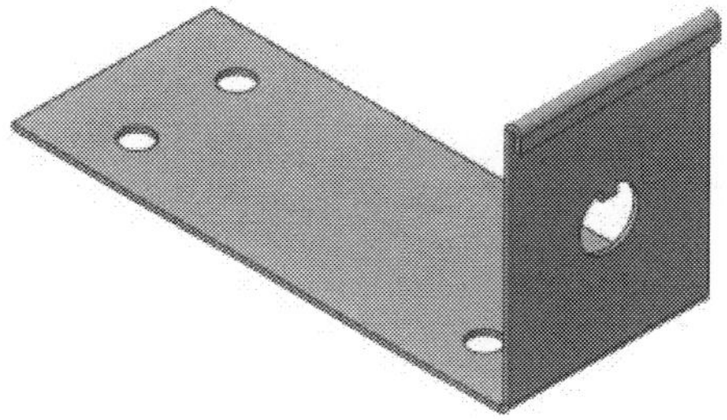

Figure 14-37 *Model with the flange wall*

Creating the Flange Walls on the Left Edge of the Base Wall

1. Create the flange walls on the left edge of the base wall by following the same procedure that was used for creating the first two flange walls on the right edge of the base wall, refer to Figure 14-38. You can also mirror the flange walls with respect to the RIGHT plane.

Creating the Cut Feature on the Left Flange Wall

1. Create the cut feature on the left flange wall by following the procedure used for creating the cut on the right flange wall. The model after creating the cut feature is shown in Figure 14-38.

Figure 14-38 *Model after creating the flange walls on the left edge of the base wall and the cut feature on the left flange wall*

Creating the Flat Wall

1. Choose the **Flat** tool from the **Shapes** group; the **Flat** dashboard is displayed and you are prompted to select the attachment edge to create the wall.

2. Select the front edge of the base wall; the **Rectangle** profile of the wall is chosen by default and the preview of the wall is displayed in the drawing area.

3. Choose the **Rectangle** option from the dashboard; a flyout is displayed. Choose the **User Defined** option from the flyout.

4. Choose the **Shape** tab from the **Flat** dashboard to display the slide-down panel. Choose the **Sketch** button from the slide-down panel; the **Sketch** dialog box is displayed.

5. Orient the model such that the top face of the base wall is at the top and draw the sketch for the user-defined flat wall, as shown in Figure 14-39. Note that the sketch drawn should be open at the top. Choose the **OK** button to exit the sketcher environment.

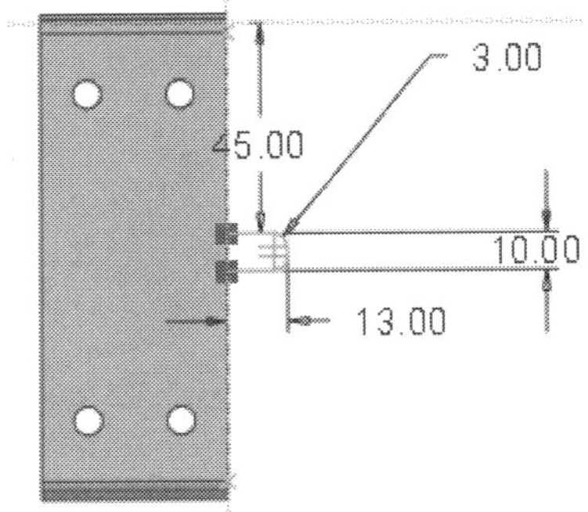

Figure 14-39 *The sketch for the flat wall*

6. Choose the **Relief** tab from the **Flat** dashboard to display the slide-down panel and then select the **Rectangular** option from the **Type** drop-down list. Accept the default parameters for the relief. You need to choose the **Flip** button to flip the direction.

7. Choose the **OK** button from the **Flat** dashboard to complete the creation of the flat wall.

8. Create a bend feature on the flat wall, refer to Figure 14-23. The model similar to the one shown in Figure 14-40 is displayed in the drawing area.

Creating the Flat Pattern of the Model

1. Choose the **Flat Pattern** tool from the **Bends** group; the flat pattern of the model is created and the model similar to the one shown in Figure 14-41 is displayed in the drawing area.

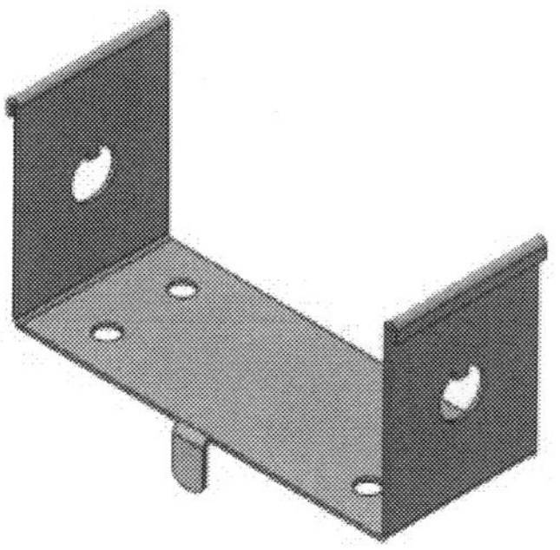

Figure 14-40 Model after creating the flat wall

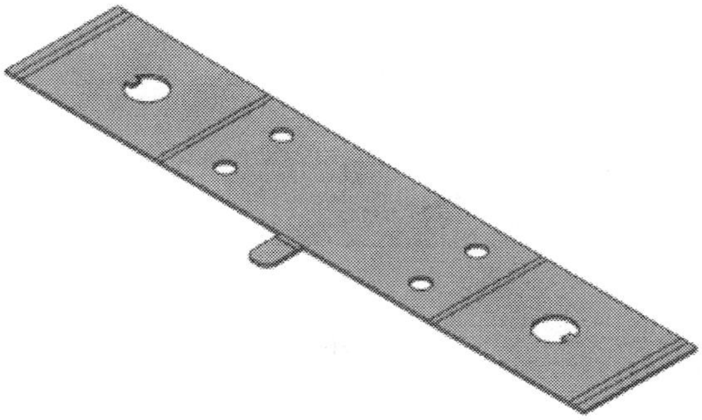

Figure 14-41 Model after creating the flat pattern

Saving the Model

1. Choose the **Save** button from the **Quick Access** toolbar to save the model.

EXERCISES

Exercise 1

Create the sheetmetal component shown in Figure 14-42. The flat pattern of the component is shown in Figure 14-43. The dimensions of the model are shown in Figure 14-44.

(Expected time: 30 min)

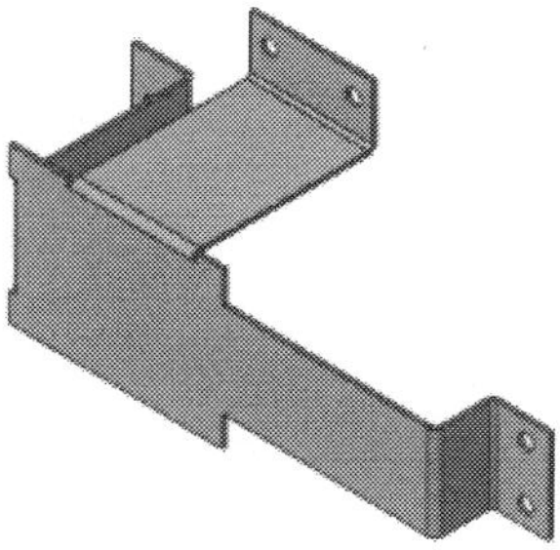

Figure 14-42 Model for Exercise 1

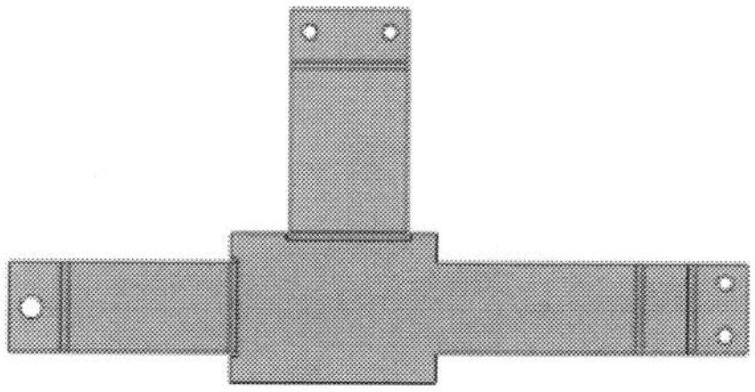

Figure 14-43 Flat pattern of the component

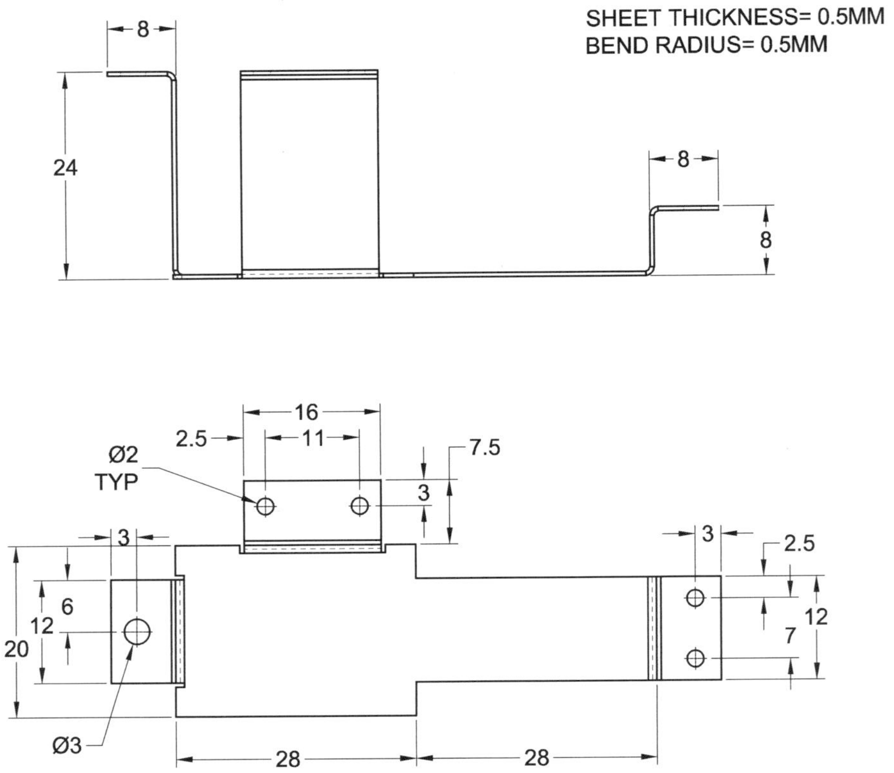

Figure 14-44 Orthographic view of the component

Exercise 2

In this exercise, you will create the sheetmetal component shown in Figure 14-45. The flat pattern of the component is shown in Figure 14-46. The dimensions of the model are shown in Figure 14-47. **(Expected time: 30 min)**

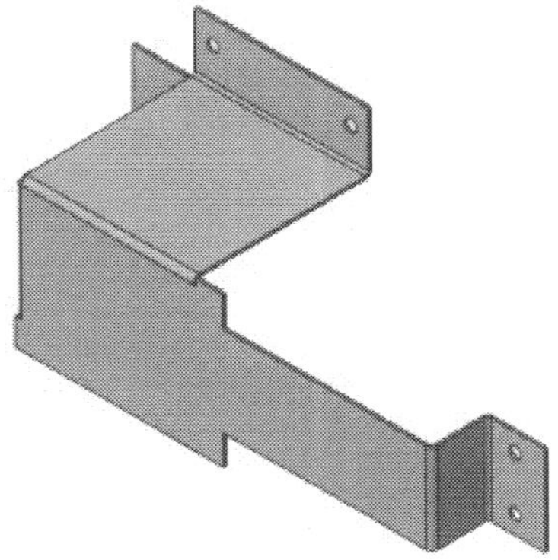

Figure 14-45 Sheetmetal part for Exercise

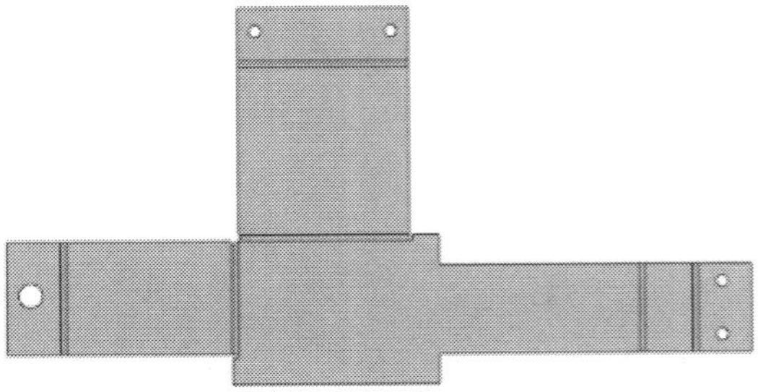

Figure 14-46 Flat pattern of the part

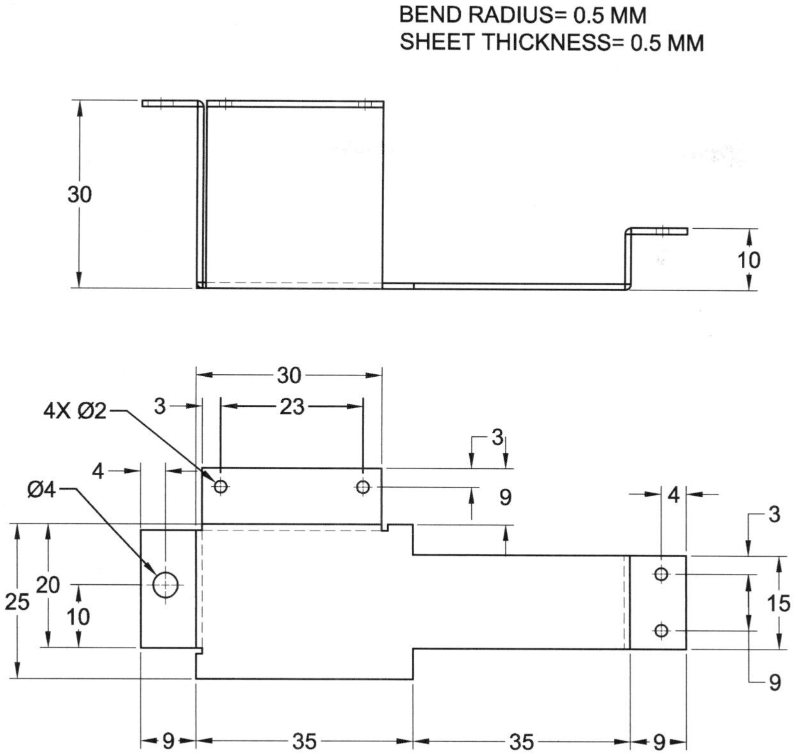

Figure 14-47 Dimensions of the sheetmetal part

This page is intentionally left blank

Index

This page is intentionally left blank

Made in United States
Orlando, FL
05 November 2026

90231089R00149